Prediction of Defects in Material Processing

Prediction of Defects in Material Processing

edited by
Mircea Predeleanu, Arnaud Poitou,
Germaine Nefussi & Luc Chevalier

First published in 2002 by Lavoisier and Hermes Science Publications
First published in Great Britain and the United States in 2004 by Kogan Page Science, an imprint Kogan Page Limited
Derived from *International Journal of Forming Processes,* Volume 5, No. 2-3-4.

First South Asian Edition 2007

Apart from any fair dealing for the purposes of research or private study, or criticism or review, as permitted under the Copyright, Designs and Patents Act 1988, this publication may only be reproduced, stored or transmitted, in any form or by any means, with the prior permission in writing of the publishers, or in the case of reprographic reproduction in accordance with the terms and licences issued by the CLA. Enquiries concerning reproduction outside these terms should be sent to the publishers at the undermentioned addresses:

Kogan Page Limited	Kogan Page India
120 Pentonville Road	4737/23 Ansari Road
London N1 9JN	New Delhi- 110002
UK	
www.koganpagescience.com	

© Lavoisier
© Kogan Page Limited

The right of Mircea Predeleanu, Arnaud Poitou, Germaine Nefussi and Luc Chevalier to be identified as the editors of this work has been asserted by them in accordance with the Copyright, Designs and Patents Act 1988.

ISBN 1-9039-9640-6

British Library Cataloguing-in-Publication Data

A CIP record for this book is available from the British Library.

Library of Congress Cataloging-in-Publication Data

Prediction of defects in material processing / edited by Mircea Predeleanu, ...[et al.]
  p. cm.
ISBN 1-9039-9640-6
  1. Materials--Defects--Congresses. 2. Manufactures--Defects--Congresses. I. Predeleanu, M.
TA409.P725 2004
620.1'12--dc22
                                2003023053

Typeset by Kogan Page
Printed in Brijbasi Art Press Ltd., I-72, Sector-9, Noida, U.P. India.

Contents

Preface ix
Introduction xi

1. Application of Extended Bernoulli's Theorem to Prediction of
Central Bursting Defects in Extrusion and Drawing 1
Sergei E. Alexandrov and Elena A. Lyamina

2. Identification of a Continuum Damage Model from Micro-hardness
Measurements 11
Gilles Arnold, Olivier Hubert, René Billardon and Martin Dutko

3. Growth and Coalescence of Non Spherical Voids in Metals Deformed
at Elevated Temperature 23
Antoine Baldacci, Helmut Klöcker and Julian Driver

4. Lattice Misorientations in Titanium Alloys 37
Nathan R. Barton and Paul R. Dawson

5. Interaction of Initial and Newly Born Damage in Fatigue Crack Growth
and Structural Integrity 51
Vladimir V. Bolotin

6. A Ductile Damage Model with Inclusion Considerations 63
Cyril Bordreuil, Jean-Claude Boyer and Emmanuelle Sallé

7. Failure Prediction in Anisotropic Sheet Metals under Forming
Operations using Damage Theory 73
Michel Brunet, Fabrice Morestin and Hélène Walter

8. Analysis and Experiment on Void Closure Behaviour Inside the
Sheet during Sheet Rolling Processes 85
Dyi-Cheng Chen and Yeong-Maw Hwang

9. Analysis of the Interply Porosities in Thermoplastic Composites Forming Processes 95
Anthony Cheruet, Damien Soulat, Philippe Boisse, Eric Soccard and Serge Maison-Le Poec

10. Predicting Material Defects in Reactive Polymeric Flows 107
Francisco Chinesta

11. Anisotropic Damage Model for Aluminium 117
Patrick Croix, Franck Lauro and Jérôme Oudin

12. Shape Defects Measurement in 3D Sheet Metal Stamping Processes 135
Leonardo D'Acquisto and Livan Fratini

13. Cavity Defects and Failure Study of Ceramic Components 147
Ioannis Doltsinis

14. Prediction of Necking Initiation during the Bending of Metal-rubber Profiles by FEM Simulations of the Forming Process 165
Monique Gaspérini, Cristian Teodosiu, David Boscher and Eric Hoferlin

15. Application of the Variational Self-consistent Model to the Deformation Textures of Titanium 175
Pierre Gilormini, Yi Liu and Pedro Ponte Castañeda

16. Influence of Strain-hardening and Damage on the Solid-state Drawing of Poly(oxymethylene) 185
Christian G'Sell and Ian M. Ward

17. Mathematical Modelling of Dynamic Processes of Irreversible Deforming, Micro- and Macro-fracture of Solids and Structures 199
Alexey B. Kiselev and Alexander A. Luk'yanov

18. Numerical Simulation of Ductile Damage in Metal Forming Processes. Part 1: Theoretical and Numerical Aspects 211
Khémaïs Saanouni, Philippe Lestriez, Abdelhakim Cherouat and Jean-François Mariage

19. Numerical Simulation of Ductile Damage in Metal Forming Processes. Part 2: Some Applications 225
Khémaïs Saanouni, Philippe Lestriez, Abdelhakim Cherouat and Jean-François Mariage

20. A Simplified Model of Residual Stresses Induced by Punching 239
Vincent Maurel, Florence Ossart and René Billardon

21. Wrinkling and Necking Instabilities for Tube Hydroforming 249
Germaine Nefussi, Noel Dahan and Alain Combescure

22. Forming Limit Curves in Blow Molding for a Polymer Exhibiting
Deformation Induced Crystallization 261
Arnaud Poitou, Amine Ammar and Germaine Nefussi

23. Modelling of Thin Sheet Blanking with a Micromechanical Approach 271
Christophe Poizat, Said Ahzi, Nadia Bahlouli, Laurent Merle and
Christophe Husson

24. Anisotropy in Thin, Canning Sheet Metals 281
David W.A. Rees

25. Progress in Microscopic Modelling of Damage in Steel at
High Temperature 293
Marc Remy, Sylvie Castagne and Anne Marie Habraken

26. Use of Laser-Doppler Velocimetry and Flow Birefringence
to Characterize Spurt Flow Instability during Extrusion of
Molten HDPE 305
Bruno Vergnes, Laurent Robert and Yves Demay

27. Induced Non Homogeneity in a Saturated Granular Media
Submitted to Slow Shearing 315
Nicolas Roussel, Christophe Lanos and Yannick Mélinge

28. *A Priori* Model Reduction Method for Thermo-mechanical
Simulations 325
David Ryckelynck

29. Fatigue Analysis of Materials and Structures using a
Continuum Damage Model 341
Omar Salomón, Sergio Oller and Eugenio Oñate

30. Influence of Initial and Induced Hardening on the Formability in
Sheet Metal Forming 353
Sébastien Thibaud and Jean-Claude Gelin

31. Computational Characterization of Micro- to Macroscopic Mechanical Behaviour and Damage of Polymer containing Second-phase Particles 369
Yoshihiro Tomita and Wei Lu

Index 379

Preface

This special issue of the International Journal of Forming Processes contains selected papers presented at the Fourth International Conference on Materials Processing Defects, held in Cachan, France, during 24–26 September 2002. As the previous conferences held in 1987, 1992 and 1997, this Conference focussed advanced methods for predicting and avoiding the occurrence of defects in manufactured products. The following topics have been considered: microstructural evolutions during processing, induced properties in some materials, damage modelling and fracture criteria, instability analysis, characterisation of formability, specific methods for the prediction of defects, influence of defects on the integrity of structures, new numerical techniques.

Major manufacturing operations are covered, conventional and new materials are considered: metal alloys, ceramics, composites and polymers.

We would like to thank the authors for their efforts and co-operation in making this conference successful.

Mircea Predeleanu
Arnaud Poitou
Germaine Nefussi
Luc Chevalier

Introduction

The prediction of defects arising in material forming processes and the evaluation of the straining effects in manufactured products are the main objectives of the theoretical and applied research in this field, with obvious economic implications. The results of these investigations are basic elements in optimisation of the material process sequences, product microstructure, mechanical properties and surface finish, tooling geometry and materials, friction, process equipment, etc.

In the past, defects in material working have tended to draw little attention from researchers. Efforts were devoted mostly to ideal materials and material characteristics, and to simplified processes which are easier to solve successfully. Tentative methods have been proposed to explain some defects arising in metal products by using slip-line field theory, upper- and lover-bound techniques of limit analysis. But, as is known, the limit analysis of plasticity theory is one alternative to the absence of precise solutions. On the other hand no prediction of defects can be provided by such methods.

Significant advances were made in the past two decades by introducing a new approach based on mathematical damage models, new bifurcation and instabilities theories in to defect analysis and formability characterisation. In addition, computer techniques based essentially on finite element methods have permitted simulation of plastic flow for more complex constitutive relations, replacing the trial-and error methods which are still commonly used in some material forming industries.

The series of conferences devoted to predictive methods of material processing defects initiated by the author in 1987, has drawn attention to these new approaches and their proceedings contain valuable contributions to this field.

A list of the most common defects arising in the processing of metals and composite materials published by Johnson and his associates is impressive (more than two hundred defects are mentioned). The defects are classified according to both the main working operations (rolling, forging, extrusion, drawing, bending, etc) and to the nature of the material (metals, fibre-reinforced plastics, metal-matrix composites, coated materials).

The nature of defects is varied, but most of them are due to internal or surface fractures (central bursts or chevrons, alligatoring, fir tree, edge cracking, etc.) and/or to instability phenomena (wrinkling, necking, earring, folding, etc.). In addition, thermomechanical treatment, spring back behaviour and tooling geometry can favour the occurrence of many other body or surface defects.

Among the microstructural rearrangements induced by extensive plastic straining during the forming processes, nucleation and growth of voids and cracks are perhaps the most important. These phenomena, called damage, lead to progressive deterioration of the material in the sense that it diminishes its resist capacity to subsequent loading. An advanced evolution of damage favours the appearance of plastic instabilities and ductile fractures, internal or external by coalescence of the microcavities.

Practical consequences of these straining effects are unsatisfactory limitations in working operations and/or unacceptable products. Apart form the determination of forming limits, another equally important interest in damage analysis is determining the "soundness" of the material undergoing a forming operation. As noted, the absence of apparent cracks on the free-stress boundary of a product does not exclude an advanced state of damage within the material itself. For the upsetting test, metallographic examinations indicate that sub-surface void formation occurs prior to the appearance of a surface crack during the same bulk forming operations. It was also noticed for less ductile materials, that fracture occurs through the full section and is not confined to the bulge equator area of the cylindrical upset specimen.

Since the internal microstructure of the material is strongly affected by damage phenomena, the mechanical and physical properties are consequently also modified. The global (macroscopical) rheological behaviour of the material will therefore be dependent both on the "initial" state of damage and on its evolution during the deformation process. Conversely, every stage of damage evolution (nucleation, growth, coalescence) is controlled by stress and strain fields at both the microscopic and the macroscopic levels. Therefore, the coupling of damage effects with plastic deformation must be included in constitutive modelling.

Two classes of continuum damage models were developed and used in defect analysis: models based on a phenomenological approach (in the Kachanov's sense) and models based on a micro-mechanical analysis (such as Gurson's model) which have gained wider acceptance. The Gurson model was extended by including various geometries and orientations of microvoids (Gologanu-Leblond model) and thus applied in defect analysis. Some results are presented at the MPD 4 Conference.

The developments of these damage models by including the initial and/or induced material anisotropy, the closure of existing microvoids, and new damage evolution laws have provided tools even more reliable for defect analysis.

Among the main accomplishments obtained and presented at the MPD Conferences can be cited:

1) Prediction of the occurrence of bulk or surface defects. For the rupture defects, appropriate ductile fracture criteria have been added to the governing equations including damage variables. The following have been predicted: central

bursts during extrusion and during drawing, failures in upsetting, sheet metal blanking and hot rolling, hot tears, scratches, sharkskin. For geometrical defects such as wrinkling, necking, etc., new bifurcation and instability criteria have been used coupled with the damage models. For the spring back and the residual stresses, also involved in the occurrence of forming defects, theoretical and experimental results have been obtained.

2) Evaluation of damage induced in a product during processing represents one of successful results of this new approach and could mark an important and useful advance for the manufacturing industries. Lacking operative technical equipments able to evaluate the microstructural deterioration of every product, computer simulation could assure much more control on its soundness, avoiding possible accidents during service life. Using different damage models cited above, iso-damage charts have been obtained for the main forming processes such as extrusion, up-setting, H-shaped cross-sectional forging, pipe-bulging, wire and strip drawing, deep-drawing as well as for deformation tests such as collar-testing, torsion-traction or extension of e notched rod. It was shown that a formability analysis by means of just the "apparent" cracking observations is not always satisfactory.

3) Formability characterisation by a theoretical analysis that couples material flow equations including damage with instability criteria must become a necessary complement to the wide experimental activity of the workers in the field. The results of the test measuring material characteristics to be deformed without cracking or other undesirable conditions cannot be useful for an engineering analysis unless are known local stress, strain and strain rate and temperature. That is the case, for example, for the determination of the forming limit diagrams: fracture f.l.d or necking f.l.d. More suitable predictions of the rupture or of the necking with respect to experimental data have been obtained by using a coupled strain-damage model and appropriate fracture or respectively localised strain criteria.

Workability tests can establish interactions among material aspects (microstructure evolutions, physical and mechanical properties, etc.) and process variables only if these are compared with the results of a theoretical and/or numerical approach of these tests.

4) The influence of processing-induced defects on the integrity of structures, introduced as topic at the last two MPD Conferences, can be treated successfully knowing not only the nature and the localisation of some imperfections, but also the straining and damaging effects supported by every manufactured part of the structure. Special methods and numerical techniques have been proposed, such as new algorithmic refinements, adaptive meshing to avoid mesh-dependency of the results due to the localisation of plastic flow, probability approach for the reliability of cast components containing flows, fatigue life prediction methods of damaged structures and applications of damage models to crack problem of cast parts.

To improve predictions of the occurrence of material processing defects and the knowledge of the micro-structural evolution and processing-induced properties of formed products, the following trends and research needs can be mentioned:

– developments of micromechanical damage models including thermo-mechanical and environment effects on the behaviour of the material;

– micromechanical analysis of fracture mechanisms and fracture criteria under general three-dimensional loading conditions, including rate effects, inertial effects, and temperature;

– damage modelling of free surface and interface elements including finishing operations and environment effects;

– efficient and reliable methods for predicting anisotropy induced by large strains occurring in forming processes;

– experimental methods for quantitative characterisation of damage and straining effects on the microscale;

– development of computer methods for the estimation of the residual strength of damaged structures or the fatigue limit.

The present volume contains important advances following some of these research trends.

Mircea Predeleanu
LMT-Cachan, CNRS/Université Pierre et Marie Curie, France

Chapter 1

Application of Extended Bernoulli's Theorem to Prediction of Central Bursting Defects in Extrusion and Drawing

Sergei E. Alexandrov
Department of Mechanical Engineering, Center for Mechanical Technology and Automation, University of Aveiro, Portugal, and Institute for Problems in Mechanics, Moscow, Russia

Elena A. Lyamina
Institute for Problems in Mechanics, Moscow, Russia

2   Prediction of Defects in Material Processing

1. Introduction

Damage to the workpiece center during extrusion and drawing invariably results in increased scrap and production costs. Such a defect is termed *central burst* or *chevron*. Several methods have been proposed to predict the formation of this defect ([ARA 86; AVI 68; RED 96]). For design purpose, it is important to develop a simplified method for quick analysis of the influence of different parameters on the initiation of the defect. Such a method based, on the upper bound technique, has been proposed and applied in [AVI 68] among others. A significant disadvantage of this method is that it is not concerned with stress distribution whereas it is well known that the hydrostatic stress is extremely important for the prediction of ductile fracture [ATK 96]. In the case of extrusion processes, this drawback has been mentioned in [ARA 86] where a numerical solution based on a theory of plasticity for porous materials has been obtained. The purpose of the present paper is to develop an approach for determining the hydrostatic stress distribution in extrusion and drawing with the use of an upper bound solution and relationships along streamlines in steady flows of general isotropic continua when the streamlines are coincident with principal stress trajectories obtained in [RIC 00]. It is also proposed to use singular velocity fields [ALE 01a] to improve upper bound solutions. Once the hydrostatic stress distribution is found, a fracture criterion based on the workability diagram proposed in [VUJ 86] can be applied to predict the initiation of central burst.

2. Approach

It has been shown in [RIC 00] that the equation

$$\frac{\rho}{2}\frac{\partial u^2}{\partial s} = \frac{\partial(\sigma_1 - w)}{\partial s} - \rho\frac{\partial \psi}{\partial s} - \frac{\sigma_1}{\rho}\frac{\partial \rho}{\partial s} + \frac{\left[(\sigma_2 - \sigma_1)\mathrm{div}(u_2\mathbf{e}_2) + (\sigma_3 - \sigma_1)\mathrm{div}(u_3\mathbf{e}_3)\right]}{u} \quad [1]$$

is valid along any streamline L in a steady flow of a general continuum where the streamline is also a trajectory of principal stress. Here σ_1, σ_2 and σ_3 are the principal stresses, ds is an infinitesimal length element of the streamline, ρ is the mass density, ψ is the potential of body forces, $\mathbf{e}_1, \mathbf{e}_2, \mathbf{e}_3$ are the orthogonal basis of eigenvectors of the stress tensor, u_1, u_2, u_3 are the components of the velocity vector in this basis, and w is the plastic work. Since L is a streamline, $u_1 = u$ is also the magnitude of the velocity vector and u_2 and u_3 vanish on L. Equation [1] has been derived assuming that the direction of velocity vector coincides with the direction of $\mathbf{e}_1$. In other words, this means that $u_1 > 0$. For many metal forming process, inertia terms and body forces may be neglected and the material may be assumed to be incompressible, $\rho = const$. In this case equation [1] simplifies to

$$\frac{\partial \sigma_1}{\partial s} - \frac{\partial w}{\partial s} + \frac{\left[(\sigma_2 - \sigma_1) \operatorname{div}(u_2 \mathbf{e}_2) + (\sigma_3 - \sigma_1) \operatorname{div}(u_3 \mathbf{e}_3)\right]}{u} = 0 \qquad [2]$$

Figure 1 shows the axisymmetric extrusion (drawing) process where the diameter of a circular bar is reduced from the initial radius of unity to the final radius b ($b < 1$). The extrusion ratio is defined by $\lambda = 1/b^2$ ($\lambda > 1$), and the semi-cone angle of the die is denoted by α. In the case of extrusion $F_{dr} = 0$ and $F_{ex} \neq 0$, and in the case of drawing $F_{dr} \neq 0$ and $F_{ex} = 0$. The axis of symmetry is a streamline and, therefore, equation [2] is valid along this axis where central bursting may occur. Assume that the material is rigid/perfectly plastic. In this case Mises yield criterion can be written in the form

$$(\sigma_1 - \sigma_2)^2 + (\sigma_2 - \sigma_3)^2 + (\sigma_3 - \sigma_1)^2 = 2\sigma_Y^2 \qquad [3]$$

where σ_Y is the yield stress in tension, a material constant. It is possible to assume, without the loss of generality, that the axis of symmetry is a trajectory of the principal stress σ_1. At the axis of symmetry $\sigma_2 = \sigma_3$ and equation [3] results in

$$\sigma_1 - \sigma_2 = \sigma_Y \quad \text{and} \quad \sigma_1 - \sigma_3 = \sigma_Y \qquad [4]$$

By definition, the circumferential velocity, say u_3, vanishes everywhere. Therefore,

$$\operatorname{div}(u_3 \mathbf{e}_3) = 0 \qquad [5]$$

The term $\operatorname{div}(u_2 \mathbf{e}_2)$ can be rewritten in the form

$$\operatorname{div}(u_2 \mathbf{e}_2) = \mathbf{e}_2 \bullet \operatorname{grad} u_2 + u_2 \operatorname{div} \mathbf{e}_2 \qquad [6]$$

The component u_2 vanishes on any streamline. Therefore, equation [6] becomes

$$\operatorname{div}(u_2 \mathbf{e}_2) = \frac{1}{r} \frac{\partial u_2}{\partial \theta} \qquad [7]$$

where $r\theta\varphi$ is the spherical coordinate system shown in Figure 1. At an arbitrary point of the plastic flow, the velocity components in the basis $\mathbf{e}_1, \mathbf{e}_2, \mathbf{e}_3$ and in the basis of the spherical coordinates, u_r and u_θ, are connected by the following transformation rule (Figure 2)

$$u_1 = -u_r \cos\gamma + u_\theta \sin\gamma \quad \text{and} \quad u_2 = u_r \sin\gamma + u_\theta \cos\gamma \qquad [8]$$

where γ is the orientation of the trajectory of the stress σ_1 relative to the axis of symmetry. It is clear (Figure 2) that $\gamma = 0$ at this axis. Substituting [8] into [7] gives

$$\operatorname{div}(u_2 \mathbf{e}_2) = \frac{1}{r}\left(-u \frac{\partial \gamma}{\partial \theta} + \frac{\partial u_\theta}{\partial \theta}\right) \qquad [9]$$

4   Prediction of Defects in Material Processing

at the axis of symmetry. In the case under consideration, the plastic work rate is given by

$$u\, \partial w/\partial s = \sigma_Y \xi_{eq} \qquad [10]$$

where ξ_{eq} is the equivalent strain rate defined by

$$\xi_{eq} = \sqrt{(2/3)\xi_{ij}\xi^{ij}} \qquad [11]$$

where ξ_{ij} are the components of the strain rate tensor in any coordinate system.

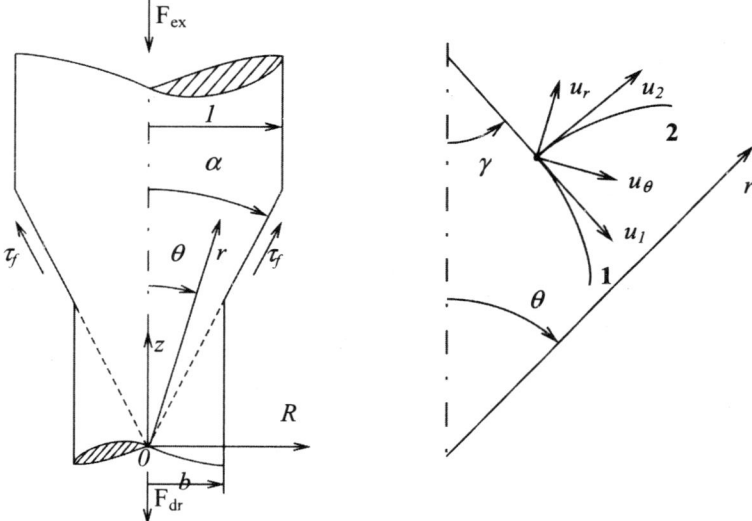

Figure 1. *Extrusion (drawing) geometry*   **Figure 2.** *Velocity components in two different bases*

Using [4], [5], [9], and [10] and taking into account that $ds = -dr$ and $\sigma_1 = \sigma_{rr}$ (σ_{rr} is the radial stress in the spherical coordinates) along the axis of symmetry, equation [2] may be transformed to

$$\frac{\partial(\sigma_{rr}/\sigma_Y)}{\partial r} = -\frac{1}{r}\frac{\partial \gamma}{\partial \theta} - \frac{1}{ur}\frac{\partial u_\theta}{\partial \theta} - \frac{\xi_{eq}}{u} \qquad [12]$$

The right hand side of this equation is solely determined by the velocity field and, therefore, may be approximated by an upper bound solution. Then, equation [12] can be integrated to find σ_{rr} and, therefore, the hydrostatic stress. To this end, it is necessary to specify one boundary condition. However, in the case of rigid/plastic materials the corresponding condition is given in integral form from which the value of σ_{rr} at the axis of symmetry cannot be found. However, it is possible to specify this value for a conservative prediction of central bursting. The exact condition in the cylindrical coordinate system Rz shown in Figure 1 is

$$\int_0^b R\sigma_{zz} dR = 0 \qquad [13]$$

at the exit section (in the case of extrusion) or

$$\int_0^1 R\sigma_{zz} dR = 0 \qquad [14]$$

at the entry section (in the case of drawing). In equations [13] and [14], σ_{zz} is the axial stress. It follows from [13] and [14] that σ_{zz} must change its sign. It also follows from physical considerations that σ_{zz} is negative at the die surface where friction occurs. Therefore, σ_{zz} should be positive at the axis of symmetry. However, $\sigma_{zz} = \sigma_{rr}$ at the axis of symmetry. Since the possibility of ductile fracture increases with increasing hydrostatic stress, a conservative prediction can be made by assuming that

$$\sigma_{rr} = 0 \qquad [15]$$

at the exit (in the case of extrusion) or at the entry (in the case of drawing). Condition [15] will be used in the present paper.

3. Upper bound solution

A review of upper bound solutions for axisymmetric extrusion (drawing) processes has been given in [ALE 01b]. In this work, a new solution has been proposed which, in contrast to other solutions, accounts for the singular behaviour of real velocity field in the vicinity of maximum friction surfaces (friction surfaces where the friction stress is equal to the shear yield stress) [ALE 01a]. In [ALE 01b], it has been shown that the singular velocity field results in a better prediction, as compared with other kinematically admissible velocity fields of the same structure, even if the friction stress is lower than the shear yield stress. In the present study, the solution found in [ALE 01b] is modified to account for the zero shear stress at the axis of symmetry (This seems to be important because the purpose of the study is to predict fracture in the very vicinity of this axis). Moreover, it is common knowledge that central burst occurs for relatively small area reductions [ARA 86]. Therefore, the present solution deals with dies of given semi-cone angle whereas the solution [ALE 01b] is mainly concerned with dead zone formation though several calculations have been done for flow with no dead zone. The solution [ALE 01b] is based on the following equation for the extrusion pressure P obtained in [OSA 75].

$$\frac{P}{\sigma_Y} = A + B\ln\lambda \qquad [16]$$

$$A = \frac{4}{\sqrt{3}\sin^2\alpha}\int_0^\alpha (g^2 + g'^2)\sin^2\theta \, d\theta, \quad B = \frac{m}{\sqrt{3}\sin\alpha}\left[g(\alpha)\cos\alpha + g'(\alpha)\sin\alpha\right] +$$

$$+ \frac{2}{\sin^2\alpha}\int_0^\alpha (g^2\cos\theta + gg'\sin\theta)\sqrt{1 + \frac{1}{12}\left[\frac{3gg'\cos\theta + (g'^2 + gg'' - g^2)\sin\theta}{g^2\cos\theta + gg'\sin\theta}\right]^2}\sin\theta \, d\theta$$

where m is the friction factor ($0 \le m \le 1$) and $g = g(\theta)$ is an arbitrary function of θ ($g' \equiv dg/d\theta$, $g'' \equiv d^2g/d\theta^2$) satisfying the following conditions

$$g(\alpha) = 1, \quad g(\theta) \ge 0, \quad 0 \le \theta \le \alpha \qquad [17]$$

In the spherical coordinate system, the velocity field is given by

$$u_r = -v(\theta)/r^2 \quad \text{and} \quad u_\theta = 0 \qquad [18]$$

where

$$v(\theta) = g(\theta)\left[g(\theta)\cos\theta + g'(\theta)\sin\theta\right]/\sin^2\alpha \qquad [19]$$

A function $g(\theta)$ defines the shape of rigid/plastic boundaries in the form

$$r = g(\theta)/\sin\alpha, \quad 0 \le \theta \le \alpha \qquad [20]$$

$$r = g(\theta)/(\sqrt{\lambda}\sin\alpha), \quad 0 \le \theta \le \alpha \qquad [21]$$

at the entry and at the exit, respectively. In [ALE 01b], the function $g(\theta)$ has been chosen in the form $g(\theta) = 1 + c(\alpha - \theta)^{3/2}$. It follows from [18] and [19] that the corresponding velocity field is singular and that its behaviour in the vicinity of the friction surface $\theta = \alpha$ coincides with the asymptotic behaviour found in [ALE 01a]

$$\partial u_r/\partial \theta = O\left(1/\sqrt{\alpha - \theta}\right) \qquad [22]$$

However, this velocity field does not satisfy the condition

$$\xi_{r\theta} = 0 \qquad [23]$$

at $\theta = 0$. Even though it is not necessary to satisfy this condition to apply the upper bound theorem, it is advantageous to take [23] into account since this condition is satisfied in the exact solution. Moreover, the purpose of the present study is to obtain the solution at $\theta = 0$ that increases the significance of condition [23]. The simplest modification of the solution [ALE 01b] satisfying [22] and [23] is

$$g(\theta) = 1 + c(\alpha^2 - \theta^2)^{3/2} \qquad [24]$$

Substituting [24] into [16] gives P as a function of c. After minimization of the right hand side of [16] with respect to c, it is possible to find c and P. Usually, the value of P is of interest. However, in our case, the value of c is more important since it enters equation [12]. At $m = 1$, the variation of c with the process parameters, α and λ, is shown in Figure 3.

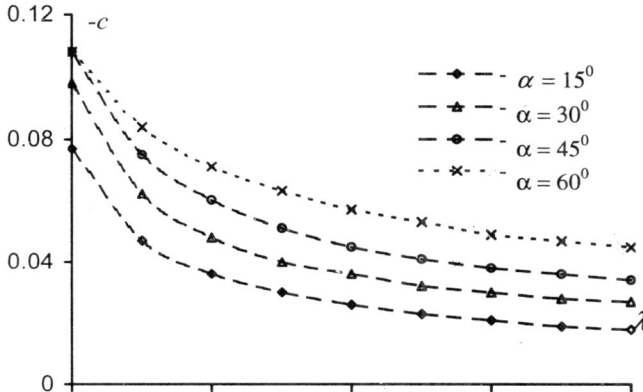

Figure 3. *Variation of the parameter c with the extrusion ratio and semi-cone angle*

4. Central burst initiation

Using the velocity field found in the previous section and the associated flow rule, the right hand side of [12] can be obtained. It follows from geometrical considerations (Figure 2) and the flow rule that

$$\tan[2(\gamma + \theta)] = 2\sigma_{r\theta}/(\sigma_{\theta\theta} - \sigma_{rr}) = 2\xi_{r\theta}/(\xi_{\theta\theta} - \xi_{rr}) \qquad [25]$$

where $\sigma_{\theta\theta}$ and $\sigma_{r\theta}$ are the stress components and ξ_{rr}, $\xi_{\theta\theta}$, and $\xi_{r\theta}$ are the strain rate components in the spherical coordinates. Differentiating this equation and using [18], [19] and [24] gives $\partial\gamma/\partial\theta$ as a function of θ. In particular,

$$d\gamma/d\theta = -2c\alpha/(1 + c\alpha^3) - 7/6 \qquad [26]$$

at the axis of symmetry, $\theta = 0$. Since $\xi_{r\theta} = 0$ at the axis of symmetry, it follows from [11] that $\xi_{eq} = \xi_{rr}$. Therefore, with the use of [18],

$$\xi_{eq}/u = 2/r \qquad [27]$$

at $\theta = 0$. Also, it follows from [18] that $\partial u_\theta / \partial \theta = 0$. Substituting this equation, [26] and [27] into [12] leads to

$$r\partial(\sigma_{rr}/\sigma_Y)/\partial r = -2c\alpha/(1+c\alpha^3) - 19/6 \quad [28]$$

This equation can be immediately integrated to give

$$\sigma_{rr}/\sigma_Y = -\left[2c\alpha/(1+c\alpha^3) + 19/6\right]\ln(r/r_0) \quad [29]$$

where r_0 is a constant of integration and c was found in the previous section (Figure 3). In the case of extrusion, it follows from [15], [21] and [24] that

$$\sigma_{rr} = 0 \quad \text{at} \quad r = r_{ex} = (1+c\alpha^3)/(\sqrt{\lambda}\sin\alpha) \quad [30]$$

Substituting [30] into [29] leads to

$$\sigma_{rr}/\sigma_Y = -\left[2c\alpha/(1+c\alpha^3) + 19/6\right]\ln(r/r_{ex}) \quad [31]$$

In the case of drawing, it follows from [15], [20] and [24] that

$$\sigma_{rr} = 0 \quad \text{at} \quad r = r_{en} = (1+c\alpha^3)/\sin\alpha \quad [32]$$

Substituting [32] into [29] leads to

$$\sigma_{rr}/\sigma_Y = -\left[2c\alpha/(1+c\alpha^3) + 19/6\right]\ln(r/r_{en}) \quad [33]$$

For the prediction of central burst initiation, a fracture criterion based on the workability diagram [VUJ 86] will be applied. To this end, in the case under consideration, it is necessary to calculate the quantity

$$\beta_{av} = 3\int_{r_{en}}^{r} \frac{\sigma \xi_{eq}}{\sigma_Y u_r} dr \quad [34]$$

where σ is the hydrostatic stress. In its general form, the criterion has been described in [VIL 87]. A particular form of this criterion is Oyane's criterion (see, for example [SHI 90]). In this case the workability diagram is a hyperbola. At the axis of symmetry, it follows from [4] that

$$\sigma/\sigma_Y = \sigma_{rr}/\sigma_Y - 2/3 \quad [35]$$

Using [27], [31], [33], and [35], equation [34] can be transformed to

$$\beta_{av} = \left(\frac{6c\alpha}{1+c\alpha^3} + \frac{19}{2}\right)\left[\ln^2\left(\frac{r}{r_{ex}}\right) - \ln^2\left(\frac{r_{en}}{r_{ex}}\right)\right] + 4\ln\left(\frac{r}{r_{en}}\right) \quad [36]$$

in the case of extrusion and to

$$\beta_{av} = \left(\frac{6c\alpha}{1+c\alpha^3} + \frac{19}{2}\right)\ln^2\left(\frac{r}{r_{en}}\right) + 4\ln\left(\frac{r}{r_{en}}\right) \quad [37]$$

in the case of drawing. In addition to β_{av}, it is necessary to find the equivalent strain

$$\varepsilon_{eq} = \int_{r_{en}}^{r}\left(\xi_{eq}/u_r\right)dr \quad [38]$$

Using [27], equation [38] can be integrated to give

$$\varepsilon_{eq} = 2\ln(r_{en}/r) \quad [39]$$

At each material point ε_{eq} is a function of β_{av}. According to the criterion [VIL 87], fracture occurs if the curve $\varepsilon_{eq}(\beta_{av})$ at any point and the workability diagram intersect. In the case under consideration, it is sufficient to apply the fracture criterion at $r = r_{ex}$. It follows from [36], [37] and [39] that

$$\beta_{av} = \mp\frac{1}{8}\left(\frac{12c\alpha}{1+c\alpha^3} + 19\right)\varepsilon_{eq}^2 - 2\varepsilon_{eq} \quad [40]$$

where the upper sign corresponds to extrusion and the lower sign to drawing. A typical application of equation [40] is illustrated in Figure 4 for $\alpha = 15^0$. One can see that, in the case of drawing, fracture is predicted for a wide range of reduction whereas no fracture is predicted in the case of extrusion. The workability diagram for a steel used in these calculations has been found in [VIL 87].

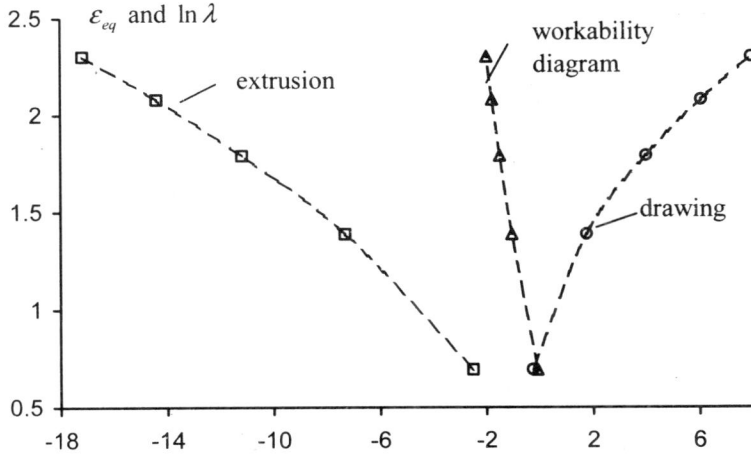

Figure 4. *Central burst prediction,* $\alpha = 15^0$

5. Conclusions

A simplified method for central burst prediction in extrusion and drawing has been developed. To apply this method, it is necessary to find an upper bound solution and, then, to solve equation [12]. This has been done in the present paper using a radial velocity field. The main result is illustrated in Figure 4. For other semi-cone angles, similar results can be obtained with the use of the value of c given in Figure 3. Thus a wide range of the process parameters can be covered and a safe domain for each process can be found.

Acknowledgements

The research was partly supported by RFFI (grants 02-01-06180, 02-01-00419).

6. References

[ALE 01a] ALEXANDROV S., RICHMOND O., "Singular plastic flow fields near surfaces of maximum friction stress", *Int J. Non-Linear Mech.*, Vol. 36, 2001, pp. 1–11.

[ALE 01b] ALEXANDROV S., MISHURIS G., MISHURIS W., SLIWA R., "On the dead zone formation and limit analysis in axially symmetric extrusion", *Int. J. Mech. Sci.*, Vol. 43, 2001, pp. 367–379.

[ARA 86] ARAVAS N., "The analysis of void growth that leads to central bursts during extrusion", *J. Mech. Phys. Solids*, Vol. 34, 1986, pp. 55–79.

[ATK 96] ATKINS A., "Fracture in forming", *J. Mater. Process. Technol.*, Vol. 56, 1996, pp. 609–618.

[AVI 68] AVITZUR B., "Analysis of central bursting defects in extrusion and wire drawing", *Trans. ASME. J. Engng Ind.*, Vol. 90, 1968, pp.79–91.

[OSA 75] OSAKADA K., NIIMI YU., "A study on radial flow field for extrusion through conical dies", *Int. J. Mech. Sci.*, Vol. 17, 1975, pp. 241–254.

[RED 96] REDDY N., DIXIT P., LAL G., "Central bursting and optimal die profile for axisymmetric extrusion", *Trans. ASME J. Manufact. Sci. Engng*, Vol. 118, 1996, pp. 579–584.

[RIC 00] RICHMOND O., ALEXANDROV S., "Extension of Bernoulli's theorem on steady flows of inviscid fluids to steady flows of plastic solids", *C.R. Acad. Sci. Paris, Ser. IIb*, Vol. 328, 2000, pp. 835–840.

[SHI 90] SHIMA S., NOSE Y., "Development of upper bound technique for analysis of fracture in metal forming", *Ingenieur-Archiv*, Vol. 60, 1990, pp. 311–322.

[VIL 87] VILOTIC D., "Fracture Properties of Steels in Bulk Cold Metal Forming", *University of Novi Sad Press, Novi Sad*, 1987.

[VUJ 86] VUJOVIC V., SHABAIK A., "A new workability criterion for ductile metals", *Trans. ASME. J. Engng Mater. Technol.*, Vol. 108, 1986, pp. 245–249.

Chapter 2

Identification of a Continuum Damage Model from Micro-hardness Measurements

Application to a Backward Extrusion Process

Gilles Arnold, Olivier Hubert and René Billardon
LMT-Cachan, Cachan, France

Martin Dutko
Rockfield Softward Ltd, Swansea, Wales, UK

12   Prediction of Defects in Material Processing

1. Introduction

The prediction of damage in an industrial forging process can be of great interest for the optimisation of the forging steps in order to avoid or limit the initiation of defects induced by plastic deformation. This paper presents some results of a study performed during the FORMAS BRITE-EURAM project [FORMAS]. The process considered is a single step backward extrusion of a tube with internal teeth (Figure 1).

Figure 1. *CAD image of the piece after forging*

The material used in the present paper is a 16 Mn Cr 5 steel. An analysis of the microstructure has been performed and shows two kinds of inclusions that could influence damage evolution: cementite nodules homogeneously distributed and MnS sulfides strongly elongated in the extrusion direction of the bar (Figure 2). Lemaitre's isotropic ductile damage model [LEM 85a, LAD 84, LEM 92] has been chosen to describe the damage evolution.

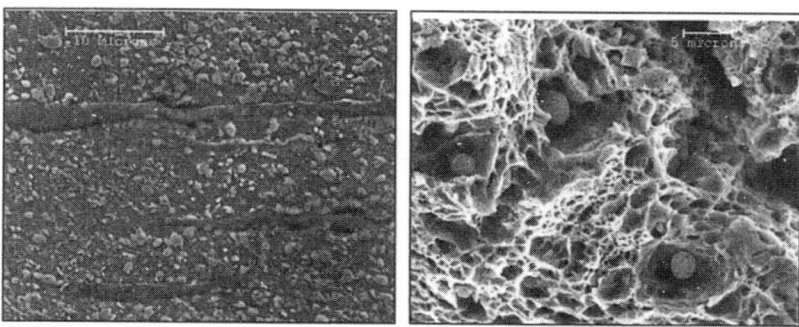

Figure 2. *a) Elongated MnS inclusions and cementite nodules as observed inside the same specimen (etched and polished), b) Nodules of cementite in the ferritic matrix as observed after ductile failure in tension*

2. Damage measurements and identification

Measurement of damage through the direct evaluation of the surface of the defects is very difficult and strongly dependent on the surface polishing conditions. Various indirect evaluations have been proposed through the measurements of the evolution of mass density, electric conductivity, ultrasonic wave propagation speed or internal damping. Hypotheses must be made to relate damage to the variations of any of the previous quantities. Equivalent strain hypothesis associated to the concept of effective stress allows one to relate damage to the variations of mechanical properties such as elastic modulus or hardness.

2.1. *Principle of damage measurement*

The procedure used in this study is based on micro-hardness measurements to identify ductile damage. This kind of analysis enables one to perform local damage evaluations.

In the following, microhardness refers to Meyer-Vickers hardness, that is equal to the average pressure applied to a pyramidal indentor with a square base so that an almost square indentation is obtained (mean diagonal of the notch d). For a given load F that corresponds to a mass of a few dozens of grams; hardness H is defined as:

$$H = \frac{2F}{d^2} \qquad [1]$$

The method, first applied on a pure copper and an annealed AISI 1010 steel [BIL 87], is based on the phenomenological model establishing that, in the case of no damage, micro-hardness is proportional to the flow stress with a shift p_H in strain induced by the hardness indentation.

$$H(p + p_H) = k_H \sigma(p) \qquad [2]$$

Parameters k_H and p_H can be identified from hardness measurements made on a tensile test specimen in regions where strains are small enough not to induce damage. For strains that are larger than a plastic strain threshold p_D, it is assumed that isotropic damage D(p) increases and relation [2] is modified as [3]:

$$H(p + p_H) = k_H (1 - D(p)) \sigma^*(p) \qquad [3]$$

$$H^*(p + p_H) = k_H \sigma^*(p) \qquad [4]$$

where σ^* denotes the flow stress that would exist without damage. This effective stress can be deduced from the flow stress measured for smaller strains by

14 Prediction of Defects in Material Processing

extrapolation to stràins larger than p_D. Hence, damage D can be derived from the following expression:

$$D = 1 - \frac{H}{H^*} = 1 - \frac{\sigma}{\sigma^*} \quad [5]$$

where H denotes the current hardness of the material, which is a function of the current values of damage D and cumulative plastic deformation ($p+p_H$), and where H^* denotes the hardness of a fictitious comparison material with the same cumulated plastic strain state but without damage.

2.2. Experimental procedure using a tensile test

A standard tensile test up to final failure has been made on a specimen that was machined in the longitudinal direction of the bar (Figure 3a). After failure, the specimen exhibits a large range of cumulative plastic deformations from 0 – in the specimen heads – to very high values close to the necking.

The hardness tests are performed using a given couple (mass, time), that corresponds to a minimal scatter of the measured values. Calibration tests are performed on a sample taken in the initial bar of the raw material that is supposed undamaged.

The tensile test specimen is coated in a resin and carefully cut in a mid-plane along its length using a micro-cutting machine (Figure 3b). The mid-plane surface of the sample is polished so that micro-hardness measurements can be made all along the axis of the specimen, as indicated in Figure 3c.

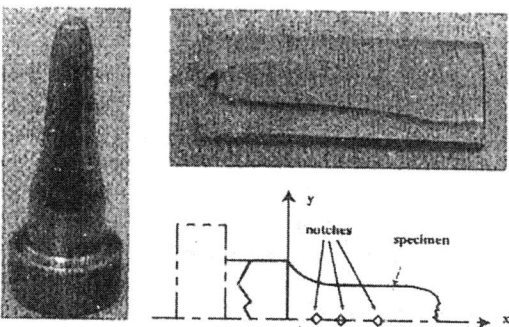

Figure 3. *a) "Half" tensile test specimen after failure, b) Coated and polished specimen, c) Micro-hardness measurements along the axis of the specimen*

It is possible to evaluate the plastic strain from the final shape of the specimen. The "local" plastic deformation – for a given initial abscissa x_i – is derived from the measurement of the "local" diameter of the tensile specimen. The following approximation for the plastic strain tensor in the specimen is used:

$$\underline{\underline{\varepsilon}}_p = \begin{vmatrix} \varepsilon & 0 & 0 \\ 0 & -\varepsilon/2 & 0 \\ 0 & 0 & -\varepsilon/2 \end{vmatrix} \quad \text{with} \quad \varepsilon = -2\ln\left[\frac{\phi_f}{\phi_i}\right] \quad [6, 7]$$

where $\phi_f(x_f)$, $\phi_i(x_i)$ and ε respectively denote the final diameter, initial diameter and longitudinal plastic deformation.

At this step, it must be noticed that initial and final diameters correspond to a section that has different initial and final abscissa. An algorithm has been developed to compute the initial axial position of all the sections where hardness measurements have been performed from the initial and final geometries. The maximal plastic strain level reached before failure is about 1.4 (Figure 4).

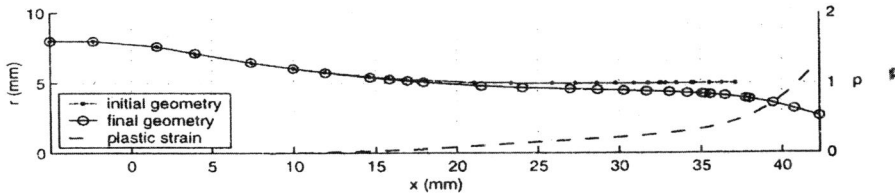

Figure 4. *Initial $r_i(x_i)$ and final $r_f(x_f)$ geometries of the specimen and cumulated plastic strain along the axis of the specimen*

2.3. *Identification of the parameters of the damage law*

2.3.1. *Constitutive equations*

The yield surface is defined by

$$\frac{\sigma_{eq}}{1-D} - R_y = 0 \qquad [8]$$

where σ_{eq} denotes the Von Mises equivalent stress, D the isotropic damage variable and R_y the size of the elastic domain that depends on cumulative plastic strain p and absolute temperature T. It is chosen herein to model hardening evolution during the tensile test by the following expression [LEM 85b].

$$R_y = g(T)\left[\sigma_y + R_0\left(1 - \exp(-p/b_{R0})\right) + C_0\left(1 - \exp(-p/b_{C0})\right)\right]$$
$$\text{with} \quad g(T) = \frac{1 - \exp(-B_0/T)}{1 - \exp(-B_0/T_0)} \quad [9]$$

Following Lemaitre's ductile damage evolution law [LEM 85a] is defined by

$$\dot{D} = \left(\frac{Y}{S}\right)^s \dot{p} H(p - p_D) \quad \text{with Heaviside function} \quad H(x) = \begin{cases} 0 \text{ if } p \leq 0 \\ 1 \text{ if } p \geq 0 \end{cases} \quad [10]$$

where S, s and p_D denote material dependent parameters and Y the thermodynamical force such that

$$Y = \frac{\sigma_{eq}^2}{2E(1-D)^2} R_v \quad \text{with} \quad R_v = \frac{2}{3}(1+\nu) + 3(1-2\nu)\left(\frac{\sigma_H}{\sigma_{eq}}\right)^2 \quad [11]$$

where σ_H denotes the hydrostatic stress: $\sigma_H = \frac{1}{3}\text{Tr}(\sigma)$.

Damage evolution may be different under dominantly positive or negative hydrostatic stress states. In order to take into account this unilateral effect, a material parameter h is introduced [LAD 84] (or see for instance [DES 00]) so that the expression of the thermodynamical force is modified to

$$Y = \frac{1+\nu}{2E}\left[\frac{\sum_j \langle \sigma_j \rangle}{(1-D)^2} + \frac{h\sum_j \langle -\sigma_j \rangle^2}{(1-hD)^2}\right] - \frac{\nu}{2E}\left[\frac{\langle 3\sigma_H \rangle^2}{(1-D)^2} + \frac{h\langle -3\sigma_H \rangle^2}{(1-hD)^2}\right] \quad [12]$$
$$\text{with} \quad \langle x \rangle = \begin{cases} x \text{ if } x \geq 0 \\ 0 \text{ if } x \leq 0 \end{cases}$$

2.3.2. *Identification from tensile tests (h=1)*

Two sets of parameters have to be previously identified. First, an optimal couple $\{k_H = 4.1, p_H = 0.05\}$ enables a good correlation between hardness and stress values, as defined in equation [2]. Second, a set of hardening parameters has to be identified in order to extrapolate the stress-strain curve measured during the tensile test before damage appears (Figure 5). The extrapolation uses exponential hardening laws [9] and the values of identified parameters are given in Table 2. It can be noticed that the maximal cumulative plastic strain measured during such a test is small (p_{max}= 0.2) compared to the high values measured in the vicinity of the necking. Hence, it can be assumed that this plot describes the behaviour of an undamaged material.

The damage is calculated using equation [5] from hardness values H and hardening extrapolation σ^*, so that parameters S and p_D of Lemaitre's damage model [10] can be identified, exponent s being set to 1.

Identification of a Continuum Damage Model   17

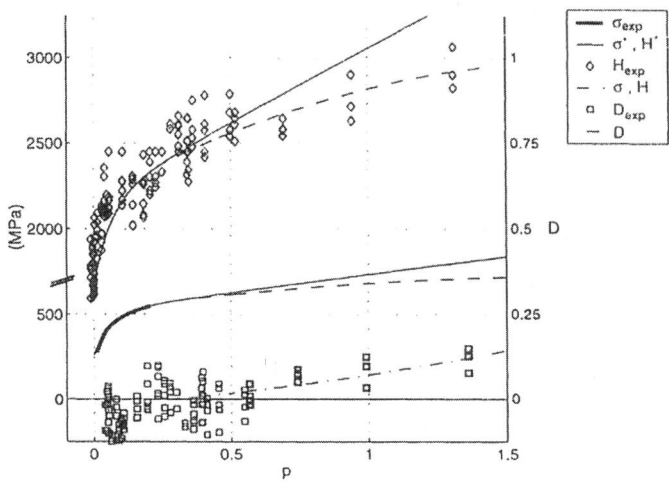

Figure 5. *Identification of the damage law from micro-hardness measurements performed along a tensile test specimen*

NOTE: For this material, the hardness is almost constant whatever the plastic strain is within the range 0.5 to 1.5, although damage increases in the same plastic strain range. As a consequence, for this material, the accuracy of the damage evaluation from the hardness measurements is low.

3. Upsetting tests and unilateral effect

3.1. *Experimental results*

Discs have been subjected to compression loadings (upsetting) as presented in Figure 6a. The discs have an initial height h_i of 20mm and an initial diameter of 60mm. Table 1 gives final height h_f, conventional deformation $\Delta h/h_i$ and the corresponding logarithmic plastic strain p. The corresponding effective hardness H^*, defined by equation [4], is deduced from the parameters that are identified from the tensile test (see section 2.3.2).

Table 1. *Final deformations and effective hardness H^* of the compressed discs*

h (mm)	h_i = 20	16	12	10	8	5
$\Delta h/h_i$ (%)	0%	−20%	−40%	−50%	−60%	−75%
p	0	0.22	0.51	0.69	0.91	1.38
H^*(MPa)	1065	2285	2580	2740	2935	3345

Each disc has been cut into two halves. The surface cut has been polished to perform hardness measurements along the diameter. The evolution of the hardness along the radius is plotted in Figure 6b.

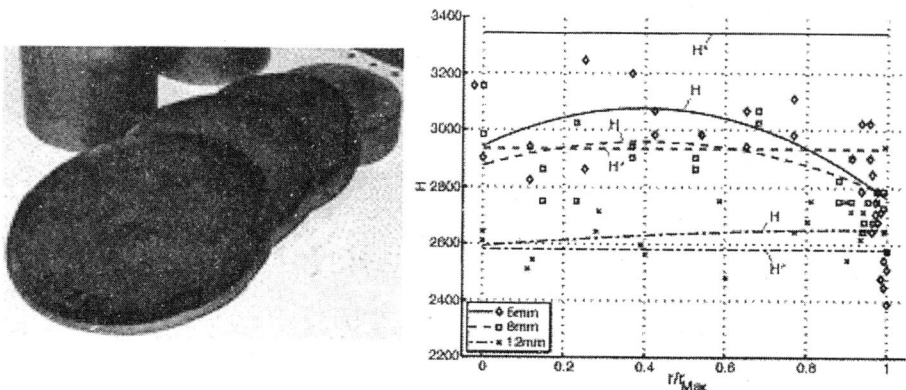

Figure 6. a) *Compressed discs obtained after upsetting tests*, b) *Hardness profiles measured along the normalised radius (r/r_{Max}) of the discs (h_f=12, 8 and 5 mm)*

3.2. Numerical results

Lemaitre's model has been implemented in Elfen FE code [ELFEN]. A coupled thermo-mechanical analysis has been performed using the parameters previously identified (Table 2). It must be noticed that parameter B_0 has been previously identified from tests performed in the temperature range 20°C to 600°C and heat transfer coefficients have been taken from the literature. The friction coefficient f has been measured from classical ring tests (f = 0,085). Because of the high deformation levels, automatic remeshing is required.

Table 2. *Values of the identified parameters*

E	180 GPa	σ_y	256 MPa
ν	0.3	Q_0	255 MPa
s	1	b_{Q0}	0.066
S	10.8 MPa	C_0	5282 MPa
P_D	0.30	b_{C0}	24.1
T_0	298°K	B_0	438°K

First simulations have been performed without taking into account unilateral effect (h = 1). In that case, the highest and lowest values of damage are respectively located in the centre of the cylinder and near the outer radius. Hence, contrary to experimental evidence (see for example [VAZ 01]), the simulation predicts fracture in the centre of the disc (D = 1) before final height is reached and very small damage at the outer radius.

In order to improve the results, unilateral effect can be introduced in the simulations. Some authors (see for instance [SAA 00]) proposed to take h = 0. Figure 7 shows the damage field obtained with h = 0.08 as proposed in [CES 02].

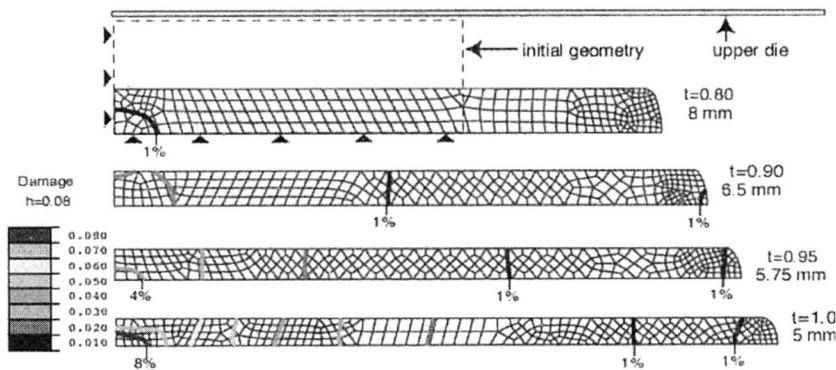

Figure 7. *FE simulation of the upsetting test with unilateral effect (h = 0.08). Damage field at time step corresponding to h_f=8, 6.5, 5.75 and 5 mm. Thanks to symmetry, one quarter of the specimen is shown*

During the upsetting of the cylinders, the stress state in the centre of the disc is close to pure hydrostatic stress state and stress triaxiality is very high. As a consequence, if no unilateral effect is taken into account the damage evolves quickly until fracture. On the opposite, the stress state near the outer radius is similar to shear stress state, which corresponds to small stress triaxiality and, as a consequence, to a slow damage evolution. The introduction of unilateral effect enables one to reduce drastically the damage rate in the centre.

From hardness profiles measured on discs, it is possible to deduce damage from the value of effective hardness H^* corresponding to the mean plastic strain in the disc (almost uniform according to FE simulations). As noted previously, hardness measurement is not very accurate for measuring damage. Nevertheless, it is possible to exhibit that damage appears in the centre for discs compressed to 8 and 5 mm and that damage seems to be higher near the outer radius (Figure 8).

The values measured in the centre of the discs can help to identify the value of h by an inverse method. A compromise has to be found to obtain reasonably good results in the centre and near the outer radius.

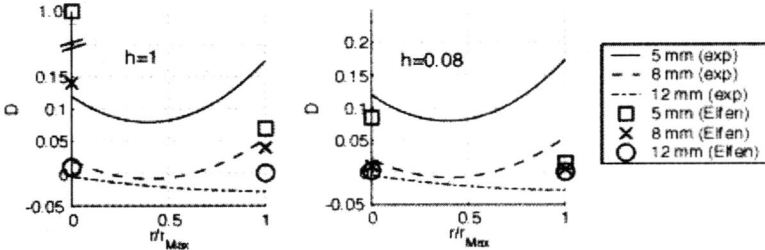

Figure 8. *Comparison of damage values obtained from hardness measurements (lines) and FE analysis (dots) with and without unilateral effect*

4. Conclusion

The identification procedure of Lemaitre's ductile damage law from microhardness tests performed on a tensile test specimen has been applied to a 16MnCr5 steel.

Figure 9. *FE simulation of the backward extrusion process – damage field*

Upsetting tests have been performed. Hardness measured on the discs enables one to deduce damage field. A procedure is proposed to identify the unilateral effect that takes into account of compressive stress states from confrontations between these hardness measurements and corresponding numerical simulations.

First simulations of the industrial process enable one to exhibit a critical area near the edge of the punch, with a good agreement with the experimental observations (Figure 9).

Acknowledgements

The authors acknowledge the support of the European Commission for the funding of the FORMAS project. Personal thanks are given to Achim Hess, Christian Walter ZFS (Zentrum Fertigungstechnik Stuttgart, Stuttgart, Germany) and Rainer

(Neher Wezel GmbH, Frieckenhausen, Germany) for performing all upsetting tests and industrial process experiments and preliminary numerical simulations.

5. References

[BIL 87] BILLARDON R., DUFAILLY J., LEMAITRE J., "A procedure based on Vickers'microhardness tests to measure damage fields", *Proc. 9th Structural Mechanics in Reactor Technology*, pp. 367–373, Balkema, 1987.

[CES 02] CÉSAR DE SÁ J.M.A. and PIRES F.M.A., "Modelling of ductile fracture initiation in bulk metal forming", *Proc. 5th ESAFORM Conference on metal forming*, M. Pietrzik, Z. Mitura and J. Kaczmar (Editors), Krakow, Poland, 2002, pp. 147–152.

[DES 00] DESMORAT R., "Strain localization and unilateral conditions for anisotropic induced damage model", *Continuous Damage and Fracture*, Benallal A. (Editor), Cachan, France, 2000, pp. 71–79.

[ELFEN] Rockfield Software Limited, Swansea, U.K.

[FORMAS] BRITE-EURAM Project - Proposal BE97-4614, Contract BRPR-CT98-0648.

[LAD 84] LADEVEZE P. and LEMAITRE J., "Damage effective stress in quasi unilateral conditions", *16th Int. Congress of Theor. And Appl. Mechanics*, Lyngby, Denmark, 1984.

[LEM 85a] LEMAITRE J., "A continuous damage mechanics model for ductile fracture", *Journal of Applied Mechanics,* Trans. ASME, pp. 83–89, 1985.

[LEM 85b] LEMAITRE J. and CHABOCHE J.-L., *Mechanics of solid materials*, Cambridge University Press, 1990 (or *Mécanique des matériaux solides*, Dunod, Paris, 1985).

[LEM 92] LEMAITRE J., *A Course on Damage Mechanics*, Springer-Verlag, 1992.

[SAA 00] SAANOUNI K. AND HAMMI Y., "Numerical simulation of damage in metal forming processes", *Continuous Damage and Fracture*, Benallal A. (Editor), Cachan, France, 2000, pp. 353–361.

[VAS 01] VASQUEZ V. and ALTAN T., "Measurement of flow stress and critical damage value in cold forging", *Inter. Journal of Forming Processes*, 2001, Vol. 4, No.1–2, pp. 167–175.

Chapter 3

Growth and Coalescence of Non Spherical Voids in Metals Deformed at Elevated Temperature

Antoine Baldacci, Helmut Klöcker and Julian Driver
Microstructure and Processing Department, Ecole Nationale Supérieure des Mines, Saint-Etienne, France

1. Introduction

In as-cast structures, complex shaped voids are observed. Deformation–induced break-up of intermetallic particles in aluminium alloys also leads to complex void shapes. Analysing the growth and coalescence of complex shaped voids is therefore of significant interest.

Tvergaard and Needleman (Tvergaard and Needleman, 1984) used a modified Gurson model for porous ductile materials to show that void coalescence implies the localisation of plastic flow in the inter-void ligament and a transition to uniaxial straining of the unit cell. Koplick and Needleman (Koplick and Needleman, 1988) used similar axisymmetric finite element cell model calculations to determine the critical void volume fraction f_c, at which the coalescence process starts. Steglich and Brocks (Steglich and Brocks, 1988) used cell model calculations to study elasto-plastic and Brocks, Sun and Hönig (Brocks, Sun and Hönig, 1995) elasto-viscoplastic damaged materials.

In the present work, the growth and coalescence of complex shaped voids embedded in an elastic-viscoplastic matrix is analysed by finite element calculations and a modified Gurson model taking into account the rate sensitivity of the matrix flow stress and the initial void shape (from a sphere to a cylinder).

2. Problem formulation

2.1. *Material model*

The matrix material was described as an isotropically hardening elastic-viscoplastic solid. The matrix flow stress σ_M is given by the following empirical relation (Koplick and Needleman, 1988):

$$\sigma_M = g(\bar{\varepsilon})\left(\frac{\dot{\bar{\varepsilon}}}{\dot{\varepsilon}_0}\right)^m \qquad g(\bar{\varepsilon}) = \sigma_0\left(1+\frac{\bar{\varepsilon}}{\varepsilon_0}\right)^n \qquad \varepsilon_0 = \frac{\sigma_0}{E} \qquad [1]$$

Here, $\bar{\varepsilon} = \int \dot{\bar{\varepsilon}} dt$, and $g(\bar{\varepsilon})$ represents the effective stress vs. effective strain response in a tensile test carried out at a strain rate such that $\dot{\bar{\varepsilon}} = \dot{\varepsilon}_0$. During hot working most metal alloys exhibit a similar behaviour with small values of the strain hardening exponent n and with values of m varying with the temperature. Throughout this work a constant value of 0.01 has been considered for n. Four different values of the strain rate sensitivity m were considered: 0.01, 0.07, 0.10 and 0.15. The lowest m value corresponds to room temperature. The ratio between the initial yield stress σ_0 and Young's modulus E was kept constant throughout this paper (E = 500 σ_0) corresponding to typical material behaviour at hot working temperature.

2.2. Cell model and numerical implementation

A cylindrical cell of initial length L_0 (Figure 1) containing a hole of radius r_0 and axis l_0 was considered. Different curvatures ($1/\rho_0$) of the void surface were considered. The void shape varies continuously from a cylinder ($\rho_0 = 0$) to a sphere ($\rho_0 = r_0$). The initial cell radius and cell height were fixed to unity ($R_0 = L_0 = 1$). The initial void radius was taken equal to the initial void axial dimension ($r_0 = l_0$). The initial cell geometry is characterised by the initial void volume fraction f_0 and the void shape $a = \rho_0/r_0$. $a = 1$ corresponds to an initially spherical void, whereas $a = 0$ corresponds to an initially cylindrical void.

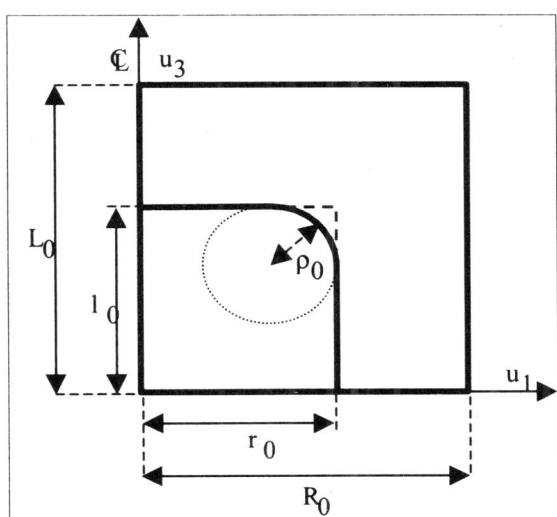

Figure 1. *Cell model geometry: ¼ of the unit cell*

The initial void radius r_0 is related to the initial volume fraction f_0:

$$f_0 = \frac{2}{3}\left(\frac{r_0}{R_0}\right)^3 \left\{ a^3 + 3(1-a)\left[3a^2\left(\frac{\pi}{4} - \frac{1}{2}\right) + \frac{1}{2}(1-a) + \frac{1}{2}\left(1 - (1-a)^2\right)\right] \right\} \quad [2].$$

The cell was subjected to homogeneous axial and radial displacements u_3 and u_1. The macroscopic principal strains and the effective strain are given by:

$$E_{11} = E_{22} = \ln\left(\frac{R}{R_0}\right) \quad E_{33} = \ln\left(\frac{L}{L_0}\right) \quad E_{eq} = \frac{2}{3}|E_{33} - E_{11}| \quad [3]$$

The corresponding macroscopic stresses $\Sigma_{11} = \Sigma_{22}$ and Σ_{33} are the average forces on the boundaries divided by the current areas. The macroscopic equivalent stress, Σ_{eq}, the mean stress Σ_m and stress triaxiality T are given by

$$\Sigma_{eq} = |\Sigma_{33} - \Sigma_{11}| \quad \Sigma_m = \frac{(2\Sigma_{11} + \Sigma_{33})}{3} \quad T = \frac{\Sigma_m}{\Sigma_{eq}} \quad [4]$$

During the loading history the overall axial strain rate $\dot{E}_{33}$ was imposed and the triaxiality ratio T was kept constant corresponding to a constant ratio between the axial and radial stress. The ABAQUS finite element code with axisymmetric 4-node elements was used. Figure 2 shows typical finite element meshes for three different initial void shapes.

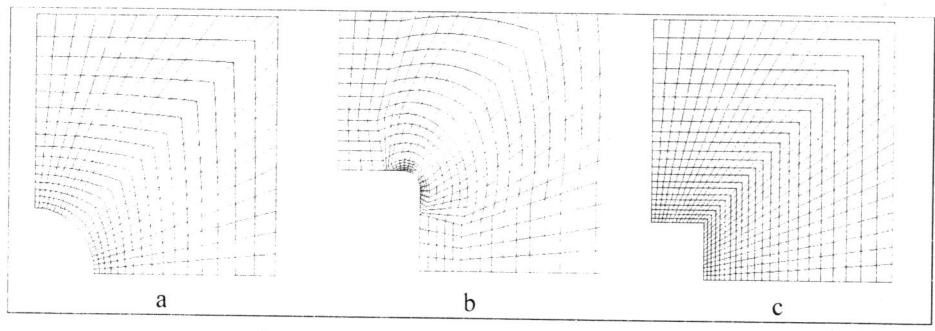

Figure 2. *Finite element meshes for voids with different shape factors. a = 1 a), a = 0.2 b) and a = 0.0 c)*

2.3. m-dependent porous material model

Leblond, Perrin, Suquet (Leblond *et al.*, 1994) have proposed an overall stress potential for a viscoplastic (Norton) material containing spherical voids. This stress potential reproduces the exact solution of a hollow sphere loaded in hydrostatic tension. Here the extension of Klöcker and Tvergaard (Klöcker and Tvergaard, 2000) to elastic-viscoplastic voided solids has been used:

$$\Phi = \left(1 + \frac{2}{3}f\right)\frac{\Sigma_{eq}^2}{\sigma_M^2} + f^* q_1 p_m\left(\frac{3}{2} q_2 \frac{\Sigma_m}{\sigma_M}\right) - \left[1 + \left(\frac{1-m}{1+m}\right)(q_1 f^*)^2\right] \quad [5]$$

$$p_m(x) = h_m(x) + \frac{(1-m)}{(1+m)} / h_m(x) \quad h_m(x) = [1 + m_x(1+m)]^{1/m}$$

Leblond et al. introduced the corrective factor $(1+2f/3)$ to respect the Hashin-Shtrikman lower bound. Without this correction, the above potential tends to the Gurson potential for an ideal-plastic (m = 0) solid. Parameters q_1 and q_2 were introduced into Gurson's model by Tvergaard (1981, 1982). The function f* was proposed by Tvergaard and Needleman (Tvergaard and Needleman, 1984) to account for the effects of rapid void coalescence at failure.

$$f^* = \begin{cases} f & f \le f_c \\ f_c + \frac{f_u - f_c}{f_f - f_c}(f - f_c) & f \ge f_c \end{cases} \qquad [6]$$

The constant f_u is the value of f* at zero stress, i.e. $f_u = 1/q_1$. As $f \to f_c$, $f^* \to f_u$ and the material loses all stress carrying capacity. Different values for f_f and f_c were proposed in the literature for corresponding low temperature analysis based on the modified Gurson criterion. Tvergaard and Needleman (1984) suggested that the values of f_c and f_f be taken as 0.15 and 0.25, respectively. Koplick and Needleman (Koplick et al., 1988), based on numerical calculations, suggested a constant value of 0.13 for f_f and f_c varying between 0.03 and 0.055 depending on the initial void volume fraction. Here, particular attention will be paid to the variation of f_c with the strain rate sensitivity (temperature) and the initial void geometry.

In general, the evolution of the void volume fraction results from the growth of existing voids and the nucleation of new voids. In the present paper only growth is considered but a void nucleation term can be added independently.

3. Results

The cell geometry has been described by the void aspect ratio $a = \rho_0/r_0$ and the initial void volume fraction f_0. Two different initial void volume fractions ($f_0 = 0.0013$ and $f_0 = 0.0104$) and five different void aspect ratios (a = 0., 0.2, 0.5, 0.8 and 1.0) have been considered. For each particular cell geometry, three different values of the triaxiality parameter (T = 1, 2 and 3) and four different values of the strain rate hardening parameter m (0.01, 0.07, 0.10 and 0.15) were considered.

3.1. Cell model results

3.1.1. Overall behaviour: stress strain curves and void volume fraction evolution

Figure 3 shows cell model results for an initially spherical (a = 1) and initially cylindrical void (a = 0). The volume fraction is $f_0 = 0.0013$. Figure 3a corresponds to a stress triaxiality of T = 1. The maximum stress is only slightly affected by the void shape. The critical strain E_c corresponding to the equivalent stress drop is smaller for initially spherical cavities than for initially cylindrical voids. The void volume fraction increases more rapidly for initially spherical voids (a = 1) than for initially cylindrical ones (a = 0). Figure 3(b) and 3(c) show similar results for an overall stress triaxiality of T = 2 and T = 3 respectively. At high values of the stress triaxiality parameter (T = 2, 3), the difference between initially spherical and initially cylindrical voids becomes negligible.

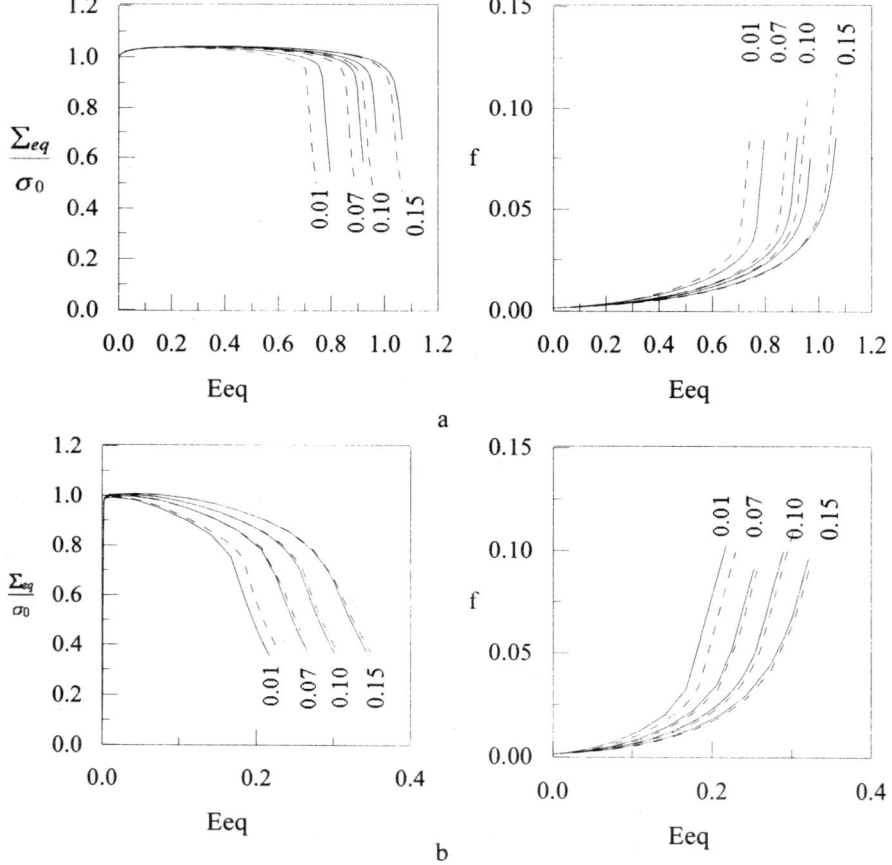

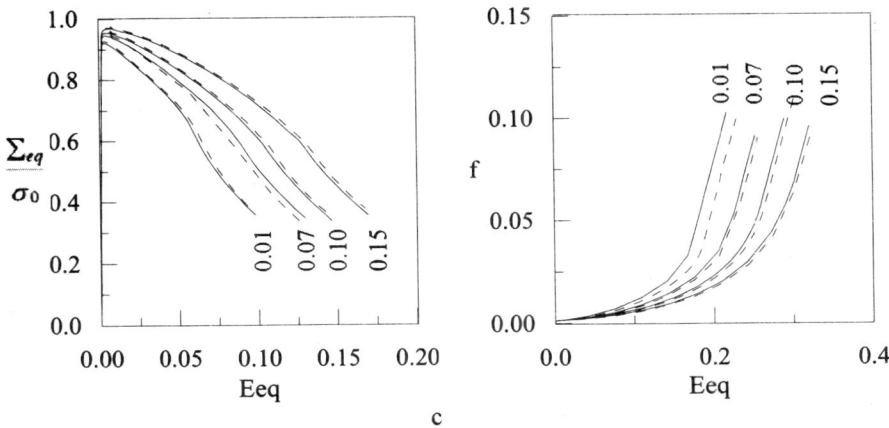

Figure 3. *Cell model results for a cylindrical void (continuous line) and a spherical void (discontinuous line) for $f_0 = 0.0013$ with a stress triaxiality $T = 1$ (a), $T = 2$ (b) and $T = 3$ (c)*

Figure 4 shows cell model results for an overall stress triaxiality of $T = 1$, a strain rate hardening parameter $m = 0.1$ and several values of the initial void volume fraction.

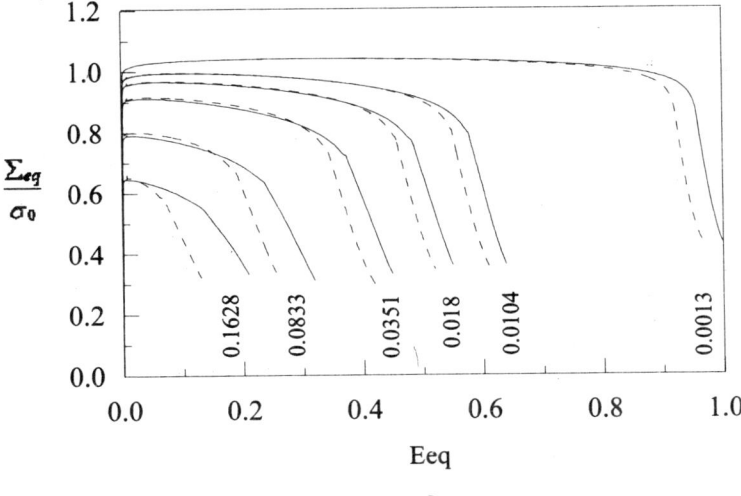

a

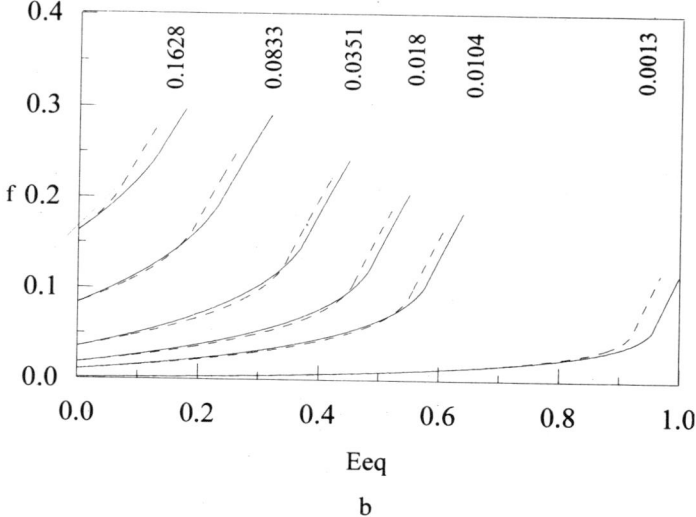

b

Figure 4. *Cell model results for a cylindrical void (continuous line) and a spherical void (discontinuous line) with a strain rate hardening exponent m = 0.10 and a stress triaxiality T = 1 for different initial void volume fractions f_0. (a) Macroscopic effective stress vs. effective strain response. (b) Void volume fraction vs. macroscopic effective strain*

Figure 4a shows equivalent stress vs. equivalent strain curves. The maximum stress decreases if the initial void volume fraction increases. The maximum equivalent stress does not depend on the initial void shape. The critical strain E_c corresponding to the drop of the overall equivalent stress is smaller for initially spherical voids than for initially cylindrical voids. The difference between the initially spherical and initially cylindrical voids increases if the initial void volume fraction increases. Figure 4b shows the evolution of the void volume fraction as a function of the overall equivalent strain. The void volume fraction increases more rapidly if the initial void volume fraction is increased. Initially, spherical voids grow more rapidly than initially cylindrical voids. The void volume fraction at the onset of uniaxial straining does not depend on the initial void shape. The difference between initially spherical and initially cylindrical cavities increases if the initial void volume fraction is increased.

3.1.2. *Void shape change*

Figure 5 shows cell model results corresponding to a stress triaxiality of T = 1 and a strain rate hardening parameter m = 0.1. The initial void shape and the corresponding void shape at the onset of uniaxial straining are shown. Figure 5a corresponds to an initial void volume fraction $f_0 = 0.0013$ and Figure 5b corresponds to an initial void volume fraction $f_0 = 0.0833$. The void shapes at the onset of uniaxial straining (corresponding to the overall stress drop) are very close independently of the initial void shape. A high initial void volume fraction (f_0 = 0.0833) leads to small differences in the inter-void ligament size between initially spherical and initially cylindrical voids. Small initial void volume fractions lead to the same inter-void ligament size for initially spherical and cylindrical cavities at the onset of uniaxial straining.

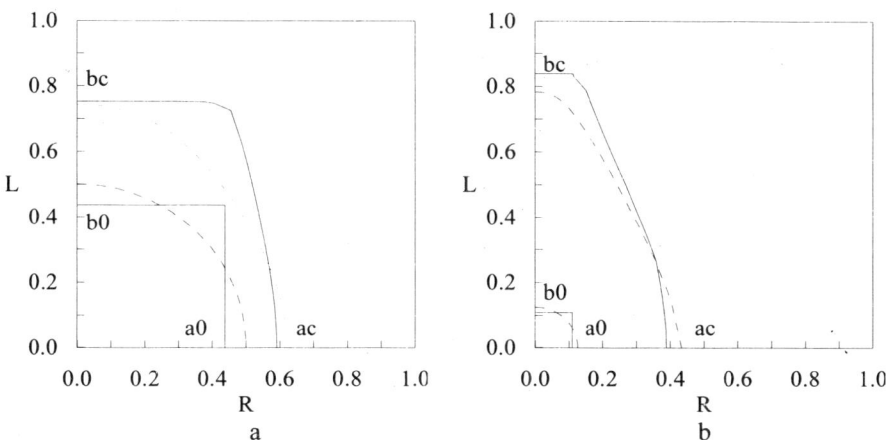

Figure 5. *Initial void shapes(a0 ,b0) and void shapes at coalescence (ac, bc) with a stress triaxiality T = 1, a strain rate hardening exponent m = 0.10 and initial void volume fractions f_0 = 0.0833 a) and f_0 = 0.0013 b)*

3.2. m-dependent porous material model

3.2.1. *m-dependent porous material model predictions*

Figures 6 and 7 show the comparison between cell model calculations and the m-dependent porous material. Figure 6 corresponds to an initial void volume fraction of 0.0013 and several values of the strain rate hardening parameter m. A small value of f_0 leads to very close results for initially spherical and cylindrical voids. Obviously, the growth of initially spherical or cylindrical voids may be described by the same m-dependent porous material model. The same optimisation parameters q_1, q_2 and f_c have been chosen for all initial void shapes. Figure 7 corresponds to a stress triaxiality of T = 1, a strain rate hardening parameter m = 0.1 and several values of the initial void volume fraction f_0. The difference between initially spherical and initially cylindrical voids increases if the initial void volume fraction increases. But the m-dependent porous material model still gives a very good approximation of the overall equivalent stress and void volume fraction.

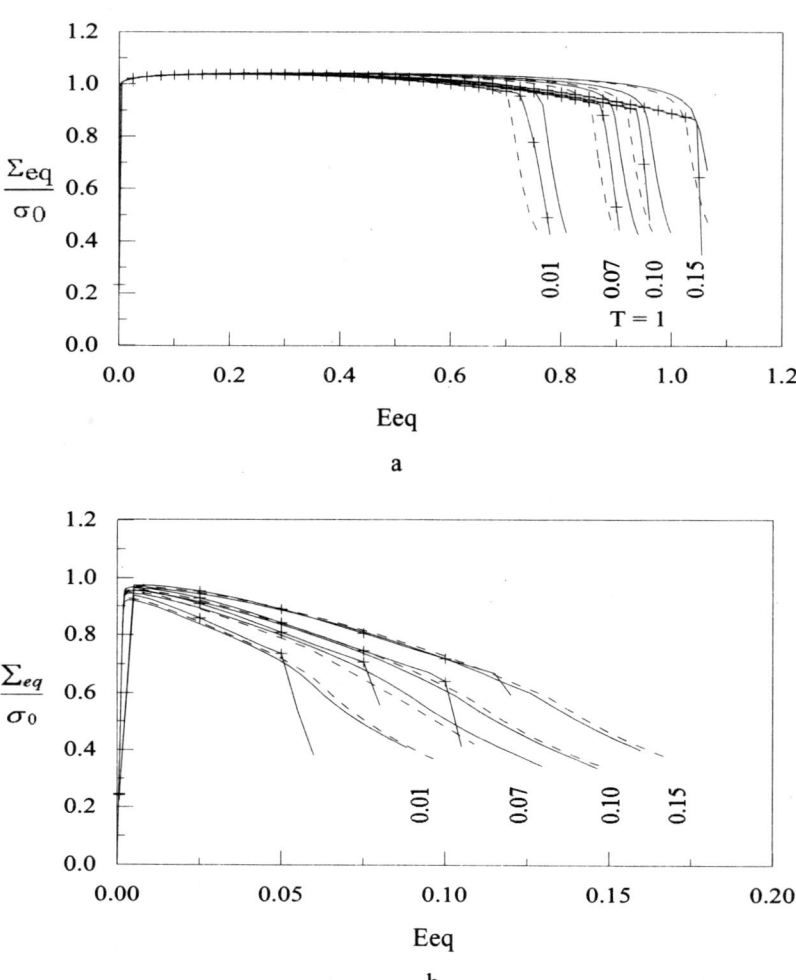

Figure 6. *m-dependent porous material model predictions (+) compared to cell model results for a spherical void (discontinuous) and a cylindrical void (continuous) with an initial void volume fraction $f_0 = 0.0013$ Macroscopic effective stress vs. effective strain response for a stress triaxiality $T = 1$ a), $T = 3$ b)*

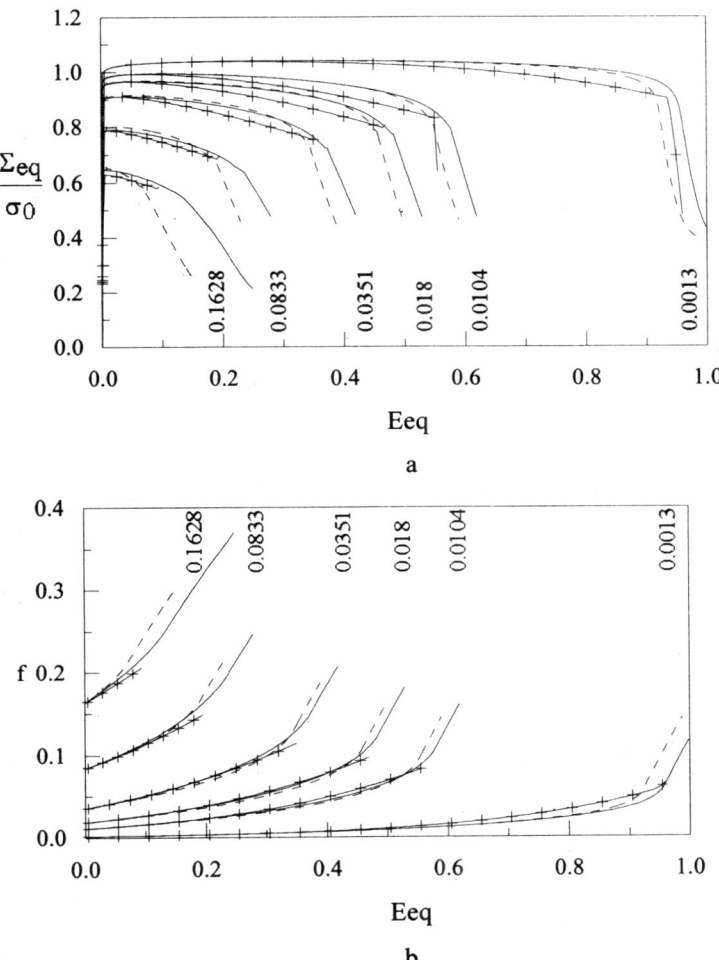

Figure 7. *m-dependent porous material model predictions (+) compared to cell model results for a spherical void (discontinuous line) and for a cylindrical void (continuous line) with a strain rate hardening exponent m = 0.10 and a stress triaxiality T = 1 for different initial void volume fraction*

3.2.2. Optimisation parameters in the m-dependent porous material model. Variation of the critical void volume fraction

Throughout this work, constant values for q_1 and q_2 and f_f have been used. The values of q_1 and q_2 were chosen in order to be close to the equivalent stress-strain curves before the onset of localization. The parameters q_1 and q_2 were chosen equal to 1.382 and 1.0 respectively. The cell model calculations have shown no major influence of void geometry, remote loading (i.e. the stress triaxiality) and the strain rate hardening exponent m on void growth after flow localization. Thus, the void volume fraction f_f at final failure was chosen equal to 0.085.

The cell model results have shown that an increase in the strain rate sensitivity leads to an increase of the void volume fraction at the onset of flow localisation in the ligaments between voids. The cell model calculations have also shown that smaller initial void volume fractions lead to smaller void volume fractions at the onset of flow localization. Thus, the critical void volume fraction f_c decreases if the initial void volume fraction f_0 decreases (Koplick and Needleman, 1988; Tvergaard, 1990; Klöcker and Tvergaard, 2000). Figure 8 shows the variation of the critical void volume fraction f_c at the onset of uniaxial straining with the remote stress triaxiality, the strain rate hardening parameter m and the initial void volume fraction f_0.

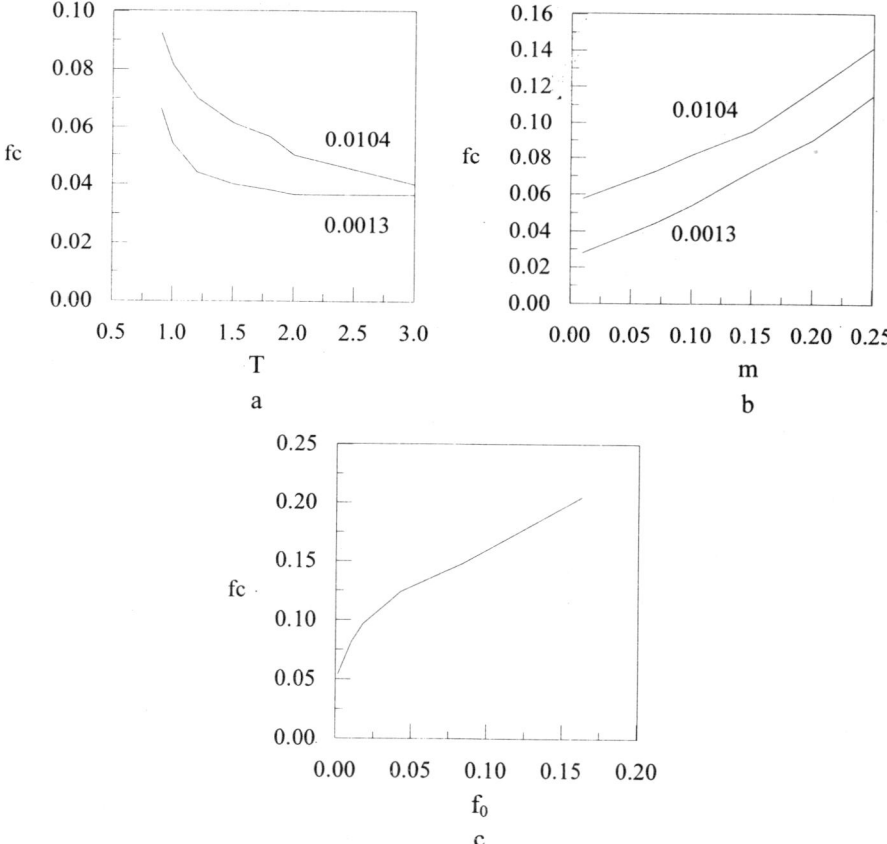

Figure 8. *Critical void volume fraction f_c vs. a) stress triaxiality T (f_0 = 0.0013 and 0.0104, m = 0.10), b) strain rate hardening exponent m (f_0 = 0.0013 and 0.0104, T = 1), c) initial void volume fraction f_0 (m = 0.10, T = 1)*

4. Discussion

The present numerical study of void growth to coalescence is analogous to that of Koplik and Needleman (Koplik and Needleman, 1988), but in the present investigation the focus is on deformation at elevated temperatures corresponding to hot working conditions (Klöcker and Tvergaard, 2000). In real hot working situations voids nucleate from second phase particles and rarely have spherical or ellipsoidal shapes. Thus, the present work goes a step further in the application of micromechanical models to real materials.

The present work analyses the effect of hot working process parameters i.e. the temperature (through the strain rate hardening parameter m) and the stress triaxiality as well as the material characteristics (through the void geometry) on void growth to coalescence.

Void growth to coalescence has been investigated by two models. *Cell model calculations* lead to deeper insight in the overall cell behaviour without any particular assumption on the velocity field. *The m-dependent porous material model* is based on an extension of Gurson's micro-mechanical approach to complex void shapes.

4.1. *Cell model results*

The cell model calculation show a delay in fracture strain if the remote stress triaxiality decreases and if the strain hardening parameter (the temperature) increases. Initially, spherical voids lead to slightly larger strains to fracture than initially cylindrical voids. The influence of the initial void shape is almost negligible at all values of the remote stress triaxiality.

4.2. *m-dependent porous material model*

As shown in Klöcker and Tvergaard (Klöcker and Tvergaard, 2000), to represent correctly the effect of rather large m-values a specially derived stress potential has to be used. When introducing the $f^*(f)$ model to approximately represent the effect of final failure by void coalescence it is found that the critical value of the void volume fraction, fc, should depend on m-value. In addition, as is also known from room temperature behaviour, the value of fc should depend on the initial void volume fraction. These results are very close to those obtained by Klöcker and Tvergaard (Klöcker and Tvergaard, 2000). A single model describing void growth to coalescence for initial void shapes reaching from a sphere to a cylinder has been presented. This point is of primary importance when trying to apply the modified Gurson model to real material with rather complicated void shapes.

5. References

Brocks W., Sun D.Z., Hönig A. "Verification of the transferability of micromechanical parameters by cell model calculations with visco-plastic materials", *Int. J. of Plasticity*, Vol. 11, 1995, pp. 971–989.

Gurson A.L., "Continuum theory of ductile rupture by void nucleation and growth – I. Yield criteria and flow rules for porous ductile media", *J. Engng. Materials Technol.*, Vol. 99, 1977, pp. 2–15.

Klöcker H., Tvergaard V., "Void growth and coalescence in metals deformed at elevated temperature", International Journal of Fracture, Vol. 106, 2000, pp. 259–276.

Koplick J., Needleman, A., "Void growth and coalescence in porous plastic solids", *Int. J. Solids Structures*, Vol. 24, 1988, pp. 835–853.

Leblond J.B., Perrin G., Suquet P., "Exact results and approximate models for porous viscoplastic solids", *Int. J. of Plasticity*, Vol. 10, 1994, pp. 213–235.

Needleman A., "Void growth in an elastic-plastic medium", *J. Appl. Mech.* Vol. 41, 1972, pp. 964–970.

Steglich D., Brocks W., "Micromechanical modelling of damage and fracture of ductile materials", *Fatigue & Fracture of Engineering Materials & Structures*, Vol. 21, 1998, pp. 1175–1188.

Tvergaard V., "Influence of voids on shear band instabilities under plane strain conditions ", *Int. J. Fracture*, Vol. 17, 1981, pp. 389–407.

Tvergaard, V., "On localisation in ductile materials containing spherical voids", *Int. J. Fracture*, Vol. 18, 1982, pp. 237–252.

Tvergaard V., "On the creep constrained diffusive cavitation of grain boundary facets", *J. Mech. Phys. Solids*, Vol. 32, 1984, pp. 373–393.

Tvergaard V., Needleman, A., "Analysis of the cup-cone fracture in a round tensile bar", *Acta Metallurgica*, Vol. 32, 1984, pp. 157–169.

Tvergaard V., "Material failure by void growth to coalescence", *Advances in Applied Mechanics*, Vol. 27, 1990, pp. 83–151.

Chapter 4

Lattice Misorientations in Titanium Alloys

Modelling the Origins of Defects

Nathan R. Barton
Department of Mechanical and Aerospace Engineering, University of California, San Diego, USA

Paul R. Dawson
Sibley School of Mechanical and Aerospace Engineering, Cornell University, New York, USA

1. Introduction

Defects can occur from hot working operations of titanium alloys in relation to several features of their microstructures. One possibility is for voids to form as a consequence of cavitation. These may be associated with grain boundaries where the lattices are highly misoriented, especially if the hexagonal close packed c-axis lies parallel to the tension direction in one grain [BIE 01]. For alloys with lamellar microstructures, the failure of the hot working process to transform the lamellae into more equiaxed grains can leave the material with what is regarded as a defect, namely the retained lamellar structure. Dynamic globularization denotes the breakup of the lamellar microstructure that is an intended goal of the hot working operations. Recent articles report on globularization and cavitation observed experimentally using orientation imaging microscopy (OIM) [BIE 02], as well as the flow softening that is apparent during the breakdown forming operations. With the purpose of better understanding the origins of processing defects in titanium alloys, finite element simulations have been conducted at the scale of the lamellar microstructure. Results from [BIE 02] and from the references contained therein in large part motivate the issues examined via the modeling reported here.

With lamellae of hexagonal close packed α and body centered cubic β, the colony microstructures exhibit considerable deformation heterogeneity. This heterogeneity arises from the difference in overall strength of the phases, with the α phase being stronger, and from the highly anisotropic plastic response of the α phase. Further, shape instabilities of plate-like microstructures drive the globularization mechanisms. Deformation helps to destabilize the plates by disrupting the planar surfaces and by introducing internal misorientation boundaries with high surface energy [MAR 80, MAL 89, COU 89]. Methods that utilize mean field assumptions, such as the lamellar-structures self consistent framework [LEB 99], allow for bulk texture predictions taking account of some microstructural interactions, but they do not capture details of the deformation heterogeneity that are of interest here. The finite element method, together with a crystallographic constitutive model, may be used to resolve the heterogeneity of the deformation field within the lamellar structure and to predict both of these factors. Such results may ultimately guide efforts, such as those reported in [PET 80] and elsewhere, to control microstructural evolution during processing of titanium alloys.

2. Simulations and results

We simulate hot isothermal compression at a strain rate of $0.1\,\mathrm{s}^{-1}$ – a rate typical of that for billet forging in a hydraulic press. Time at temperature is short enough that microstructural coarsening is not a significant issue and need not be included in the model. Under these conditions, dynamic globularization is not appreciable until strains of roughly 0.5 [SHE 99]. For Ti-6Al-4V, the α/β transus temperature is near 995° C and at 815° C the composition is 80%α and 20%β. All simulations reported here are at 815° C. Preliminary simulations conducted at 900° C, and thus having a

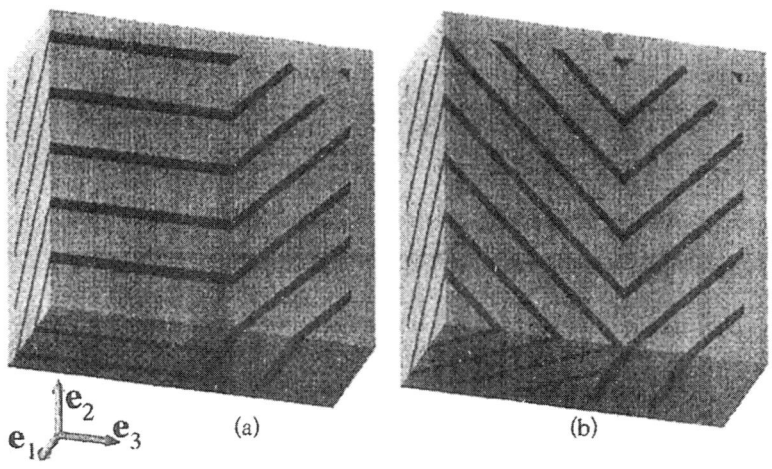

Figure 1. *Initial meshes, showing α (lighter) and β (darker) phase distributions*

larger volume fraction of β, suffered severe mesh distortion due to localization of plastic flow at strains less than 1%.

Descriptions of the finite element and constitutive model used here may be found in [MAR 98, BAR 01a]. Given the temperature and strain rates involved, the hot forming operations of interest lie in the dislocation glide dominated regime of deformation [SEM 99]. A multiplicative decomposition of the deformation gradient allows for plasticity by dislocation glide (slip) on a specified set of slip systems and for anisotropic elasticity. Anisotropic elastic moduli at 815° C are the same as those used in [BAR 01a]. In the α phase, strengths of the prismatic, basal, and pyramidal slip systems remain in the relationship $\tau_c^{pri} = 0.7\tau_c^{bas} = 0.22\tau_c^{pyr}$. Initially, $\tau_c^{pri} = 50$ MPa and $\tau_c^{pg} = 15$ MPa with τ_c^{pg} being the strength of the pencil glide slip systems in the β phase. A Voce law determines the slip system hardening rate, and parameters are chosen so that the hardening saturates quickly. Saturation values of τ_c^{pri} and τ_c^{pg} are 70 MPa and 20 MPa, respectively. The model does not distinguish between slip systems based on their relationship to the habit plane or their slip line length.

Figures 1a and 1b show the initial microstructures, each composed of nearly $210 \cdot 10^3$ ten-noded tetrahedral finite elements. In this idealization of colony microstructure, each lamella traverses its colony. Lattice orientations in the lamellae within a colony follow the Burgers orientation relationship [BUR 34, NEW 53, MCQ 63] – $(0001)_\alpha \parallel (110)_\beta$ and $[\bar{2}110]_\alpha \parallel [\bar{1}1\bar{1}]_\beta$. Under this orientation relationship, twelve α variants may form within a given parent β grain. The habit plane for the phase boundary is taken to be $\{334\}_\beta$.

All of the β in the mesh depicted in Figure 1a has the same initial orientation and the domain is meant to correspond to a portion of a parent β grain. Parts of three co-

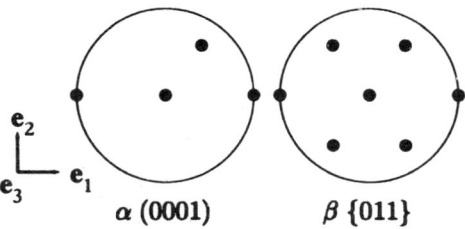

Figure 2. *Initial pole locations*

lonies are contained within the domain, and the domain is bordered by "primary" α on three sides. On the other three sides, those visible in the figure, mirror boundary conditions are applied. The initial β orientation is chosen so that it is invariant under the mirror boundary conditions. Thus one eighth of a parent β grain is simulated. Ideally such a domain would be embedded within a larger aggregate, and would involve higher aspect ratios of lamellae, but available computer resources limit the size of the mesh. Note that we do not include the thin layer of β experimentally observed to lie between the primary (grain boundary) α and the lamellae [SEM 99]. Compression occurs along the e_3 direction. As noted above, the three visible faces in Figure 1à have mirror boundary conditions. The top and back faces are traction free, and on the remaining face (at the right) we also apply a mirror boundary condition.

The initial pole figures in Figure 2 illustrate the $(0001)_\alpha \parallel (110)_\beta$ part of the Burgers orientation relationship for the α variants in Figure 1à. In the α pole figure, the primary α corresponds to the upper right pole, with the other poles being the colonies. The colony with its habit plane normal to the loading axis (e_3) has its (0001) pole (c-axis) along the loading axis. Due to the higher strength of the pyramidal slip systems, this colony has comparatively high resistance to plastic deformation.

We construct the mesh depicted in Figure 1b in similar fashion. However, the colony at the bottom left has been buckled before the start of the simulation. Lattice orientations in both the α and β lamellae in that colony have been rotated about an axis in the habit plane to approximate a buckling mode of deformation. We rotate the habit plane in the mesh by a corresponding amount. This is done to investigate the flow softening produced by such a buckling mode.

2.1. *Globularization*

The discretized microstructure was loaded in compression and the deformations accumulated over incremental steps in strain. A strain of 6.5% was achieved in the non-buckled (reference) mesh before mesh distortion became too severe to continue the simulation. Although the final accumulated strain is small in comparison to experimental strain levels, it is illustrative to discuss the results of this type of finite element simulation and how they reveal trends in the evolution of microstructure.

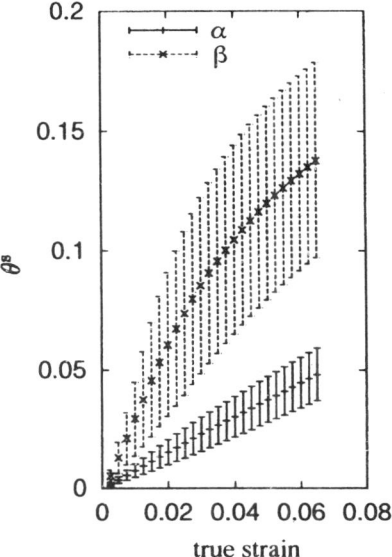

Figure 3. *Average value by phase of θ^s versus true strain; bars show standard deviation of θ^s*

Over the deformation sequence, the lattice orientation within each finite element is updated according to the equations for crystal plasticity. This provides data for the spatial variation in lattice orientation over each lamellae as well as misorientations across phase boundaries. Using the methods described in [BAR 01b], lattice orientation within the lamellae is characterized by spread about an average value and by the spatial correlation of this spread. The typical angular deviation from the average orientation, in a volume average sense, within the lamella is then denoted θ^s. Figure 3 shows the evolution of θ^s with strain. The results in [BAR 01a] include θ^s evolution for an equiaxed titanium alloy with similar ratios of strength among the slip systems but with no hardening of the slip systems. The general trend in [BAR 01a] was toward more rapid orientation spread (increase in θ^s), particularly in the β phase, with increasing volume fraction of α. In [BAR 01a], the 70%α/30%β alloy most closely matches the composition here. Higher initial rate of θ^s accumulation seen for the lamellar microstructure may be due to the difference in microstructure or to the higher volume fraction of α. Rate of θ^s accumulation in the α phase is nearly the same in the lamellar and equiaxed simulations.

For globularization, not only the degree of lattice orientation spread but also the ordering of this spread is of importance. Correlation between spatial position in the lamellae and the lattice misorientation provides an indication of the ordering of the spread. A second order correlation tensor may be constructed such that its three sin-

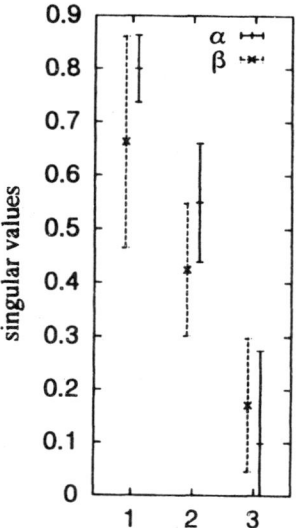

Figure 4. *Singular values of* **X** *at* 6.5% *true strain ; bars show standard deviation*

gular values range strictly between zero and unity, with values near unity indicating perfect correlation between a spatial direction and a misorientation axis [BAR 01b]. The distributions of these singular values over the lamellae of both phases are shown in Figure 4. Misorientation structure within the α lamellae is found to be more ordered than in the β lamellae, so that for the same amount of spread θ^s the β lamellae have more internal misorientation boundaries.

Figure 5a shows elements in which the lattice orientation is misoriented by at least 24°, which is half of the observed maximum value. This more direct observation of the misorientation structure indicates that high misorientations develop near the colony boundary having its c-axis aligned with the compression axis. Relatively high misorientations also develop around the boundary of the mesh due to the influence of the boundary conditions. These observations are supported by Figures 6 and 7 which show angle of misorientation from the lamella average orientation.

All factors indicate that deformation is concentrated in the β phase and that it develops internal misorientation boundaries at a more rapid rate than the α early in the deformation. If such trends continued at larger strains, shape instability would be greater in the β lamellae. However, the overall strain accumulation achieved in the simulations was relatively small and only a small sampling of colonies was investigated. Thus, the data available from these simulations do not warrant any stronger conclusions regarding lattice orientation breakup in the lamellar microstructure simulations.

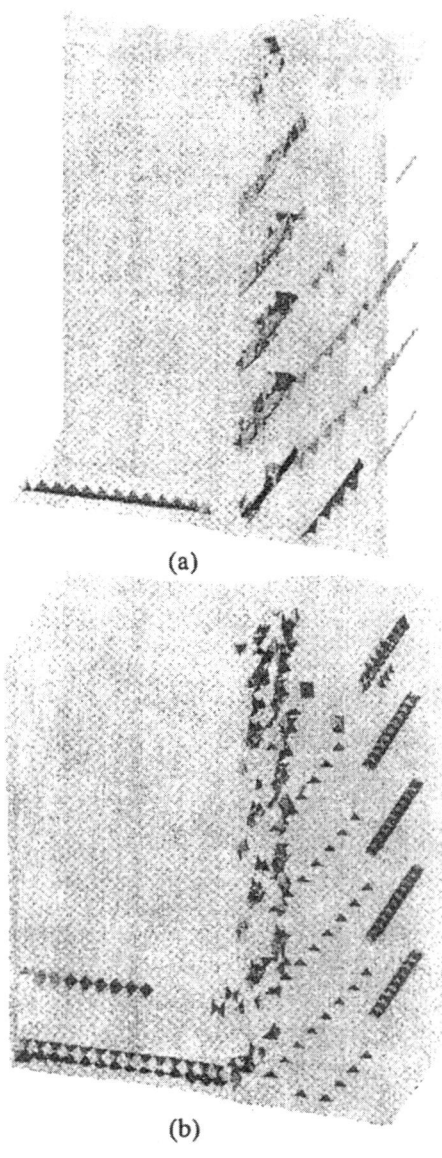

Figure 5. At 6.5% strain, (a) elements misoriented more than 24° and (b) elements having at least one quarter of the maximum tensile pressure

44   Prediction of Defects in Material Processing

Figure 6. *Misorientation angle from lamella average, with lighter shades representing higher misorientation angles*

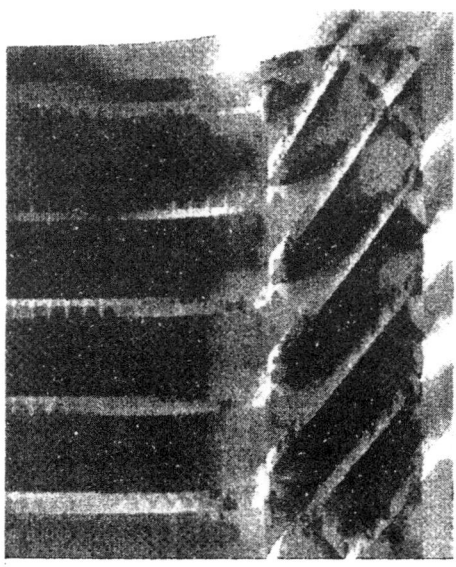

Figure 7. *Misorientation angle from lamella average, on a slice through the mesh interior*

2.2. Cavitation

Workability in lamellar titanium alloys is frequently limited by the onset of cavitation, which is observed experimentally even in compression tests. It is known that compressive deformation modes in materials with highly anisotropic microstructural constituents can produce substantial tensile hydrostatic stresses (tr$(\sigma)/3$) [DAO 96]. Elements in which the tensile hydrostatic stress has reached at least 58 MPa, one quarter of the maximum value at 6.5% strain, are shown in Figure 5b. Notice that both the tensile hydrostatic stress and the amount of misorientation have high values along the interface of the c-axis colony. All elements appearing in both Figures 5a and 5b lie along the c-axis colony interface. There are very few such elements in comparison to the total numbers of elements in this microstructure (over 200,000).

The combined action of misorientation structure and stress concentration increase the likelihood of cavitation, with misorientation structure associated with disorder in the material and lower packing density. Lattice curvature and greater disorder in the material result in a lower mean free path for dislocation motion. The corresponding increase in pinning rate drives an increase in dislocation density [BEA 00]. Reduction in dislocation mean free path also results in a higher resistance to deformation and elevated flow stress, so that the material is able to support larger stresses. Higher dislocation densities and any localization of flow would also increase dislocation annihilation rates. In the presence of a tensile hydrostatic stress, this could lead to vacancy accumulation and to voids of dimension comparable to the length of a Burgers vector. Hydrostatic stress driven vacancy migration and concentration may be regarded as an incipient damage mechanism. High disorder in the material would feed such a mechanism, with dislocations moving to a void surface contributing to the volume of the void [DAV 77].

A number of variants on the above scenario result in microscale damage to the material. In general a reduction in order in the material diminishes the ability of dislocations to glide and reduces ductility. It is significant that zones with high misorientation appear in the same region as tensile hydrostatic stresses – at the boundary of a colony with its c-axis aligned with the compression direction. Experimentally observed cavitation sites often involve such boundaries [BIE 02].

2.3. Stress response and flow softening

Over the strain range simulated using the reference mesh, the compression stress, formed from a volume average over the domain, increases monotonically. Over the same strain range an upper bound mean field assumption (iso-strain, or extended Taylor) shows a peak stress at roughly 3% strain followed by a slight stress drop-off due to lattice reorientation. In addition to the initial orientations of the volume fractions found in the finite element mesh, the upper bound simulation includes the orientations induced under the mirror boundary conditions.

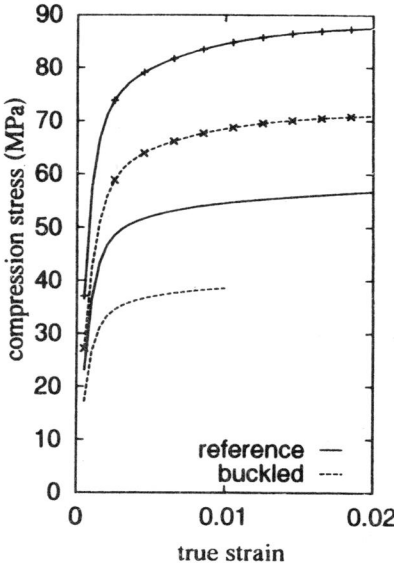

Figure 8. *Upper bound (marked with symbols) and finite element stress response*

Figure 8 shows the stress response in the reference and buckled cases. The buckled finite element simulation was conducted only to a strain sufficient for flow stress determination. In proportion to the reference flow stress, the finite element simulation predicts a larger drop in the flow stress for the buckled microstructure. While the drop in flow stress is comparable to that observed experimentally, it results from rotating a large volume fraction of material away from a stiff orientation. Even if such a change in the orientation distribution were experimentally justified, flow softening is typically observed in the strain range covered by the reference finite element simulation conducted here. Thus it seems that the factors most directly responsible for the observed flow stress drop are not captured by these simulations. While buckling does result in a softer response, the fact that the simulations do not predict buckling and corresponding flow softening at smaller strains suggests that other mechanisms are responsible for the experimentally observed softening.

3. Discussion

Materials with highly anisotropic microstructural constituents and with phases of disparate strength exhibit heterogeneous deformation fields. Heterogeneity helps to drive some desirable microstructural evolution, such as globularization, but may also contribute to cavitation defects. Even though the model and parameters used here do not develop the observed flow softening, they do reveal trends that have important implications relative to the initiation of defects. Namely, we see a strong correlation between the occurrence of high lattice misorientation near grain boundaries and high tensile stress. These zones possess the driving elements for void nucleation: vacancies in the lattice structure and high levels of hydrostatic tension. Introduction of a coupling between lattice curvature and dislocation accumulation, such as that used in [BEA 00], should be considered for any further investigation of damage driven by misorientation structure.

Some relatively simple avenues are available that might enable the simulations to capture the flow softening behavior. For example, it is possible to initialize the slip system strength to be larger than the saturation value appearing in its evolution equation. This could be justified if the prior processing history resulted in a dislocation density that is higher than specified for the strain rate in compression loading. However, rate sensitivity of the saturation value is not typically large, and it is doubtful one could obtain the desired result with reasonable initial values of slip system strength. A "loss of Hall-Petch strength" [SEM 01] has been suggested to explain the drop in flow stress, and phenomenological models might capture the macroscopically observed trends. Phenomenological constitutive models have been proposed that do not relate the drop in macroscopic flow stress to a particular microstructural feature [BLA 98]. However, at the length scales simulated here, it may be necessary to introduce a state variable model that is more directly connected to dislocation motion and accumulation [MEC 81, EST 84, ARS]. Significant questions remain concerning the most appropriate models at the scale of lamella within colonies.

Localization and heterogeneity of the deformation also continue to present sizable hurdles. Three dimensional remeshing constitutes one possible solution. This approach is especially challenging, however, in the face of material with internal boundaries and detailed state variable descriptions. Eulerian finite element methods offer another solution to the problem of large localized deformations. Sufficient progress has been made on boundary tracking methods in micromechanics simulations [BEN 98, BEN 02]. In this case, advection of the variables used in the state descriptions remains a serious issue.

Acknowledgements

This work has been supported by the Office Naval of Research (contract NOOO 14-95-1-0314) through funding from DARPA. The authors are also grateful for the use of computer resources at the Cornell Theory Center.

4. Bibliography

[ARS] ARSENLIS A., PARKS D. M., « Modeling the evolution of crystallographic dislocation density in crystal plasticity », To appear in *J. Mech. Phys. Solids*.

[BAR 01a] BARTON N. R., DAWSON P. R., « On the spatial arrangement of lattice orientations in hot-rolled multiphase titanium », *Modelling Simul. Mater. Sci. Eng.*, vol. 9, 2001, p. 433–463.

[BAR 01b] BARTON N. R., DAWSON P. R., « A methodology for determining average lattice orientation and its application to the characterization of grain substructure », *Metall. Mater. Trans. A*, vol. 32, 2001, p. 1967–1975.

[BEA 00] BEAUDOIN A. J., ACHARYA A., CHEN S. R., KORZEKWA D. A., STOUT M. G., « Considerations of grain-size effect and kinetics in the plastic deformation of metal polycrystals », *Acta Mater.*, vol. 48, 2000, p. 3409–3423.

[BEN 98] BENSON D. J., « Eulerian finite element methods for the micromechanics of heterogeneous materials : dynamic prioritization of material interfaces », *Compt. Methods Appl. Mech. & Eng.*, vol. 151, 1998, p. 343–360.

[BEN 02] BENSON D. J., « Volume of fluid interface reconstruction methods for multimaterial problems », *Adv. Appl. Mech.*, vol. 55, 2002.

[BIE 01] BIELER T. R., SEMIATIN S. L., « The effect of crystal orientation and boundary misorientation on tensile cavitation during hot deformation of Ti-6Al-4V », *Lightweight Materials for Aerospace Applications*, Warrendale, PA, 2001, TMS.

[BIE 02] BIELER T. R., GLAVICIC M. G., SEMIATIN S. L., « Using OIM to Investigate the Microstructural Evolution of Ti-6Al-4V », *JOM-J. Min. Met. Mat. S.*, vol. 54, n° 1, 2002, p. 31–36.

[BLA 98] BLACKWELL P. L., BROOKS J. W., BATE P. S., « Development of microstructure in isothermally forged Nimonic alloy AP1 », *Mater. Sci. Technol.*, vol. 14, 1998, p. 1181–1188.

[BUR 34] BURGERS W. G., « On the process of transition of the cubic-body-centered modification into the hexagonal-close-packed modification of zirconium », *Physica*, vol. 1, 1934, p. 561–586.

[COU 89] COURTNEY T. H., MALZAHN KAMPE J. C., « Shape instabilities of plate-like structure – II. Analysis », *Acta Metall.*, vol. 37, 1989, p. 1747–1758.

[DAO 96] DAO M., KAD B. K., ASARO R. J., « Deformation and fracture under compressive loading in lamellar TiAl microstructures », *Phil. Mag. A*, vol. 74, n° 3, 1996, p. 569–591.

[DAV 77] DAVISON L., STEVENS A. L., KIPP M. E., « Theory of spall damage accumulation in ductile metals », *J. Mech. Phys. Solids*, vol. 25, n° 1, 1977, p. 11–28.

[EST 84] ESTRIN Y., MECKING H., « A unified phenomenological description of workhardening and creep based on one-parameter models », *Acta Metall.*, vol. 32, n° 1, 1984, p. 57–70.

[KIM 80] KIMURA H., IZUMI O., Eds., *Titanium '80, science and technology : proceedings of the Fourth International Conference on Titanium*, Kyoto, Japan, 1980, Metallurgical Society of AIME.

[LEB 99] LEBENSOHN R., « Modelling the role of local correlations in polycrystal plasticity using viscoplastic self-consistent schemes », *Modelling Simul. Mater. Sci. Eng.*, vol. 7, 1999, p. 739–746.

[MAL 89] MALZAHN-KAMPE J. C., COURTNEY T. H., LENG Y., « Shape instabilities of plate-like structure – I. Experimental observations in heavily cold worked *in situ* composites », *Acta Metall.*, vol. 37, 1989, p. 1735–1745.

[MAR 80] MARGOLIN H., COHEN P., « Evolution of the equiaxed morphology of phases in Ti-6Al-4V », Kimura, Izumi [KIM 80], p. 1555–1561.

[MAR 98] MARIN E. B., DAWSON P. R., « On modelling the elasto-viscoplastic response of metals using polycrystal plasticity », *Compt. Methods Appl. Mech. & Eng.*, vol. 165, n° 1–4, 1998, p. 1–21.

[MCQ 63] MCQUILLAN M. K., « Phase transformations in titanium and its alloys », *Met. Rev.*, vol. 8, n° 29, 1963, p. 41–104.

[MEC 81] MECKING H., KOCKS U. F., « Kinetics of flow and strain-hardening », *Acta Metall.*, vol. 29, 1981, p. 1865–1875.

[NEW 53] NEWKIRK J. B., GEISLER A. H., « Crystallographic aspects of the beta to alpha transformation in titanium », *Acta Metall.*, vol. 1, 1953, p. 370–374.

[PET 80] PETERS M., LUETJERING G., « Control of microstructure and texture in Ti-6Al-4V », Kimura, Izumi [KIM 80], p. 925–935.

[SEM 99] SEMIATIN S. L., SEETHARAMAN V., GHOSH A. K., « Plastic flow, microstructure evolution, and defect formation during primary hot working of titanium and titanium aluminide alloys with lamellar colony microstructures », *Phil. Trans. R. Soc. Lond. A*, vol. 357, 1999, p. 1487–1512.

[SEM 01] SEMIATIN S. L., BIELER T. R., « The effect of alpha platelet thickness on plastic flow during hot working of Ti-6Al-4V with a transformed microstructure », *Acta Mater.*, vol. 49, n° 17, 2001, p. 3565–3573.

[SHE 99] SHELL E. B., SEMIATIN S. L., « Effect of initial microstructure on plastic flow and dynamic globularization during hot working of Ti-6Al-4V », *Met. Trans. A*, vol. 30A, 1999, p. 3219–3929.

Chapter 5

Interaction of Initial and Newly Born Damage in Fatigue Crack Growth and Structural Integrity

Vladimir V. Bolotin
Mechanical Engineering Research Institute, Russian Academy of Sciences, Moscow, Russia

1. Introduction

Material defects occur both during the processing and the in-service stages of the life of any structural component. Very often process-induced defects become the origin of cracks and crack-like defects developing in the future history of an engineering system, affecting its load-carrying capacity, integrity, and durability. The objective of this paper is to discuss the influence of defects born during the manufacturing stage on the propagation of fatigue cracks under cyclic loading in service life.

The initiation of fatigue cracks growth occurs, as a rule, at the micromechanical level. Up till a certain size a crack is considered as a microdefect without its influence on integrity and load-carrying capacity of a structural component. For example, cracks of which depth is less than 1 mm are usually considered in mechanical engineering and related domains of industry as insignificant. The fail-safe and damage-tolerance philosophy in design allows one to neglect such defects if proper procedures are applied to predict the stable in-service crack growth or to follow this growth by instrumental means [ATL 97].

The theory of fatigue crack growth [BOL 85] is based on the synthesis of damage mechanics and fracture mechanics. Namely, the crack growth process is interpreted as a result of the interaction of generalised forces of fracture mechanics in the process of damage accumulation due to cyclic loading and other, not necessary mechanical, actions. This approach is discussed in detail in [BOL 99] with the applications to fatigue in elastic, elastic-plastic and visco-elastic-plastic solids, to corrosion fatigue, and stress corrosion cracking. Along with traditional materials, fatigue of composites of fibrous and laminar structure is considered.

If process-induced defects are involved, they enter in the model of fatigue crack growth as initial conditions. Namely, it is assumed that a body is initially damaged, and the distribution of damage in the body is known. Then the damage acquired during service life is added to the initial damage that affects all the patterns of fatigue crack growth.

Damage mechanics is an advanced domain of the up-to-date mechanics of solids [KAC 86, CHA 88, KRA 96, LEM 95, MUR 82]. It is developed in two versions, continuum damage mechanics and micromechanics. The first version is based on the notion that damage in a body is described by introducing additional field variables, namely the damage measures are continuously distributed at least in certain parts of the body. The second version considers damage on the level of material's microstructure. Continuum damage mechanics is compatible with all the branches of common solid mechanics (elasticity theory, plasticity theory, etc.), whereas micromechanics of damage is related to solid state physics going to the intrinsic mechanisms of damage nucleation and development. From the viewpoint of fracture and fatigue, continuum damage mechanics is suitable for common structural materials whereas the microstructural approach becomes necessary when we turn to composite materials with their large-scale microstructure. In this paper the models of fatigue crack growth are discussed using the concepts of continuum damage mechanics. Moreover, its simplified version based on the scalar description of damage [KAC 86] is used here. One of the reasons to use such an approach instead of more consistent (tensor) models is that macroscopic cracks, as a rule, are space-oriented formations. For example, considering a planar opening mode crack, one does not need to know the angular distribution of damage around the crack tip. In addition, the scalar damage measures are easier to interpret and easier to evaluate experimentally than tensor measures.

In this paper the concepts of the mechanics of fatigue crack growth are generalised on the case when initial damage is present. Using a few numerical examples, it is demonstrated how initial damage affects the process of fatigue crack growth. Two types of initial damage are considered: the near-surface damage and the damaged strip situated ahead of an initial position of the crack tip. The final results are discussed in the terms of fatigue crack growth rate diagrams referring the crack growth rate to the range of the corresponding stress intensity factor.

2. Analytical fracture mechanics

A generalised version of fracture mechanics based on the notions of analytical mechanics is useful to develop mathematical models of fatigue crack growth and related phenomena. Let a body contain a number of cracks given with parameters $a_1, \ldots, a_m$. Following the recently introduced terminology [MAU 95], [KIE 00], one can name these parameters *configurational coordinates*. Treating the cracks in structural materials as irreversible, we choose the crack parameters satisfying the conditions

$$\delta a_j \geq 0, \quad j = 1, \ldots m \qquad [1]$$

The principle of virtual work for mechanical systems with unilateral constraints requires that in any equilibrium state the condition $\delta W \leq 0$ holds. Here δW is the amount of work performed on any admissible (isochronic) variations of the displacement field and the configurational coordinates. If only the equilibrium states in the common sense are considered, and the Lagrangian generalised coordinates are not subjected to unilateral constraints, the Lagrangian component of the virtual work vanishes. Then the condition of equilibrium of the system "cracked body-loading" with respect to configurational coordinates is presented in the form

$$\delta W_H = \sum (G_j - \Gamma_j)\delta a_j \leq 0 \qquad [2]$$

Here $H_j = G_j - \Gamma_j$ are the generalised configurational forces. In equation [2] and later these forces are split in two parts: the generalised driving forces G_j and the generalised resistance forces Γ_j.

Equation [2] results in the following conditions of system's equilibrium with respect to configurational coordinates:

$$G_j \leq \Gamma_j \quad (j=1,\ldots,m) \qquad [3]$$

There are various ways to satisfy these conditions. We say that the system is in a subequilibrium state if

$$G_j < \Gamma_j \quad (j=1,\ldots,m) \qquad [4]$$

In this case all the cracks in the body are non-propagating. If the equality holds for a part of configurational coordinates such that

$$G_j = \Gamma_j \quad (j=1,\ldots,\mu), \quad G_k < \Gamma_k \quad (k=\mu+1,\ldots,m) \qquad [5]$$

we say that the system is in an equilibrium state with respect to configurational coordinates $a_1,\ldots,a_\mu$. This state may be stable, unstable, or neutral depending on the following terms in the expansion of virtual work with respect to variations δa_j [BOL 85, BOL 99]. In the so-called linear fracture mechanics the attainment of the equality $G_j = \Gamma_j$ usually signifies brittle fracture. In fatigue mechanics we expect, as a rule, the start of the stable crack growth with respect to the coordinate a_j. If only one of inequalities [3] is violated, for example,

$$G_j > \Gamma_j \qquad [6]$$

the system "cracked body-loading" becomes unstable. Then a crack begins to propagate dynamically with respect to coordinate a_j either until a new subequilibrium state or the final fracture.

It is evident that for single-parameter cracks in elastic materials equations [4]–[6] are similar to the relationships between the J-integral and its critical value, material parameter J_C. It is, in general, not valid for cracks given with more than one parameter. The approach based on the principle of virtual work (named analytical fracture mechanics in [BOL 99]) is applicable far beyond the domain of linear fracture mechanics being applicable to the cracks in elastic-plastic, visco-elastic-plastic, ageing, etc. materials.

3. Theory of fatigue crack growth

One of the applications of analytical fracture mechanics is the theory of fatigue crack growth. To develop this theory, the equations of fracture mechanics are to be complemented by equations describing the process of damage accumulation due to cyclic loading. This means that additional field variables are to be introduced, namely, damage measures and additional parameters characterising the conditions along the crack fronts. These new variables may be of scalar or tensor nature. The simplest examples are the scalar damage measure ω and the effective curvature radius ρ at the crack tip. In the general case the functional equations are to be given for the damage field $\omega(\mathbf{x}, t)$ and the tip characteristics $\mathbf{\rho}(t)$:

$$\omega(\mathbf{x},t) = \underset{\tau=t_0}{\overset{\tau=t}{\Omega}} \{\mathbf{s}(\tau), \mathbf{a}(\tau), \mathbf{\rho}(\tau)\}$$

$$\mathbf{\rho}(t) = \underset{\tau=t_0}{\overset{\tau=t}{R}} \{\mathbf{s}(\tau), \mathbf{a}(\tau), \omega(\tau)\}$$

[7]

Here $\mathbf{s}(t)$ is the set of variables describing the loading process, $\mathbf{a}(t)$ is the set of configurational coordinates $a_1, \ldots, a_m$, and t_0 is a certain initial time moment at which the values of ω and $\mathbf{\rho}$ are treated as known. Since damage and crack tip conditions affect the material properties and the stress-strain field, the generalised forces in equations [2]–[6] depend on $\omega(\mathbf{x}, t)$ and $\mathbf{\rho}(t)$. A fatigue crack does not grow if the corresponding configurational forces satisfy equation [4], and begins to grow if the condition given in equation [5] is attained. A jump-wise crack growth or the final fracture occurs when the system "cracked body-loading" attains unstable equilibrium or non-equilibrium state.

In the fatigue analysis (both in the theory and engineering practice), the cycle number N is used as an independent variable instead of time t. It is assumed that any characteristic cycle numbers are large compared with unity; therefore, one may treat N as a continuous variable. Both the driving and resistance forces are considered as functions of N. As applied loads are varying in time, the maximal values of the driving forces enter equations [2]–[6]. The resistant forces are defined as follows:

$$H_j(N) = \sup_{t_{N-1} < t \leq t_N} \{G_j[\mathbf{s}(t), \mathbf{a}(t), \mathbf{\rho}(t), \omega(t)] - \Gamma_j[\mathbf{s}(t), \mathbf{a}(t), \mathbf{\rho}(t), \omega(t)]\}$$

[8]

Here $(t_{N-1}, t_N]$ is the duration of the N-th cycle of loading. Later on, the reference to the supremum in equation [8] is omitted with the proper interpretation of the generalised forces $G_j(N)$ and $\Gamma_j(N)$ as the functions of the cycle number. A crack begins to grow with respect to a_j when at a certain cycle number $N = N_{j*}$ the equality $G_j(N) = \Gamma_j(N)$ is attained. The final fracture with respect to a_j occurs at the cycle number $N = N_{j**}$ at the first attainment the inequality $G_j(N) > \Gamma_j(N)$.

4. Edge cracks in a plate

Consider an opening mode planar edge crack in a thick plate of linearly elastic material under remotely applied tensile stresses $\sigma_\infty(t)$. Treating the problem as two-dimensional, the crack depth a plays the role of the only configurational coordinate. Similar to equation [1] we assume that $\delta a \geq 0$, i.e. the crack is irreversible. The material damage is given by a scalar measure ω varying both in time and in a plate's volume. The interpretation of this measure will be given later. The stress range $\Delta\sigma_\infty(N) = \sigma_\infty^{max}(N) - \sigma_\infty^{min}(N)$ is the most important loading parameter. The stress ratio $R(N) = \sigma_\infty^{min}(N)/\sigma_\infty^{max}(N)$ is considered positive and sufficiently near to unity to avoid crack closure and related effects making the analysis more complicated [BOL 99]. For simplification, it is assumed that damage does not affect the elastic properties given with the Young modulus E and the Poisson ratio ν. Since we deal with a single configurational coordinate, equation [2] takes the form $\delta W_H = (G - \Gamma)\delta a$. The crack does not propagate at $G < \Gamma$, starts to propagate in a stable pattern at $G = \Gamma$, $\partial G/\partial a < \partial \Gamma/\partial a$, and propagates unstable at $G = \Gamma$, $\partial G/\partial a > \partial \Gamma/\partial a$ or $G > \Gamma$.

The next step is to evaluate the generalised forces $G(N)$ and $\Gamma(N)$. We relate them to the unit size across the crack direction. Under the above assumptions, the driving force is equal to the Irwin strain energy release rate depending on the maximal value of the stress intensity factor K within a cycle. The virtual work of fracture in this case is equal to $-\gamma \delta a$ where γ is the specific fracture work equal to the amount of energy spent on the formation of the unit crack area (without counting both crack surfaces). Hence $\Gamma = \gamma$.

It is well known that the specific fracture work is much more sensitive to damage than other material properties, elastic ones in particular. Thus, one may assume that $\gamma = \gamma(\omega)$ where ω is the damage measure. Following the simplest version of continuum damage mechanics, we introduce this measure as a scalar value $0 \leq \omega \leq 1$. The lower boundary correspond to the non-damaged material, the upper one to the completely damaged material. Then the relationship $\gamma = \gamma(\omega)$ gives a physical interpretation of the damage measure indicating a direct way to its experimental assessment. In particular, let us assume that

$$\gamma = \gamma_0(1 - \chi\omega) \qquad [9]$$

Here γ_0 is the specific fracture work in the absence of damage, and $0 < \chi \leq 1$. Then $\gamma = \gamma_0(1 - \chi)$ means the specific fracture work for the completely damaged material.

It is essential that equation [9], as any other equation $\gamma = \gamma(\omega)$, allows one to estimate the current damage by the standard fracture toughness tests applied to damaged specimens. Briefly, ω is the measure of the fracture toughness deterioration due to fatigue.

To close the system of constitutive equations, the equations for the damage accumulation and for the variation of conditions at the crack tip are needed. In this analysis, these equations are taken as in [BOL 85]. The equation of damage accumulation contains three material parameters $\sigma_f > 0$, $\Delta\sigma_{th} \geq 0$, $m > 0$, and the equation of conditions at the tip contains three additional material parameters $\rho_s > 0$, $\rho_b > 0$, $\lambda_\rho > 0$. Here σ_f and $\Delta\sigma_{th}$ characterise the resistance to damage accumulation, ρ_s and ρ_b are the lower and upper bounds for the effective tip radius, i.e. $\rho_s \leq \rho \leq \rho_b$. The length λ_ρ characterises the tendency of the crack tip to sharpening. From the viewpoint of experimenters, the magnitudes $\sigma_f \rho^{1/2}$ and $\sigma_{th} \rho^{1/2}$ are of the order of the material parameters K_f and ΔK_{th} in semi-empirical equations of fatigue crack growth. It is shown [BOL 99] that the exponent m takes the values near the Paris-Erdogan exponent m_P [CAR 94]. Thus, some of the parameters entering the equations may be experimentally assessed, at least by the order of magnitude.

5. Numerical analysis

The following numerical data are used in computations: $E = 200$ MPa, $\nu = 0.3$, $\gamma_0 = 10$ kJ/m², $h = 100$ mm, $a_0 = 1$ mm, $\sigma_f = 5$ GPa, $\Delta\sigma_{th} = 125$ MPa, $m = 4$, $\chi = 1$, $\rho_s = 10$ μm, $\rho_b = \lambda_\rho = 100$ μm. The loading is given by the applied stress range $\Delta\sigma_\infty = 100$ MPa at $R = 0.5$ to avoid crack closure and related effects.

Compared with the numerical analysis presented in [BOL 99], non-zero initial damage as a result of the processing is assumed. The near-surface damage is modelled as exponentially decaying in depth with damage measure at the surface ω_0 and the characteristic length λ_0. In the terms of equation [9] this means the decreasing of the specific fracture work at the surface till $\gamma = \gamma_0(1 - \chi\omega_0)$. Four initial values are considered: $\omega_0 = 0$ (non-damaged material), $\omega_0 = 0.25$, $\omega_0 = 0.5$, and $\omega_0 = 0.75$ at the length parameter $\lambda_0 = 5$ mm.

The numerical results are presented in Figures 1–3. The damage Ω at the crack tip is plotted in Figure 1 with respect to the cycle number N. Curves 1, 2, 3, 4 correspond to the initial damage $\omega_0 = 0, 0.25, 0.5$, and 0.75, respectively. The crack size growth histories numerated analogously are given in Figure 2.

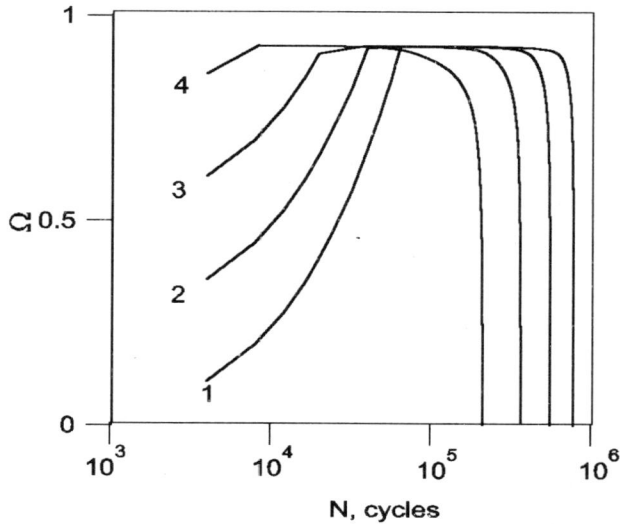

Figure 1. *Damage measure at the crack tip versus the cycle number for the near-surface initial damage: $\omega_0 = 0, 0.25, 0.5, 0.75$ (curves 1, 2, 3, 4 respectively)*

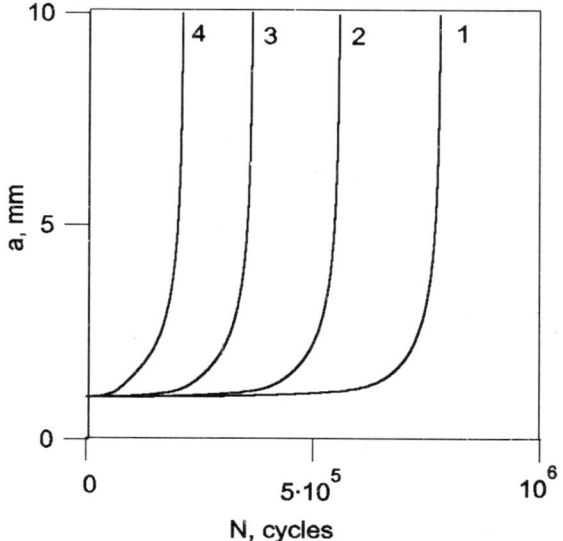

Figure 2. *Fatigue crack depth versus the cycle number (notation as in Figure 1)*

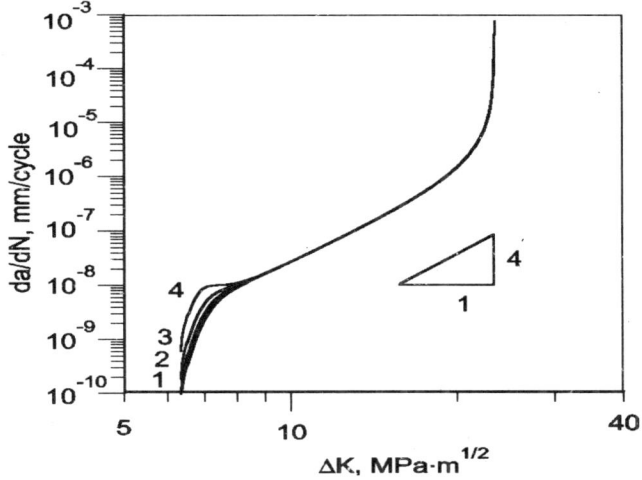

Figure 3. *Fatigue crack growth rate versus the range of the stress intensity factor (notation as in Figure 1)*

These diagrams demonstrate the influence of initial damage on the cycle number N_* at the crack growth start and the cycle number N_{**} at the final failure. Until there is no growth, the damage measure at the tip is increasing. At the attainment the equality $G = \Gamma$ the damage measure at the tip varies very slowly and then begins to decrease. This process accelerates with the approach of the final failure when the damage essentially drops. Note that the measure $\Omega(N)$ relates to the moving crack tip, i.e. corresponds to various material particles. The start of crack growth is not so observable in Figure 4 as the initial growth rate is rather low. In any case, the computations give the cycle numbers at the start of crack growth equal approximately $6.4 \cdot 10^4$, $4.0 \cdot 10^4$, $2.1 \cdot 10^4$, and $0.7 \cdot 10^4$ for the initial damage measure at the surface equal to 0, 0.25, 0.5, and 0.75, respectively. Thus, for the assumed numerical data the fatigue life measured in the cycle numbers up till the start differs within an order of magnitude. But the main contribution to the total fatigue life is made by the stage of crack growth. The cycle numbers at the final fracture for the assumed numerical data are approximately equal to $7.8 \cdot 10^5$, $5.6 \cdot 10^5$, $3.6 \cdot 10^5$, and $2.1 \cdot 10^5$. The difference is less than four times. The shortening of the fatigue life due to initial damage is, however, significant, especially in application to the damage-tolerance maintenance of structures [ATL 97].

The results of fatigue tests are usually presented in the form of crack growth rate diagrams. The diagrams predicted for the assumed numerical data are shown in Figure 3. The rate da/dN is plotted against the range ΔK of the stress intensity factor K. The numeration of curves is the same as in Figures 1 and 2. The general view of the predicted diagrams is the same as of the experimental ones [CAR 94].

The curves may be divided in three stages: the initial, near-threshold stage, the middle stage of the stable growth with a linear (in log-log plot) relationship between da/dN and ΔK, and the final stage at the approach of the final fracture. The above discussed experimental parameters K_f and ΔK_{th} are present in Figure 3 as well as the exponent m from the equation of damage accumulation. Namely, this exponent occurs very close to the exponent m_p in the Paris-Erdogan equation used in the engineering analysis of fatigue. Note that in the middle stage the damaged zone is already crossed by the crack tip. Therefore all the curves merge for every level of initial damage. The rate curves diverge essentially only near the threshold $\Delta K_{th} \approx 6.3 \, \mathrm{MPa} \cdot \mathrm{m}^{1/2}$ differing in an order of magnitude when the curves 1 and 4 are compared.

An example of a damage strip situated ahead of the initial crack tip is given in Figures 4–6. The strip of a uniformly distributed damage is situated at $2 \, \mathrm{mm} \leq x \leq 3 \, \mathrm{mm}$. Curves 1, 2, 3, 4 correspond, as in Figures 1–3, to the damage levels $\omega_0 = 0, 0.25, 0.5,$ and 0.75, respectively. As it is seen in Figures 4 and 5, the damage accumulation and crack growth histories are rather close. The cause is that the crack tip enters the damaged zone at a "mature" stage of the fatigue life. The cycle number at the crack growth start is the same in all the cases, namely $N_* = 6.4 \cdot 10^4$. The total fatigue life varies from $N_{**} = 7.78 \cdot 10^5$ in the absence of initial damage to $N_{**} = 7.50 \cdot 10^5$. The effect on the crack growth rate is more obvious. The local increase of da/dN attains two orders of magnitude. However, these anomalous deviations are rather short (Figure 6).

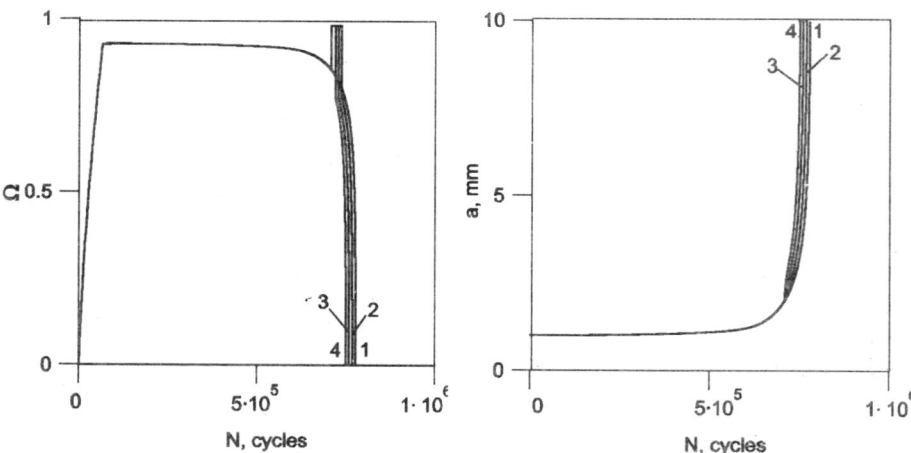

Figure 4. *Damage measure at the crack tip versus the cycle number for the strip-wise initial damage: $\omega_0 = 0, 0.25, 0.5, 0.75$ (curves 1, 2, 3, 4)*

Figure 5. *Fatigue crack depth versus the cycle number (notation as in Figure 4)*

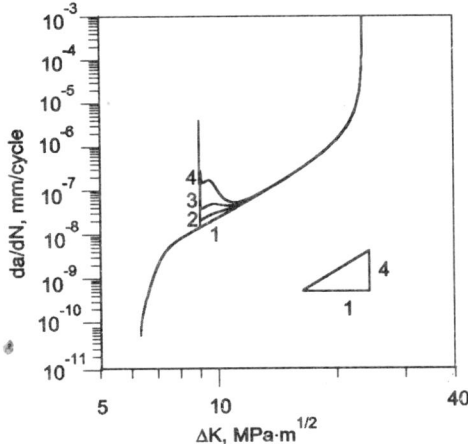

Figure 6. *Fatigue crack growth rate versus the range of the stress intensity factor (notation as in Figure 4)*

The above numerical examples are of illustrative character. Nevertheless, because of the properly chosen numerical data, the final results look like real experimental diagrams, and the leading material parameters such as ΔK_{th}, K_f, ΔK_{fc}, and m_p (as well as the fatigue life measured in cycle numbers) are of the order of magnitude typical for common carbon steels.

6. Summary

It was shown that the theory of fatigue crack growth based on the synthesis of fracture mechanics and continuum damage mechanics is applicable to the prediction of fatigue crack growth in the presence of initial damage due to processing. The effect of initial damage on the cycle numbers till the start of crack growth and till the final failure was demonstrated. For the near-surface initial damage it was shown that, although the characteristic cycle numbers are essentially affected by initial damage, discrepancy of the crack growth diagrams except their initial sections is not significant. The presence of an initial damaged zone ahead of the crack tip does not produce a significant effect on the total fatigue life if this zone is located not too close to the initial crack tip position.

Acknowledgements

This research was partially supported by the Russian Foundation of Basic Research (grant 00-01-00282) and the Alexander von Humboldt Foundation (Germany). Author is grateful to Dr. A.A. Shipkov for the performed numerical computations.

7. References

[ATL 97] Atluri S.N., *Structural integrity and durability*, Tech. Science Press., Forsyth, Georgia, 1997.

[BOL 85] Bolotin V.V., "A unified approach to damage accumulation and fatigue crack growth", *Engng Fracture Mech.*, Vol. 22, 1985, pp. 387–398.

[BOL 99] Bolotin V.V., *Mechanics of fatigue*, CRC Press, Boca Raton, Florida, 1999.

[CAR 94] Carpinteri A. (ed.), *Handbook of fatigue crack propagation in metallic structures*, Elsevier Sciences, Amsterdam, 1994.

[CHA 88] Chaboche J.-L., "Continuum damage mechanics, Parts I and II", *J. Appl. Mech.*, Vol. 55, 1988, pp. 59–72.

[KAC 86] Kachanov L.M., *Introduction in continuum damage mechanics*, Martinus Nijhoff, Dordrecht, 1986.

[KIE 00] Kienzler R. and Herrmann G., *Mechanics in material space with application to defect and fracture mechanics*, Springer, Berlin, 2000.

[KRA 96] Krajcinovic D., *Damage mechanics*, Elsevier, Amsterdam, 1996.

[LEM 95] Lemaitre J., *A course on damage mechanics*, Springer-Verlag, Berlin, 1995.

[MAU 95] Maugin G.A., "Material forces: concepts and applications". *Appl. Mech. Rev.*, Vol. 48, 1995, pp. 213–245.

[MUR 82] Mura T., *Micromechanics of defects in solids*, Martinus Niihoff, The Hague, 1982.

Chapter 6

A Ductile Damage Model with Inclusion Considerations

Three-dimensional Prediction of Cavity Growth in Finite Transformations

Cyril Bordreuil, Jean-Claude Boyer and Emmanuelle Sallé
Laboratoire de Mécanique des Solides, Insa de Lyon, Villeurbanne, France

1. Introduction

The prediction of void growth during the metal forming process is of concern for the reliability of safety parts. Classical potentials for porous materials exist for empty voids, as for the Gurson-Tvergaard yield surface [GUR 77], and underestimate the plastic dilatation under loading with a triaxiality ratio (σ_m / σ_0) of less than 1.5. Few works, dealing with ductile damage, take into account the presence of an inclusion in a void, at the macroscopic level, during large plastic straining. In this paper, a specific plastic potential, fulfilling the mass balance, and including new improvements based on a three-dimensional void growth law is proposed. The development of the present model follows previous work [BOR 01] where the influence of inclusion in voids on damage change has been shown. The shape of the defect is assumed to be aproximated by an ellipsoid, the spin of which as well as the rate of its radii are calculated for the damage change with or without internal inclusion. The real orthotropic behaviour is modelled with special attention to frame indifference.

2. Transformation of an ellipsoid

The geometry of an ellipsoid is defined by the radii of its semi-axes, the orientation in a selected frame, and its center chosen at the center of the frame, for the sake of simplicity. The set of the ellipsoid points M is obtained with the following equation for their coordinates $^0\vec{x}$:

$$(^0\vec{x})^T \mathbf{Q}^T [\mathbf{R}] \mathbf{Q} (^0\vec{x}) = 1 \qquad [1]$$

where $\mathbf{Q}$ is an orthogonal tensor which rotates the eigen directions of the ellipsoid in the working frame and $[\mathbf{R}] = Diag(\frac{1}{r_1^2}, \frac{1}{r_2^2}, \frac{1}{r_3^2})_{eigendirection}$ defines the ellipsoid with its radii r_1, r_2, and r_3.

2.1. *Deformed shape of the ellipsoid after the transformation*

Under homogeneous transformation, the deformed ellipsoid is found to be:

$$^t\vec{x}^T (^t_0\mathbf{F}^{-1})^T \mathbf{Q}^T [\mathbf{R}] \mathbf{Q} (^t_0\mathbf{F}^{-1})^t\vec{x} = 1 \qquad [2]$$

With the theorem of the singular value decomposition, this last expression becomes:

$$^t\vec{x}^T (^t_0\mathbf{F}^{-1})^T \mathbf{Q}^T [\mathbf{R}] \mathbf{Q} (^t_0\mathbf{F}^{-1})^t\vec{x} = {^t\vec{x}^T} (\mathbf{Q}')^T [\mathbf{R}'] \mathbf{Q}' \vec{x}^t \qquad [3]$$

where $\mathbf{Q}'$ and $[\mathbf{R}']$ are respectively equivalent to $\mathbf{Q}$ and $[\mathbf{R}]$ in the new frame. This important result allows one to consider the kinematics of the eigen directions of the cavity different from the eigen directions of loading.

2.2. Rate of the characteristics of the ellipsoid

The rate form of the previous equation is discussed for a formulation of the damage model. In each actual configuration $\mathcal{B}_t$ which changes with time t, the ellipsoid is defined by :

$$^t\vec{x}^T(\mathbf{Q}_t)^T[\mathbf{R}_t]\mathbf{Q}_t(^t\vec{x}) = 1 \qquad [4]$$

From derivation with respect to time, the rate form is found to be :

$$^t\vec{x}^T\mathbf{Q}_t^T[\mathbf{Q}_t\mathbf{L}_t^T\mathbf{Q}_t^T[\mathbf{R}_t] + \mathbf{W}_Q^T[\mathbf{R}_t] + [\dot{\mathbf{R}}_t] + [\mathbf{R}_t]\mathbf{W}_Q + [\mathbf{R}_t]\mathbf{Q}_t\mathbf{L}\mathbf{Q}_t^T]\mathbf{Q}_t(^t\vec{x}) = 0 \qquad [5]$$

with $^t\dot{\vec{x}} = \mathbf{L}(^t\vec{x})$ where $\mathbf{L} =_0^t \dot{\mathbf{F}}(_0^t\mathbf{F})^{-1}$ and $\dot{\mathbf{Q}}_t = \mathbf{W}_Q\mathbf{Q}_t$ where $\mathbf{W}_Q$ is the spin of the ellipsoid in the working frame and $\mathbf{Q}_t\mathbf{L}\mathbf{Q}_t^T$ the gradient of the velocity expressed in the frame of the ellipsoid. As $[\mathbf{R}_t]$ is diagonal and is expressed within the ellipsoidal frame, the previous equation is equivalent to :

$$^t\vec{x}^T\mathbf{Q}_t^T[2([\mathbf{R}_t][\mathbf{L}]_{ellipse})^s + \mathbf{W}_Q^T[\mathbf{R}_t] + [\dot{\mathbf{R}}_t] + [\mathbf{R}_t]\mathbf{W}_Q)\mathbf{Q}_t(^t\vec{x}) = 0 \qquad [6]$$

A first set of differential equations is obtained with the diagonal part of equation 6 :

$$[\dot{\mathbf{R}}_t] = -2Diag(([\mathbf{R}_t][\mathbf{L}]_{ellipse})^s) \qquad [7]$$

so the rate of the radii is only a function of the actual radii and of the diagonal component of the strain tensor $\mathbf{D}$ expressed in the frame of the ellipsoid. The solution for the ellispsoid spin is more complicated to obtain because this rotation is sensitive to material spin, strain rate and shape ratio. The next section deals with this important problem.

3. Constitutive equations

In this section, first, the concept of plastic spin is adapted to the rotational change of the ellipsoidal defect. Then, a plastic potential is built up including a dilatational dissipation.

3.1. Plastic spin

3.1.1. Theory

In this model, the ellipsoid is considered as a substructure with the radii and orientation of the ellipsoid. These developments are based on the representation theory of tensorial isotropic function. First, Wang [WAN 70], gave the conditions that a function has to fulfil in order to be invariant under an orthogonal transformation. Following Liu [LIU 82], if $\vec{n}_1$, $\vec{n}_2$ and $\vec{n}_3$ define the three directions of orthotropy, an invariant function must be written as :

$$\hat{f} = f(\sigma, \vec{n}_1 \otimes \vec{n}_1, \vec{n}_2 \otimes \vec{n}_2) \qquad [8]$$

The most important issue of the present work is the constitutive description of orientation change of the ellipsoid. A skew symmetric tensor is involved in the solution, and as any constitutive equation has to be invariant under rigid body, the spin tensor has to be expressed as a function of generators. This present work is limited to linear generators but the method can be applied to higher degree generators.

The spin is considered as a function of the strain rate, the directions of orthotropy and the shape of the ellipsoid. It can be written as :

$$\mathbf{W}^{ellipsoid}_{material} = \eta_1(\bar{\mathbf{D}}\mathbf{A}_1 - \mathbf{A}_1\bar{\mathbf{D}}) + \eta_2(\bar{\mathbf{D}}\mathbf{A}_2 - \mathbf{A}_2\bar{\mathbf{D}}) + \eta_3(\mathbf{A}_1\bar{\mathbf{D}}\mathbf{A}_2 - \mathbf{A}_2\bar{\mathbf{D}}\mathbf{A}_1) \quad [9]$$

with $\mathbf{A}_1 = \vec{n}_1 \otimes \vec{n}_1$ and $\mathbf{A}_2 = \vec{n}_2 \otimes \vec{n}_2$. $\vec{n}$ are the eigen directions of the ellipsoid and $\bar{\mathbf{D}}$ is the macroscopic plastic strain rate. η_1, η_2 and η_3 are isotropic scalar functions of the invariants of the previous arguments of the function. $\mathbf{W}^{ellipsoid}_{material}$ is the spin of the frame of the material with respect to the frame of the ellipsoid. The spin of the material with respect to the spatial frame can be split into a spin of the ellipsoid with respect to the spatial frame and the spin of the material with respect to the ellipsoid frame.

$$\mathbf{W}^{spatial}_{material} = \mathbf{W}^{spatial}_{ellipsoid} + \mathbf{W}^{ellipsoid}_{material} \quad [10]$$

The quantity of interest is $\mathbf{W}^{spatial}_{ellipsoid}$ and the following relation has to be time integrated:

$$\dot{\mathbf{Q}}^{spatial}_{ellipsoid} = \mathbf{W}^{spatial}_{ellipsoid} \mathbf{Q}^{spatial}_{ellipsoid} \quad [11]$$

with an initial condition $\mathbf{Q}^{spatial}_{ellipsoid} = \mathbf{I}$.

3.1.2. Identification of η

This subsection deals with the identification of the isotropic scalar functions. As a first model, the simplest case $\eta_1 = \eta_2 = \eta_3 = \eta$ is considered. An approximation of η is found by considering a finite element model of a unit cell of a rigid plastic material including an ellipsoidal void. A tensile test of the periodic cell with a tilted ellipsoid in the $1-2$ plane with respect to the reference frame is performed. It is assumed that η is not sensitive to : the excentricity of the ellipsoid, the void volume fraction, and the directions of the loading. Under tensile loading, no rotation of the material occurs, so the spin of the material with respect to the working frame is zero. Thus, the orientation of the ellipsoid in the working frame is identical to the orientation of the ellipsoid in the material. In the particular case of tensile stress, equation (10) becomes :

$$\mathbf{W}^{spatial}_{ellipsoid} = -\mathbf{W}^{ellipsoid}_{material} \quad [12]$$

It describes only the rotation of the ellipsoid in the plane of the 1-2 axes and can be written as

$$\dot{\theta} + \omega_{12} = 0 \quad [13]$$

where $\dot{\theta}$ is for the change of the orientation of the ellipsoid in the working frame and ω_{12} is the constitutive spin to be determined, from the constitutive equation (9) :

$$\omega_{12} = \eta \hat{D}^p_{12} \quad [14]$$

where $\hat{D}^p_{12}$ is the plastic strain expressed in the frame of the ellipsoid. If θ is the orientation of the ellipsoid in the working frame, $\hat{D}^p_{12}$ could be simply expressed by :

$$\dot{\theta} = \frac{3}{2}\eta D_1 \cos\theta \sin\theta \qquad [15]$$

The unknown is η and it is assumed that the change of the ellipsoid direction is sensitive to the equivalent plastic strain. The rate is plotted versus the equivalent plastic strain and fitted with a least squares procedure for the determination of η. This first simple model needs a constant η equal to 2.05. Better approximations of the rate could be achieved with more complicated assumptions about the scalar function η.

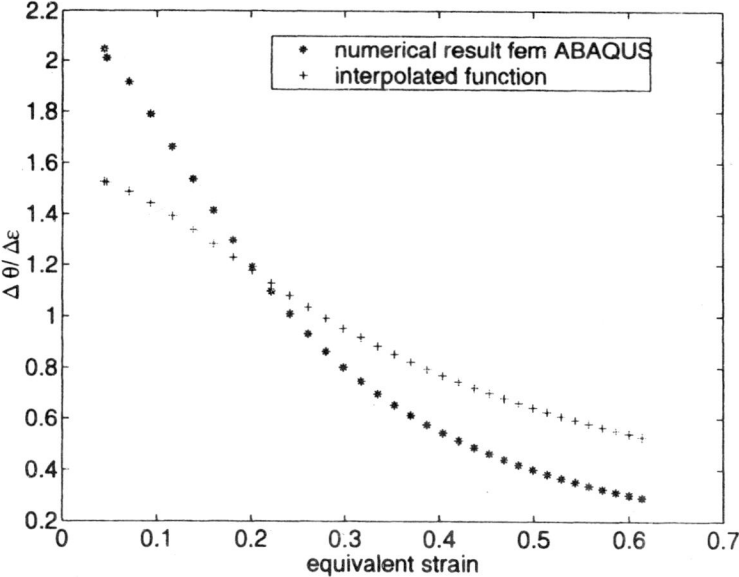

Figure 1. *Identification of η*

3.2. Compressible potential

Bordreuil, Boyer and Sallé [BOR 01] proposed a model for a porous material with inclusions inside the matrix voids. This model assumed that the eigen directions of the ellipsoid followed the stress eigen directions. Improvements of this approach are now discussed with the different concepts presented in the previous section.

3.2.1. *Foundations*

This specific potential, expressed by unit mass, is a function of stress and internal variables : scalar (equivalent plastic strain) and tensorial (orientation of the ellipsoid). The potential $\hat{\phi}$ is a modified von Mises one with a dilatational dissipation part :

$$\hat{\phi} = \frac{J_2^2}{\rho} - \frac{\sigma_0^2}{3\rho_0} + \frac{\hat{\phi}^v}{\rho} \quad [16]$$

For an ellipsoid of radii r_i, the normality rule is linked to the mass balance so that :

$$\frac{f}{1-f} \sum \frac{\dot{r}_i}{r_i} = \dot{\lambda} \frac{\partial \hat{\phi}^v}{\partial \sigma_m} \quad [17]$$

where r_i are the radii of the ellipsoid and f the void volume fraction. A model of radius rate of the ellipsoid is necessary for the volume and excentricity changes of the ellipsoid. The model proposed by Thomason [THO 88] which is a modified version of the Rice and Tracey model is selected as it can be suited to a cavity with inclusion. The radius rate is the following function of the strain rate $\bar{D}_i$ in the equivalent sound material, and of $\bar{D}$ an equivalent energetical strain rate :

$$\dot{r}_i = (\frac{5}{3}\bar{D}_i^p + \frac{1}{2}\sinh(\frac{3}{2}\frac{\sigma_m}{\sigma_0})\bar{D})\bar{r} \quad [18]$$

$\bar{r}$ is the equivalent radius of the ellipsoidal cavity. It is the radius equivalent of a spherical cavity with the same volume as the ellipsoidal cavity. In this expression, the original coefficient of the hyperbolic sine of the Rice and Tracey is set equal to 0.5 in order to obtain a simpler expression of the potential.

3.2.2. *Plastic potential*

Rice and Tracey [RIC 69] modify the strain rate field in an homogeneous material in order to take into account the presence of a spherical cavity. In our case, the eigen strain rate of equation (18) is substituted by the diagonal components of the velocity gradient in the ellipsoid frame:

$$\dot{r}_i = (\frac{5}{3}[\bar{D}_{ii}^p]_{ellipsoid} + \frac{1}{2}\sinh(\frac{3}{2}\frac{\sigma_m}{\sigma_0})\bar{D})\bar{r} \quad [19]$$

This radius rate, the normality rule, and the dissipation in the unit mass cell lead to the following dilatational potential :

$$\hat{\phi}^v = \frac{f}{1-f}\sigma_0^2(\frac{10}{9}\bar{n} : \alpha\frac{\sigma_m}{\sigma_0} + \frac{2}{9}\alpha : \mathbf{I}\cosh(\frac{3}{2}\frac{\sigma_m}{\sigma_0})) \quad [20]$$

where $\bar{n}_{ij} = \bar{D}_{ij}/\bar{D}_{eq}$ is the plastic strain rate direction and $\alpha_{ii} = \bar{r}/r_i$ is a diagonal tensor, both expressed in the ellipsoid frame. This expression satisfies the objective requirement. The compressible specific potential can be expressed as :

$$\hat{\phi} = \frac{3\sigma_{ij}^D\sigma_{ij}^D}{2\sigma_0^2} - (1-f) + \frac{3f}{1-f}(\frac{10}{9}\bar{n} : \alpha\frac{\sigma_m}{\sigma_0} + \frac{2}{9}\alpha : \mathbf{I}\cosh(\frac{3}{2}\frac{\sigma_m}{\sigma_0})) \quad [21]$$

In the case of spherical voids, this potential has the same shape as the Gurson-Tvergaard one.

3.2.3. Inclusion consideration

The approach is quite similar to the previous one, but the smoothness of the potential has to be maintained for any unilateral contact conditions between the void and the inclusion. The conditions of contact are defined by the following equations :

$$r_i^{inclusion} = r_i^{cavity}, \dot{r}_i^{cavity} \leq 0 \qquad [22]$$

if these conditions are achieved the rate $\dot{r}_i^{cavity}$ is fixed at zero. In order to fulfil the smoothness of the yield surface an additional constant $\mathbf{K}_i$ must be introduced in the expression of the plastic potential. The new potential has the form :

$$\hat{\phi}^v = \frac{f}{1-f}\sigma_0^2(\frac{10}{9}\bar{\mathbf{n}}:\alpha\frac{\sigma_m}{\sigma_0}+\frac{2}{9}\alpha:\mathbf{I}\cosh(\frac{3}{2}\frac{\sigma_m}{\sigma_0})+\mathbf{K}_i:\beta_i) \qquad [23]$$

with β_i set to 1 if contact takes place and to 0 when there is no contact between the void and the inclusion. As the plastic potential must be the same before and after contact, this is achieved when the triaxiality ratio reaches the following particular value deduced from equation (19)

$$\frac{\sigma_m^i}{\sigma_0} = \frac{2}{3}asinh(-\frac{10}{3}\frac{[\bar{D}_{ii}^p]_{ellipsoid}}{\bar{D}}) \qquad [24]$$

Equating the values of the plastic potential (19) and (23) for this particular condition leads to the constant:

$$K_i = \frac{10}{9}\alpha_i\frac{[\bar{D}_{ii}^p]_{ellipsoid}}{\bar{D}}\frac{\sigma_m^i}{\sigma_0}+\alpha_i\frac{2}{9}\cosh(\frac{3}{2}\frac{\sigma_m^i}{\sigma_0}) \qquad [25]$$

In Figure 2, the represented yield locii including the effects of an inclusion are drawn for two different assumptions about the eigen direction of the ellipsoidal void, the first case considers that they are coincident with the eigen direction of the stress tensor when the ellipsoidal void is titled in the second case.

The curve of the yield loci in Figure 2 represented in the plane of the mean stress and of the pure cylindrical shear stress can be divided in three different areas according to the contact conditions. The first one is in the range of the high positive triaxiality ratio, no point verifies the conditions of contact defined by equation (22), the plastic potential is similar to the one without inclusion in the void. The second area is the case of simultaneous contact of two symmetrical points of the void on the inclusion. Under such a condition, a plastic dilatation occurs although the mean stress is zero or negative. This is the originality of the proposed damage model. The third area is in the range of high negative triaxiality ratio, the void is contracted back on the inclusion, no more plastic dilatation takes place. The yield locus falls within the classical Von Mises criterion with a strengthened effect induced by the inclusion.

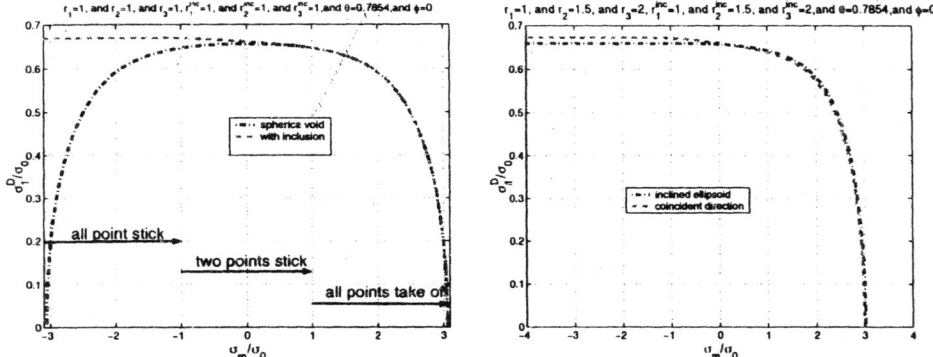

Figure 2. *Specific yield potential for different characteristics of the ellipsoid for a void volume fraction of 0.1% in case of an inclusion bound overall by the matrix and in the case of a titled ellipsoid*

4. Numerical implementation

In this section, the guide lines of the implementation of the proposed model in finite element software are outlined, with special attention to the different spins in an implicit scheme.

4.1. Stress rate

Kirchoff stress and Cauchy stress are identical in their eulerian formulation when considering the Prandtl-Reuss equation, but their rates differ. In the case of compressible plasticity, the plastic strain is no longer isochoric and the difference between the Kirchoff and Cauchy stress rates must be considered with the following expression.

$$\tau^J = \sigma^J + \sigma D_{kk} \qquad [26]$$

The constitutive matrix **H** has to be modified; it is not equal to the jacobian fourth order tensor **C** but calculated with the following expression:

$$H_{ijkl} = C_{ijkl} + \sigma_{ij}\delta_{kl} + C_{ijkl}D_{nn} \qquad [27]$$

The additional terms are of the same order as the stress when the original components are the elastic constants. This improvement acts only on the rate of convergence of the solution but has no influence on the accuracy of the results. Both stress rates have been implemented in ABAQUS/*Standard*© and similar results were found.

4.2. Integration of the internal variables

During the deformation process, internal variables evolve. Their changes have to be taken into account when $\phi = 0$ is solved at the end of each time increment. In the case of the new damage model, the orientational variables of the defect must be time integrated too.

The equivalent plastic strain is calculated with the equivalence of the specific dissipation in the sound material :

$$(1 - f)\sigma_0 \Delta \bar{\varepsilon}^p = \Delta\lambda \sigma_{ij} n_{ij}, \Delta \bar{\varepsilon}^p = \Delta\lambda \frac{\sigma_{ij} n'_{ij}}{(1-f)\sigma_0} \quad [28]$$

The void volume fraction is obtained with the incremental form of the mass balance : $\dot{f} = (1-f)D_{kk}$:

$$\Delta f = (1-f)\Delta\lambda n_{ij}\delta_{ij} \quad [29]$$

The radii are time integrated with the incremental expression of the radii rate defined by equation 18 :

$$\Delta r_i = (\frac{5}{3}\bar{n}_{ii} + \frac{1}{2}\sinh(\frac{3}{2}\frac{\sigma_m}{\sigma_0}))\Delta\bar{\varepsilon}^p \bar{r} \quad [30]$$

As isotropic elasticity is assumed, the directions of the plastic strain rate are considered constant during the increment if no contact occurs between the inclusion and the material matrix.

The orientation of the cavity is specified in the working frame with the tensor $Q_{ellipsoid}^{spatial}$. This orientation comes from the integration with respect to time of the constitutive spin of the void. The orientation of the ellipsoid with respect to the working frame is linked to the ellipsoid spin by the equation (11) where $W_{ellipsoid}^{spatial}$ has been defined by equations (10) and (9).

$$W_{ellipsoid}^{spatial} = W_{material}^{spatial} - \eta[(\bar{D}A_1 - A_1\bar{D}) + (\bar{D}A_2 - A_2\bar{D}) + (A_1\bar{D}A_2 - A_2\bar{D}A_1)] \quad [31]$$

where all W variables are skew symmetric tensors. Following Hughes and Winget [HUG 80], a relation between the incremental rotation and the incremental spin is found. The increment of rotation of the material with respect to the ellipsoid is deduced to be :

$$\Delta Q_{material}^{ellipsoid} = (I - \frac{1}{2}\omega_{material}^{ellipsoid})^{-1}(I + \frac{1}{2}\omega_{material}^{ellipsoid}) \quad [32]$$

so, the total incremental rotation of the ellipsoid with respect to the working frame is :

$$\Delta Q_{ellipsoid}^{spatial} = \Delta Q_{material}^{spatial}(\Delta Q_{material}^{ellipsoid})^{-1} \quad [33]$$

Finally, the orientation of the ellipsoid with respect to the spatial frame is given by :

$$^{t+\Delta t}Q_{ellipsoid}^{spatial} = {}^t Q_{ellipsoid}^{spatial} \Delta Q_{ellipsoid}^{spatial} \quad [34]$$

As ABAQUS/*Standard*© calculates the objective incremental rotation of the material in reference to the working frame, the use of this last expression is preferred rather equation (31).

5. Conclusion

First, the mesoscopic Rice and Tracey model is extended to an ellipsoid under finite transformation. Special attention has been paid to the spin of the ellipsoid considered as different from the plastic spin of the sound material. A simple constitutive equation of the orientation changes of the orthotropic directions of the ellipsoid is proposed and used as well as the corresponding void volume fraction rate in a new plastic potential. The discussed compressible plasticity intended for the void growth step of ductile damage includes the case of porosities and of voids with inclusion. The unilateral contact condition of the void with the inclusion acts on the yield locus and on the direction of the plastic strain rate. The proposed plastic potential leads to results similar to the Gurson-Tvergaard yield surface in the case of empty voids whereas its predictions are very different for voids in contact with inclusion as plastic dilatation occurs under zero and negative mean stress. This new plastic potential also predicts the changes of the eigen directions of the ellipsoidal defects. It is implemented in Abaqus software for the three-dimensional modelling of bulk forming processes.

6. Bibliography

[BOR 01] BORDREUIL.C, BOYER.J-C, SALLE.E, « A Ductile Damage Model with Inclusion Consideration », *Proceedings of AMPT'01*, vol. II, 2001, p. 1053-1060.

[GUR 77] GURSON.A.L, « Continuum Theroy of Ductile Rupture by Void Nucleation and Growth and Flow Rules for Porous Ductile Media », *Journal of Engineering Materials and Technology*, , 1977, p. 2-15.

[HUG 80] HUGHES.T, WINGET.D, « Finite Rotation Effects in Numerical Integration of Rate constitutive Equations Arising In Large Deformations Analysis », *International Journal of Numerical Methods in Engineering*, vol. 15, 1980, p. 1862-1867.

[LIU 82] LIU.I, « On the representation of anisotropic invariants », *International Journal of Engineering Science*, vol. 10, 1982, p. 1099-1109.

[RIC 69] RICE.J.R, TRACEY.D.M, « On the Enlargement of Voids in Triaxial Stress Fields », *Journal of the Mechanics and Physics of Solids*, , 1969, p. 201-217.

[THO 88] THOMASON.P.F, *Ductile Fracture of Metals*, Pergamon Press, 1988.

[WAN 70] WANG.C.C, « New theorem for isotropic function part 1 and part 2 », *Archive for rational Mechanics and Analysis*, vol. 36, 1970, p. 166-223.

Chapter 7
Failure Prediction in Anisotropic Sheet Metals under Forming Operations using Damage Theory

Michel Brunet, Fabrice Morestin and Hélène Walter
Laboratoire de Mécanique des Solides, INSA, Villeurbanne, France

1. Introduction

Plastic deformation in anisotropic sheet-metal consists of four distinct phases, namely, uniform deformation, diffuse necking, localized necking and final failure. The last three phases are commonly known as *non-uniform deformation*. New sheet-metals such as aluminium alloys, titanium alloys and Ni-based super-alloys, display from experimental evidence necking-failure behaviour where localised thinning is hardly visible. Plastic instability of these sheet-metals has been found to suffer material degradation which confirmed the need to properly characterize their forming limit using a theory of damage mechanics. Coupling the incremental theory of plasticity with damage and a plastic instability criterion, the enhanced criterion proposed in this paper can be used to predict not only the forming limit but also the fracture limit under proportional or non-proportional loading and is suitable for sheet-metal forming simulation by finite-element analysis.

Another case is the sheet bending process where the outer layer of the bending sheet is unstable but the sheet as a whole is not unstable. So the formation of a neck is impossible, and the boundary in bending is the strain to fracture. This important phenomenon is not related in the FLD predicted using the conventional approaches based only on plastic-instability theories. In view of these shortcomings, a new unified approach to predict the FLD is presented. This approach enables the prediction of the three stages which lead to final rupture: diffuse, localized necking and failure. In addition, the new approach takes into account the effects of micro-mechanism of voids on the FLD. This enables the introduction of the condition of plastic limit-load failure of the intervoid matrix at incipient void coalescence.

2. Damage model

Most metallic materials contain different sizes and degrees of particles, including precipitates and inclusions, which may cause micro-defects including micro-voids and micro-cracks. In the literature, one conclusion in the analysis of void growth and coalescence is that the void aspect ratio affects void coalescence, even when the void is initially spherical. Employing a rigorous micromechanical analysis, considering both prolate and oblate spheroidal voids, Gologanu et al. [GOL 97] have proposed a constitutive model retaining a form similar to that of the Gurson model. For such a purpose, the void aspect ratio, S, comes into play, for which an evolution law is derived. The Gologanu-Leblond-Devaux (GLD) model is extended here for prolate voids, plane-stress state and non-linear kinematic hardening including initial quadratic anisotropy of the matrix or base material such as:

$$\Phi = C\frac{q^2}{\sigma_y^2} + 2q_1 f^* \cosh\left(\frac{\kappa \sigma_H}{\sigma_y}\right) - (1 + q_3 f^{*2}) = 0 \qquad [1]$$

where f* is the effective void volume fraction. Consider x, y to be the "rolling" and "cross" directions in the plane of the sheet and based on Hill's quadratic yield function, the macroscopic effective stress q in equation [1] is defined as:

$$q^2 = \{\sigma-\alpha\}^T [M]\{\sigma-\alpha\} \qquad [2a]$$

and:

$$M = \begin{bmatrix} a^2 & a(b-h) & 0 \\ a(b-h) & b^2 - 2hb + f + h & 0 \\ 0 & 0 & 2n \end{bmatrix} \qquad [2b]$$

where the relative macroscopic stress tensor with respect to the center of the current yield surface is defined as:

$$\{\sigma-\alpha\} = \begin{Bmatrix} \sigma_{xx} - \alpha_{xx} \\ \sigma_{yy} - \alpha_{yy} \\ \sigma_{xy} - \alpha_{xy} \end{Bmatrix} \qquad [3]$$

In equation [2b], $a = 1 + \eta(1 - 2\alpha_2)$, $b = \eta\alpha_2$ are damage functions, f, h and n are the dimensionless Hill parameters. In the case of spherical voids (Gurson model):

$$\sigma_H = (1 - 2\alpha_2)(\sigma_{xx} - \alpha_{xx}) + \alpha_1(\sigma_{yy} - \alpha_{yy}) \qquad [4]$$

reduces to the hydrostatic stress of the relative stress tensor of equation [3]. This particular model extended here for plane stress state and non-linear kinematic hardening contains nine state variables: the 2x3 components of stress and back-stress tensor, the porosity f, the void aspect ratio S and an average yield stress for the matrix or base material σ_y. The void aspect ratio is defined by:

$$S = \ell n(R_z / R_x) \qquad [5]$$

where R_X, R_Z are the minor and major radii of the prolate ellipsoidal void in its local frame, $S \geq 0$. It is assumed here that the major axis of the cavity follows the rolling direction. The expressions of the parameters $C, \eta, \kappa, \alpha_1, \alpha_2$ in term of S and f and the evolution law of S can be found in detail in [GOL 97] and [PAR 00]. The eccentricities e_1 and e_2 of the inner and outer ellipsoidal volumes are deduced from the void shape parameter S and porosity f as:

$$e_1 = \sqrt{1-e^{-2|S|}} \qquad [6]$$

$$\frac{1-e_2^2}{e_2^3} = \frac{1}{f}\frac{(1-e_1^2)}{e_1^3} \quad \text{for prolate void} \qquad [7]$$

$$\alpha_1 = \frac{1}{2e_1^2} - \frac{1-e_1^2}{2e_1^3}\tanh^{-1}(e_1) \ , \ \alpha_1^G = \frac{1}{3-e_1^2} \qquad [8a]$$

$$\alpha_2 = \frac{1+e_2^2}{3+e_2^4} \qquad [8b]$$

$$\kappa^{-1} = \frac{1}{\sqrt{3}} + \frac{1}{\ell n f}\left[(\sqrt{3}-2)\ell n \frac{e_1}{e_2} - \frac{1}{\sqrt{3}}\ell n\left(\frac{3+e_1^2+2\sqrt{3+e_1^4}}{3+e_2^2+2\sqrt{3+e_2^4}}\right) + \ell n\left(\frac{\sqrt{3}+\sqrt{3+e_1^4}}{\sqrt{3}+\sqrt{3+e_2^4}}\right)\right] \qquad [9]$$

$$\eta = -\frac{\kappa sh(1-f)f}{1+f^2+f(\kappa H sh - 2ch)} \qquad C = \frac{\kappa f sh}{(1-f+\eta H)\eta} \qquad [10]$$

where: $sh = \sinh(\kappa H)$, $ch = \cosh(\kappa H)$, $H = 2(\alpha_1 - \alpha_2)$

For spherical voids, Gurson's model is recovered such that:

$\alpha_1 = \alpha_1^G = \alpha_2 = \frac{1}{3}$, $C = 1$, $\eta = 0$ and $\sigma_H = \sigma_M$

The equation of the evolution of the internal shape parameter S is given by:

$$\dot{S} = \frac{3}{2}\left\{1+\frac{9}{2}h_T(T,\zeta)(1-\sqrt{f})^2\frac{\alpha_1-\alpha_1^G}{1-3\alpha_1}\right\}\left\{\dot{\varepsilon}_{11} - \frac{\dot{\varepsilon}_{kk}}{3}\right\} + \left\{\frac{1-3\alpha_1}{f} + 3\alpha_2 - 1\right\}\dot{\varepsilon}_{kk} \qquad [11]$$

with $h_T(T,\zeta)$ a function dependent on the triaxiality $T = \sigma_{kk}/(3\sigma_{eq})$ according to the sign of $\zeta = \sigma_{kk}\sigma'_{ii}$ as $h_T = 1-T^2$ for $\zeta > 0$ and $h_T = 1-T^2/2$ for $\zeta < 0$.

3. Constitutive parameters

The combined isotropic-kinematic hardening of the matrix or base material is an extension of the linear model for the induced anisotropy. It provides a much more accurate approximation to the stress-strain relation of the matrix material than the linear model. It also models other phenomena such as ratchetting, relaxation of the mean stress, cyclic hardening, that are typical of materials subjected to cyclic hardening. The size of the elastic range σ_y is defined as a function of the equivalent plastic strain $\bar{\varepsilon}^p$ (and other field variables such as temperature if necessary). In this paper, this dependency is provided with a simple exponential law for materials that either cyclically harden or soften as:

$$\sigma_y = \sigma_0 + Q_\infty \left(1 - e^{-b\bar{\varepsilon}^p}\right) \qquad [12]$$

where σ_0 is the yield surface size at zero plastic strain, and Q_∞ and b are material parameters that must be calibrated from cyclic test data. The evolution of the kinematic components of the model is defined as:

$$\{d\alpha\} = C \frac{d\bar{\varepsilon}^p}{\sigma_y} \{\sigma - \alpha\} - \gamma\{\alpha\}d\bar{\varepsilon}^p \qquad [13]$$

where C and γ are additional material parameters to be adjusted. This equation is the basic Ziegler's law where the last term is the "recall" term which has been added and which introduces the non-linearity in the evolution law. It is worth noticing that introducing the non-linear kinematic hardening of the matrix material in the damage model, the effect of the bending/unbending on the predicted failure curves can be taken into account.

The initial anisotropy parameters (the r-values) are first determined independently using our digital image correlation method (DIC), [BRU 01], by mean of uniaxial tests. In order to obtain the test data for the kinematic hardening parameters identification, a bending-unbending apparatus has been built, [BRU 01].

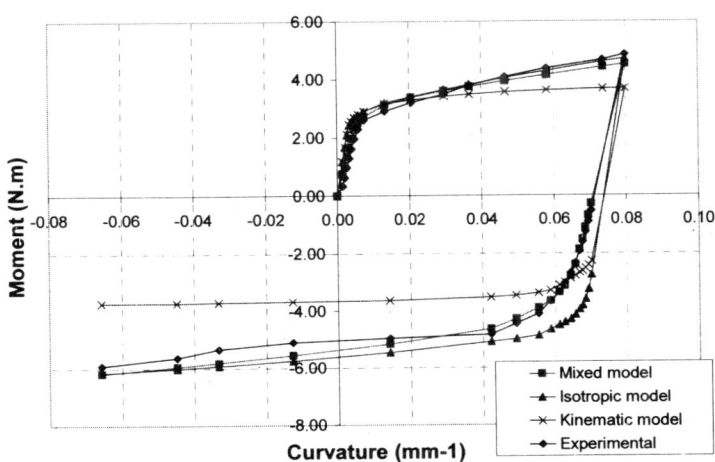

Figure 1. *Experimental and theoretical moment-curvature*

As an example, Figure 1 depicts the moment versus curvature for one loading and reverse loading. The material is an aluminium alloy of the 5182 series of strip thickness 1 mm, E = 68000Mpa, σ_0 = 139Mpa, R_0 = 0.44 and R_{90} = 0.61. It can be seen that very substantial agreement of experimental and simulated data is obtained with the converged values: C = 1895, γ = 119.4, Q = 334.5 and b =.062 for the mixed hardening model.

4. Damage parameters

The damage model can take into account the three main phases of damage evolution: nucleation, growth and coalescence. An optimisation procedure could also be performed to match the experimental and numerical finite element results as regards the load vs. displacement curve in a tensile test. However, the critical void volume fraction is not unique due to the fact that the void nucleation parameters are difficult to monitor in experiments and are usually arbitrarily chosen. To overcome this shortcoming, the void coalescence failure micro-mechanism by internal necking is considered by using a modified 3D Thomason's plastic limit-load model [PAR 00, BEN 99, THO 90]:

$$\left\{ F/\left(\frac{R_Z}{X-R_X}\right)^N + G/\left(\frac{R_X}{X}\right)^M \right\} A_n' = \frac{\sigma_1}{\sigma_y} \qquad [14]$$

where F = 0.1 and G = 1.2 are constants or adjustable coefficients, N = 2 and M = 0.5 are exponents, R_X, R_Z are the radii of the ellipsoidal void and X denotes half the current length of the reference volume element (RVE). What is interesting in the plastic limit-load criterion is that void coalescence is not only related to void volume fraction but also to void-matrix geometry, stress triaxiality and initial void spacing. By means of a void spacing ratio parameter λ_0, the anisotropic nature of rolled sheet is better accounted for in the coalescence micro-mechanism; moreover this effect is more pronounced at low stress triaxiality, [BEN 99]. Consistent with equation [14], the modified Gologanu's model is used to characterise the macroscopic behaviour and to calculate the void and matrix geometry changes using the current strain, void volume fraction f and shape factor S, such that:

$$R_X = \left(\frac{3fV}{4\pi e^S}\right)^{1/3}$$

with: $X = X_0 e^{\varepsilon_X} \qquad Z = Z_0 e^{\varepsilon_Z} \qquad Y = Y_0 e^{\varepsilon_Y}$ and:

$$Z_0 = \lambda_0 X_0 \qquad\qquad V_0 = 8 X_0 Y_0 Z_0 = 1 \qquad [15]$$

and the neat area fraction of intervoid matrix:

$$A_n = \left[1 - \frac{\pi R_X^2}{4XY}\right] \quad [16]$$

Once the equality equation [14] is satisfied, the void coalescence starts to occur and the void volume fraction at this point is the critical value f_c.

The original 3D plastic limit load model of Thomason has been modified here by changing A_n, the neat area fraction of intervoid matrix in the maximum principal stress direction σ_1, such that in equation [14]:

$$A'_n = A_n\left[1 - T\frac{\sigma_1}{\sigma_y}\right] \quad [17]$$

where T is the current triaxiality. This approximate and average correction factor has been found from several numerical 3D. FE analyses of a cube (the reference volume element RVE) containing an ellipsoidal void and spherical rigid inclusion as shown in Figure 2. A one-eighth cube model has been adopted to properly take account of plastic anisotropy. The outer faces of the unit cell remain plane and mutually orthogonal to each other during the deformation due to the periodic nature of the unit cell. Pure shear, uniaxial tension and equal biaxial tension have been considered for plane stress conditions. However, a triaxial loading with high mean stress and a pure hydrostatic tension loading could also be applied to complete the analysis. Under different loading conditions, the macroscopic stresses are calculated by averaging the surface tractions acting on the faces of the unit cell.

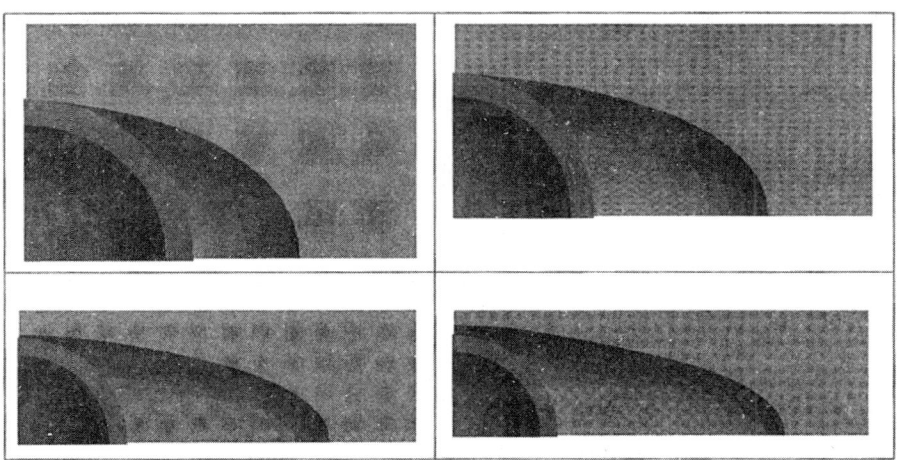

Figure 2. *Deformed RVE with rigid inclusion T = 2/3*

This improvement in the original Thomason's formula for low stress triaxiality can be seen as an average way to introduce the effects of an inclusion in the void and void spacing. Apart from the plastic limit load at coalescence, the FE analysis also enables calibration of the modified anisotropic Gologanu model with the selection of the fitting parameters $q_1 = 1.44$ and $q_3 = 2.1$ for $R_0 = 0.44$, $R_{90} = 0.61$.

5. Necking-failure criterion

The strain ratio $\beta = \Delta\varepsilon_2/\Delta\varepsilon_1$ has an evident influence on the internal damage of sheet metals. At the same level of deformation, it is generally noted that the damage increment is the greatest at plane strain such that $\Delta\varepsilon_{22} = 0$ when the localized necking occurs, which requires a drift to the plane strain state and then an additional hardening. The formulation follows our previous works, [BRU 01], the unified necking-failure criterion is formulated in terms of the principal stresses and their orientation with respect to the orthotropic axes leading to an intrinsic formulation including damage:

$$\frac{q}{\sigma_y}\left[\frac{\partial\sigma_1}{\partial q}\frac{\partial q}{\partial\sigma_y}\frac{\partial\sigma_y}{\partial\bar{\varepsilon}}\frac{\partial\bar{\varepsilon}}{\partial\varepsilon_1} + \frac{\partial\sigma_1}{\partial\beta}\frac{d\beta}{d\varepsilon_1}\right] \leq \sigma_1 \qquad [18]$$

where an analytical form of the left-hand side can be found in [BRU 01]. As an example, the comparison between the proposed necking-failure criterion and our experiments with Marciniack's tests has been obtained from the previous 5182 aluminium alloy.

The damage model takes into account the three main phases of damage evolution: nucleation, growth and coalescence with the proposed coalescence limit load model, equation [14].

$$df = df_N + df_G + df_C \qquad [19]$$

where a continuous nucleation model with one constant has been chosen:

$$df_N = A_0 d\bar{\varepsilon}^p \qquad [20]$$

Growth of existing voids is based on the apparent volume change and law of conservation of mass and is expressed as:

$$df_G = (1-f)\left(d\varepsilon_{11}^P + d\varepsilon_{22}^P + d\varepsilon_{33}^P\right) \qquad [21]$$

Figure 3 shows the experimental points and predicted curves for localized necking-failure with and without damage. For this 5182 aluminium alloy, the forming limit is a ductile-fracture forming limit, as can be seen also for an INCO 718 Ni-based alloy in Figure 4.

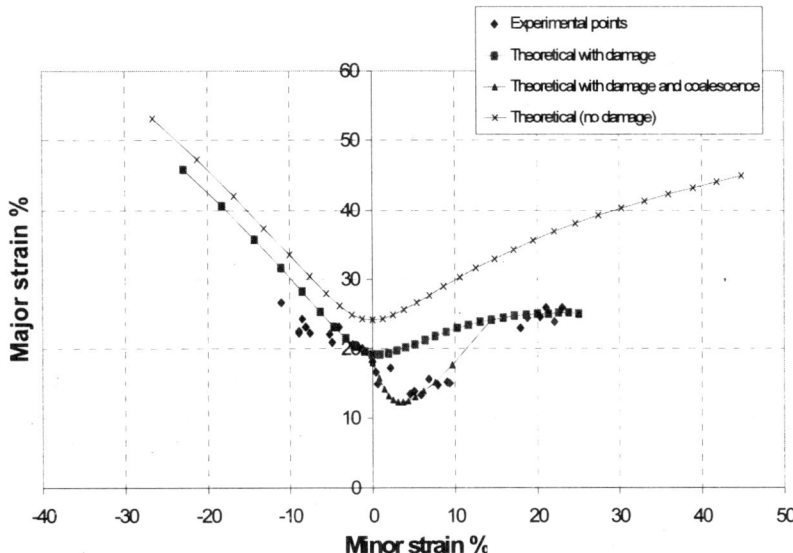

Figure 3. *Experimental and theoretical forming limit curves 5182 Al Alloy*

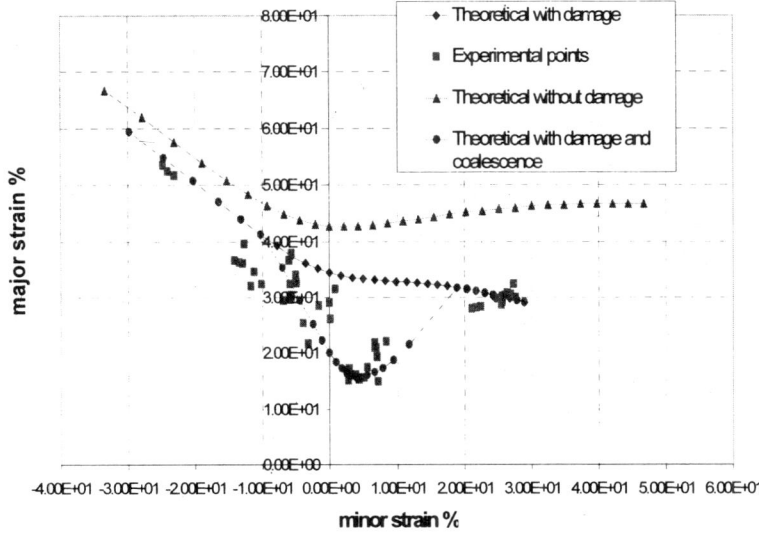

Figure 4. *Experimental and theoretical forming limit curves INCO 718 alloy*

In both cases, the initial damage factors are $f_0 = 0.001$, $S_0 = 0.01$, and the nucleation parameter $A_0 = 0.08$. The void spacing ratio parameter λ_0 in the coalescence formulation has been chosen as 1.2 for the aluminium alloy and 1 for the INCO 718.

The digital image correlation (DIC) with flat-headed or hemispherical punch have been used for the experimental points where the width of the specimens was modified in order to achieve failure in different strain states ranging from uniaxial to biaxial tension. Due to the digital image processing to identify the onset of incipient necking and crack, the scatter commonly observed in the determination of FLDs is diminished but still occurs. This is especially true for aluminium and Ni-based alloys that tend to develop highly localized deformation rapidly followed by a crack which is confirmed by the damage analysis. The concept of "incipient neck" is used as the criteria for failure but is still subjective. The area analysis procedure with the DIC allows one to reduce the variability in the determination of the FLDs because the grid size effect disappears. The local strain measurements are readily obtained by interpolation across the crack by limiting the operator variability. For the strain measurement on a typical specimen, a pattern is constituted by 8 to 12 pixels.

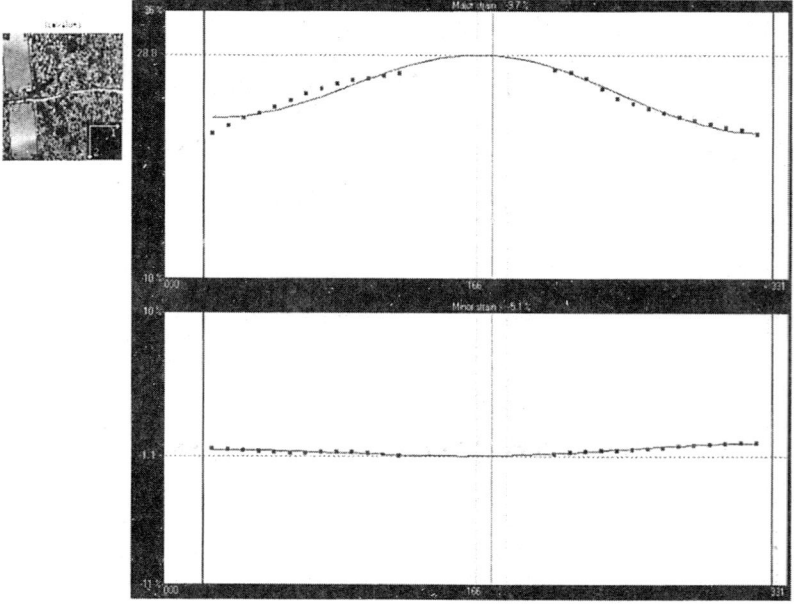

Figure 5. *Measure and interpolation of strain across a crack by DIC*

In this configuration, a pixel is around 0.07 mm, so the grid size can vary from 0.59 mm to 0.88 mm. The grey level between four pixels is interpolated in order to reach sub-pixel accuracy. With a spline interpolation method, an accuracy of 1/100 pixels can be obtained. From the strain distribution across the neck, a mathematical smoothing allows one to find the limit strain (major and minor). In the software developed, the smoothing is made automatically with a sinusoidal interpolation as shown in Figure 5.

6. Conclusion

In this work, a unified failure approach has been presented based on the theory of damage mechanics including the non-linear kinematic hardening of the matrix material and void coalescence by internal necking of the inter-void ligament. In sheet metals, the development of damage makes the strain state gradually drift to plane strain, this fact leading to a proposal for a unified instability criterion for localized necking and rupture.

7. References

[GOL 97] GOLOGANU M., LEBLOND J.B., DEVAUX J., "Recent extension of Gurson's model for porous ductile metals", in *Continuum Micromechanics*, Suquet P. Ed., Springer Verlag, Berlin, 1997, pp. 61–130.

[PAR 00] PARDOEN T., HUTCHINSON J.W., "An extended model for void growth and coalescence", *J. Mech. Physics of Solids*, 48, 2000, pp. 2467–85.

[BRU 01] BRUNET M., MORESTIN F., GODEREAUX S., "Non-linear Kinematic Hardening Identification for Anisotropic Sheet-Metals with Bending-Unbending Tests", *J. Eng. Mat. Tech.*, Vol. 123, 2001, pp. 378–384.

[BRU 01] BRUNET M., MORESTIN F., "Experimental and analytical necking studies of anisotropic sheet-metals", *Journal of Material Processing Technology*, Vol. 112, 2001, pp. 214–226.

[BEN 99] BENZERGA A.A., BESSON J., PINEAU A., "Coalescence-controlled Anisotropic Ductile Fracture", *J. Eng. Mat. Tech.*, Vol. 121, 1999, pp. 221–229.

[THO 90] THOMASON P.F., *Ductile Fracture of Metals*, Pergamon Press, 1990.

Chapter 8

Analysis and Experiment on Void Closure Behaviour Inside the Sheet during Sheet Rolling Processes

Dyi-Cheng Chen
Department of Mechanical Engineering, Institute of Technology, Taiwan, Republic of China

Yeong-Maw Hwang
National Sun Yat-Sen University, Taiwan, Republic of China

1. Introduction

When metals are made by casting processes, voids are generally formed in the workpiece due to material shrinkage or gas entrapment during solidification. If any void remains in the final metal product, it will have an unfavorable effect on the performance of the material in service. Thus, it is highly desirable to use the rolling process to eliminate the voids.

A number of papers have been published discussing the closure of voids during hot rolling of slabs. For example, Wallero (Wallero, 1985) analyzed the closure of a central longitudinal pore during hot rolling process with a square cross-sectional workpiece in experiments. His work showed the rolling mills with large rolls were advantageous in pore closure. Keife et al. (1980) studied the closure of a void in homogeneous plane strain forging between parallel dies with the upper bound approach and carried out experiments when voids of the same size and square shapes were homogeneously and squarely distributed throughout the specimen. Stahlberg (1986) has proposed a three-dimensional theoretical model of upper-bound solutions for the closure of a central longitudinal hole during hot rolling between flat parallel rolls. The above papers concluded that large roll radii and heavy reductions per pass were favorable for the closure of a central longitudinal hole. Wang et al. (Wang et al. 1996) used a FE software "ABAQUS" to investigate the effects of stress and strain on pore closure in a steel sheet during sheet rolling. They concluded that increasing the coefficient of friction between the roll and workpiece was able to eliminate the pores effectively.

Dudra and Im (Dudra and Im, 1990) used the finite element method to predict the deformation of a central void when round and square ingots were side-pressed in a plane-strain condition. They said that the effective strain was a better indicator for void closure than the hydrostatic stress. Shah (Shah, 1986) showed that voids could be closed by proper choice of forging practices including such factors such as die geometry, amount of reduction, reduction speed, and workpiece temperature. Kiuchi and Hsiang (Kiuchi and Hsiang, 1981) discussed analytically the behaviour of internal porosity in a plate rolling process using the method of limit analysis. The slab and/or the sheet in the roll gap were divided into three zones and different kinematical admissible velocity fields were assumed for each zone. Pietrzyk et al. (Pietrzyk et al., 1995) used a rigid-plastic thermal-mechanical finite-element program to examine the behaviour of voids inside a steel slab during hot rolling and to assess the tendency of void closure.

The works mentioned above just discussed the void closure behaviour fragmentarily. There were few papers involved in the investigation of the effects of various dimensions and shapes of internal voids upon critical reductions for complete void closure during sheet rolling processes so far. This paper will investigate the void closure behaviour inside a sheet during sheet rolling processes by the finite element method of continuum mechanics, and discuss systematically the effects of various rolling conditions such as the dimension of internal void, thickness reduction, etc., on the final dimensions of the void after rolling. Experiments will also be conducted using aluminum as the sheet with a round void inside. The theoretical calculations will be compared with experimental results.

2. Finite element modelling

A commercial finite element code "DEFORM" is applied to simulate void closure behaviour and stress-strain distributions around internal voids inside the sheet during sheet rolling processes. The finite element code is based on the flow formulation approach using an updated Lagrange procedure. The basic equation for the finite element formulation from the variational approach is

$$\delta \pi = \int_v \overline{\sigma} \delta \dot{\overline{\varepsilon}} dV + C \int_v \dot{\varepsilon}_v \delta \dot{\varepsilon}_v dV - \int_{S_F} \tau_i \delta v_i dS = 0 , \quad [1]$$

where

$$\overline{\sigma} = \sqrt{\frac{3}{2}} (\sigma'_{ij} \sigma'_{ij})^{1/2}, \quad \dot{\overline{\varepsilon}} = \sqrt{\frac{3}{2}} (\dot{\varepsilon}_{ij} \dot{\varepsilon}_{ij})^{1/2}, \quad \dot{\varepsilon}_v = \dot{\varepsilon}_{ii} .$$

C is a very large positive number which is called the penalty constant and is interpreted as the bulk modulus. $\delta \dot{\overline{\varepsilon}}$ and $\delta \dot{\varepsilon}_v$ are the variations in strain rate and volumetric strain rate, respectively, derived from an arbitrary variation δv_i. τ_i is the surface traction stress. By the finite-element discretization procedure, equation [1] can be converted to non-linear algebraic equations. The iteration methods adopted for solving the nonlinear equations in "DEFORM" are Newton-Raphson and the direct iteration methods. The direct iteration method is used to generate a good initial guess for Newton-Raphson method, whereas Newton-Raphson method is used for speedy final convergence. The convergence criteria for the iteration are the velocity error norm $\|\Delta \mathbf{v}\|/\|\mathbf{v}\| \leq 0.001$, and the force error norm $\|\Delta F\|/\|F\| \leq 0.01$, where $\|\mathbf{v}\|$ is defined as $(\mathbf{v}^T \mathbf{v})^{1/2}$.

For the detailed descriptions of the finite element theory and the modelling formulations, please refer to reference (Kobayashi, 1989). The schematic illustration of sheet rolling processes with an internal void inside is shown in Figure 1, where R and V are the radius and peripheral velocity of the roll, respectively; L the contact length between the sheet and roll; X the distance from the internal void to the exit of the roll gap; and t_0 and t_1 are the initial and final thickness of the sheet, respectively. The internal void is assumed to have circular or rectangular cross-sections, and its dimensions are denoted by d_0, h_0 and l_0 as shown in Figure 1. The final height and length of the internal void are h_1 and l_1, respectively. During simulations, some more assumptions are made as below:

1) The rolls are rigid;

2) The sheet is rigid-plastic material;
3) The deformation of the sheet is plane strain;
4) Anisotropy is not considered and
5) A constant friction factor at the interface between the roll and sheet is considered.

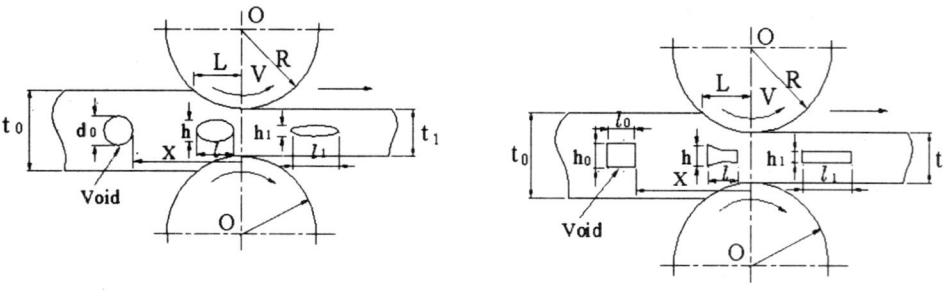

(a) Round internal void   *(b) Rectangular internal void*

Figure 1. *Schematic illustration of rolling process of sheet with an internal void inside the sheet*

3. Mesh configurations of the sheet

This model consists of two objects: the roll and sheet as shown in Figure 2. Due to symmetry, only the upper part is shown in the figure. The internal void is assumed to be located at the center of the sheet in the thickness direction. In this case, isoparametric four-node elements are used.

R=100mm, V=12mm/s, m=0.6, sheet:*Al* 1050, t_0 =10mm

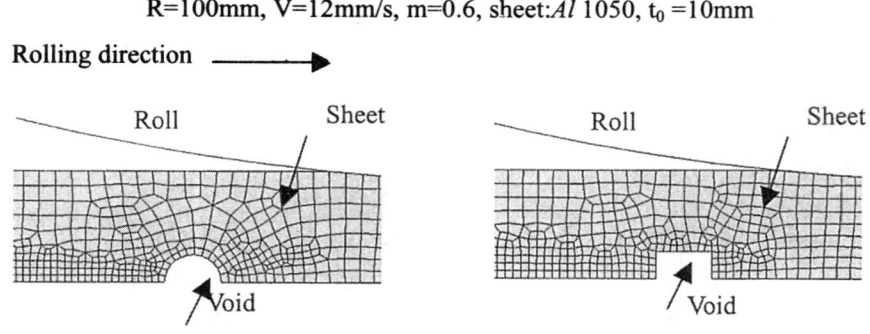

(a) Round internal void - d_0 = 2.5 mm   *(b) Rectangular internal void - $h_0 = l_0$ = 2.5 mm*

Figure 2. *The mesh layout of the sheet including a void before rolling*

The sheet is divided into about 500 elements. Two kinds of mesh densities in the deformation zone are set to make the deformation region undergoing larger strains have larger element density. The friction factor between the sheet and rolls, m, is set to be 0.6 to correspond to the dry friction condition. The material of the sheet used here is aluminum $Al1050$, the flow stress which is assumed to be a power law of the strain and its strength coefficient, C, and strain-hardening exponent, n, being 187.7 MPa and 0.079, respectively. The other parameters for FE-simulations are the time increment of 0.01 sec and the maximum strain in the workpiece for remeshing being 0.025

Figure 3 shows the effective strain distributions of the sheet at various thickness reduction during rolling for the distances from the round void to the exit being $X/L = 0.13$. The round void is located at the center of the sheet in the thickness direction. The rolling conditions are $t_0 = 10$ mm, $R = 100$ mm, $m = 0.6$, $d_0 = 2.5$ mm. It is apparent that larger thickness reductions lead to larger effective strain. Also, the largest effective strain occurs at the front and back of the void, because the sheet there undergoes more plastic deformation.

Figure 4 shows the effective strain distribution of the sheet at the roll gap during rolling. The rolling conditions are $t_0 = 10$ mm, $R = 100$ mm, $m = 0.6$, $d_0 = 2.5$ mm, $r = 30\%$. Obviously, the effective strain becomes larger when the void flows from the entrance to exit of the roll gap. The height of the void, h, becomes 0.23 mm when $X/L = -0.1$. Clearly, the volume of the void after rolling is mush smaller than that before rolling.

4. Results and discussion

The rolling conditions used in the following numerical simulations are summarized in Table 1.

Figure 5 illustrates the critical reductions, r_c, for complete void closure under different shapes of voids. The region at the hatched side stands for the void being closed completely. From this figure, it is known that the critical reduction is smaller for a round void. That is because the cross-sectional area of a round void is smaller than that of a rectangular one under the same height; accordingly, it takes smaller reductions to make the void closed completely. On the other hand r_c increases with increasing d_0/t_0 (or h_0/t_0) due to voids with a larger area needing larger reductions to make it closed completely. The critical reductions obtained for $0.15 < d_0/t_0$ (or h_0/t_0) < 0.35 are $22\% < r_c < 44\%$.

R = 100mm, V = 12mm/s, m = 0.6, sheet:*Al* 1050, t_0 = 10mm
Round void:d_0 = 2.5mm

Rolling direction ⟶

B = 0.1, C = 0.2, D = 0.3, E = 0.4, F = 0.5, G = 0.6, H = 0.7, I = 0.8

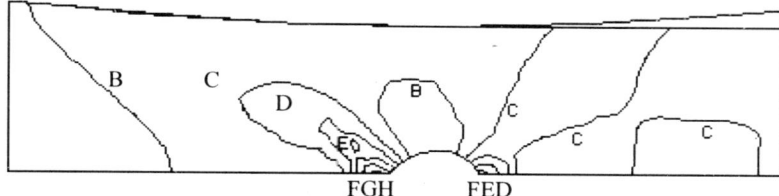

a) Reduction 15% (X/L = 0.13, h = 1.40 mm, l = 2.61 mm)

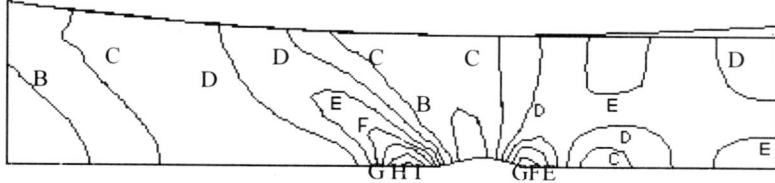

b) Reduction 25% (X/L = 0.13, h = 0.52 mm, l = 2.88 mm)

Figure 3. *Effective strain distributions of the sheet including a round void at various thickness reduction during rolling*

5. Experiments on sheet rolling with an internal void

5.1. *Experiments for determination of friction factor and properties of sheet*

In sheet rolling processes, the lubricity at the interface between the sheet and rolls is an important factor, which affects the rolling results, such as the void closure behaviour. To express the lubricity at the interface between the roll and sheet, the constant shear friction $\tau_s = mk$ is adopted, where k is the yield shear stress of the deforming material, and m is the friction factor ($0 \leq m \leq 1$). The friction factor m between the sheet and rolls is determined by the so called ring compression test. The aspect ratio of the test piece is 6:3:2. The surface of the test piece is slightly polished with sand-paper first, then cleaned by using a cloth with acetone to ensure dry friction is present. The reductions in the internal diameter of the ring by an analytical method are compared with those from experimental measurements carried out by the authors. The analytical value of the friction factor that best fits with the experimental value for the dry friction surface is about 0.6. The tensile test of specimens (*Al* 6061) was carried out at an average strain rate of 0.2 s^{-1}. The strength coefficient, C, and strain-hardening exponent, n, obtained for the sheets are 609.1 MPa and 0.128, respectively.

Analysis and Experiment on Void Closure Behaviour 91

R = 100mm, V = 12mm/s, m = 0.6, sheet:Al 1050, t_0 = 10mm, r = 30%
Round void:d_0 = 2.5mm

Rolling direction ⟶

B = 0.15, C = 0.30, D = 0.45, E = 0.60, F = 0.75, G = 0.90

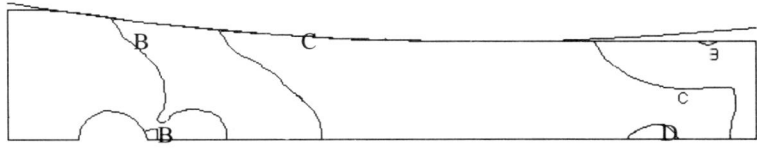

a) X/L = 0.8, h = 2.08 mm

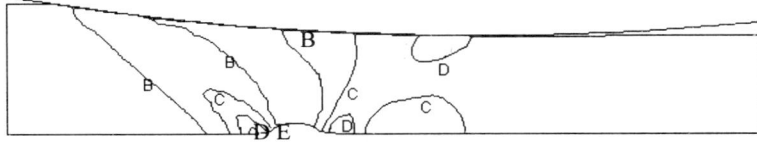

b) X/L = 0.4, h = 0.65 mm

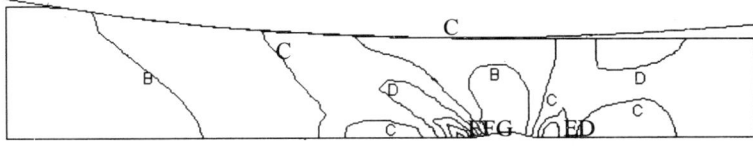

c) X/L = − 0.1, h = 0.23 mm

Figure 4. *Effective strain distributions with a round void at the roll gap during rolling*

Table 1. *Rolling conditions for simulations*

Void shape	Void dimension d_0/t_0 or h_0/t_0	Friction factor m	R/t0	Reduction r (%)
Round void	0.25	0.6	100/10	15~30
	0.15 ~ 0.35	0.6	100/10	22~44
Square void	0.15 ~ 0.35	0.6	100/10	22~44

5.2. Observations of void closure behaviour

Experiments on sheet rolling have been performed using a two-high test rolling mill. The radii of the rolls are both 100 mm. The peripheral speeds of the rolls are both 12 mm/s. Specimens for sheets are *Al* 6061 aluminum. The surfaces of the sheets are treated in the same way as in the ring compression test. After rolling, a micrometer is used to measure the length and height of the voids.

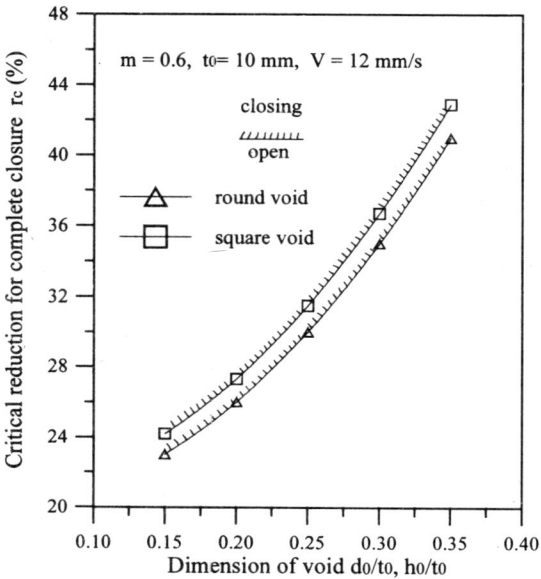

Figure 5. *Effects of void dimension on critical reductions*

$R = 100$mm, $V = 12$mm/s, $m = 0.6$, sheet: *Al* 6061, $t_0 = 10$mm, $r = 28\%$
Round void: $d_0 = 2.5$mm

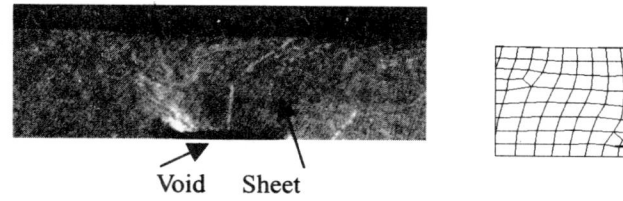

a) *Experimental results*   b) *Finite element simulation*
$[(h_1-d_0)/d_0 = -0.80, (l_1-d_0)/d_0 = 0.35]$ $[(h_1-d_0)/d_0 = -0.85, (l_1-d_0)/d_0 = 0.17]$

Figure 6. *Configurations of the sheet and internal void after rolling*

Figure 6 exhibits the comparison of the configurations of the sheet and internal void after rolling between analytical results and experimental observation. The friction factor, m, used during the FE simulations is 0.6 to correspond to the dry friction condition. It is clear that the analytically obtained void shape is quite similar to that in the experimental observation. The predicted magnitude of the variation of the final height $(h_1-d_0)/d_0$ is slightly larger than the experimental values, whereas the variation of the final length $(l_1-d_0)/d_0$ is smaller than the experimental values

6. Conclusions

In this the plastic deformation of the sheet with an internal void inside during sheet rolling paper is simulated using a FE code DEFORM^{TM} 2D. The influences of various rolling conditions, such as thickness reduction, the dimension of the internal void, etc., on the dimension of voids at the exit were discussed. From a series of simulation results, it can be concluded that (1) the largest effective strain occurs at the front and back of the void; (2) the effective strain becomes larger when the void flows from the entrance to exit of the roll gap; and (3) the critical reduction is smaller for a round void than a rectangular one with the same height. The validity of this model was verified by comparing the analytical results with the experimental measurements.

7. References

Dudra S.P., Im Y.T., Analysis of void closure in open-die forging, *Int. J. Mach. Tools Manufact*, Vol. 30, 1990, pp. 65–75.

Keife H., Stahlberg U., Influence of pressure on the closure of voids during plastic deformation, *Journal of Mechanical Working Technology*, Vol. 4, 1980, pp. 133–143.

Kiuchi M., Hsiang S.H., Two-dimensional analysis of closing behaviors of internal porosity study on application of limit analysis to rolling process, 1st Report, *J. Jap. Soc. Technol. Plasticity*, Vol. 22, 1981, pp. 927–934.

Kobayashi S., Oh S. I., Altan T., Metal Forming and the Finite Element Method, *Oxford*, New York, 1989.

Pietrzyk M., Kawalla R., Pircher H., Simulation of the behavior of voids in steel plates during hot rolling, *Steel Research*, Vol. 66, 1995, pp. 526–529.

Shah K.N., Finite element simulation of internal void closure in open-die press forging, *Advanced Manufacturing Processes*, Vol. 1, 1986, pp. 501–516.

Stahlberg U., Keife H., Lundberg M., A study of void closure during plastic deformation, *Journal of Mechanical Working Technology*, Vol. 4, 1980, pp. 51–63.

Stahlberg U., Influence of spread and stress on the closure of a central longitudinal hole in the hot rolling of steel, *Jnal. of Mechanical Working Technology*, Vol. 13, 1986, pp. 65–81.

Wallero A., Closing of a central longitudinal pore in hot rolling, *Journal of Mechanical Working Technology*, Vol. 12, 1985, pp. 233–242.

Wang A., Thomson P. F., Hodgson P. D., A study of pore closure and welding in hot rolling process, *Journal of Materials Processing Technology*, Vol. 60, 1996, pp. 95–102.

Chapter 9

Analysis of the Interply Porosities in Thermoplastic Composites Forming Processes

Anthony Cheruet, Damien Soulat and Philippe Boisse
Ecole Supérieure de l'Energie et des Matériaux, University of Orleans, France

Eric Soccard and Serge Maison-Le Poec
EADS CCR, Suresnes, France

1. Introduction

Composites with continuous fibres and thermoplastic matrices (CFRTP) are a promising alternative to the use of thermoset composites. Their use is increasing in the manufacture of new planes, for instance in the A340/500-600 (Maison *et al.*, 1998; Pora, 2001). Compared with thermoset composites their main advantage is the shorter processing cycle for equivalent mechanical properties. Additionally, their stocking, recycling and repair are easier. After heating at a temperature above melting (Figure 1a), the forming is made with a punch and die process, usually using a rubber on the die (Figure 1b). Finally, reconsolidation is obtained by applying pressure on the punch (Figure 1c). The objective of this last stage is to avoid all residual porosity at the interface of the plies. This is important and critical because the health requirements for loaded aeronautical parts are severe.

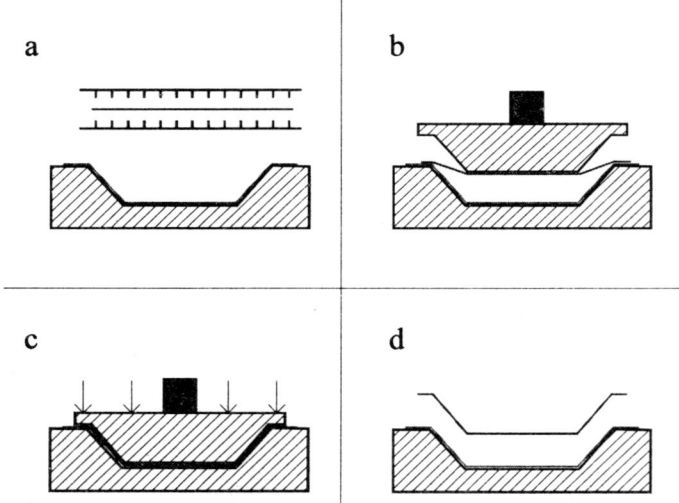

Figure 1. *Differents stages of forming, a) Heating of the CFRTP, b) Forming with punch and die, c) Reconsolidation phase, d) Final part*

The present paper is focused on the experimental analysis of the compaction stage and its numerical simulation. The evolution of the porosity is studied using micrographies of specimens formed at different ratios of complete forming. They show the importance of the porosities and consequently of the compaction stage that must ensure that all the voids disappear. The finite element analysis of the forming is made using one shell set per ply. There is a viscous contact between the plies and they can slide relative to reach other. Concerning the reconsolidation, porosities only disappears if sufficient normal stress is achieved in the ply. In order to model this compaction, a new shell element is defined in which a through-the-thickness strain degree of freedom is introduced. The stress through-the-thickness is not zero and this kind of finite element can describe correctly the compression during the recompaction stage without changing the general type of finite element analysis. Examples of the numerical simulation of the forming and the compaction stage will be shown in the case of a Z profile.

2. Sheet forming simulations

2.1. *Dynamic explicit scheme*

The simulation is made in the PAM-FORM® code (De Luca et al., 1998). An explicit dynamic approach is used to simulate the forming process. At each time step, the acceleration nodal vector an is calculated by solving the discrete equation of the motion at t_n:

$$Ma_n + F^{int}(u_n) = F^{ext} \qquad [1]$$

where M is the diagonal mass matrix, F^{int} and F^{ext} are the nodal interior and external loads:

$$M = A \int_{\Omega^e_t} N^t N \rho dV, \qquad F^{int} = A \int_{\Omega^e_t} B^t \sigma dV \qquad [2]$$

$$F^{ext} = A \int_{\Omega^e_t} N^t \overline{f} dV + A \int_{\Gamma^e_t} N^t \overline{t} dV \qquad [3]$$

N and B are the displacement and strain interpolation matrices, A is the assembling operator, σ is the Cauchy stress and $\overline{t}$ and $\overline{f}$ are the surface and volume exterior loads. An explicit scheme is used to compute u_{n+1}.

$$a_n = M^{-1}\left[F^{ext}_n - F^{int}(u_n)\right] \qquad [4]$$

$$v_{n+\frac{1}{2}} = v_{n-\frac{1}{2}} + \Delta t\, a_n \qquad [5]$$

$$u_{n+1} = u_n + \Delta t\, v_{n+\frac{1}{2}} \qquad [6]$$

Consequently, there is no need to solve a system of equations and the method is very effective. The time step size has to ensure the stability condition of the integration scheme accounting for the element size and the material data (Belytschko, 1983).

This explicit approach is used in many forming simulations and has been used in the case of woven composite reinforcement forming in (Boisse et al., 2001).

2.2. Application to CFRTP plate forming

One of the main forming modes is the relative sliding between the plies. In order to permit this sliding each ply is modelled by a set of shell element. An example is shown in Figure 2 in the case of a Z reinforcement. It is made of ten plies that slide during the bending of the initially flat plate (Figure 2). The relative sliding between the plies (Figure 3) is in agreement with experiment (Cheruet, 2001).

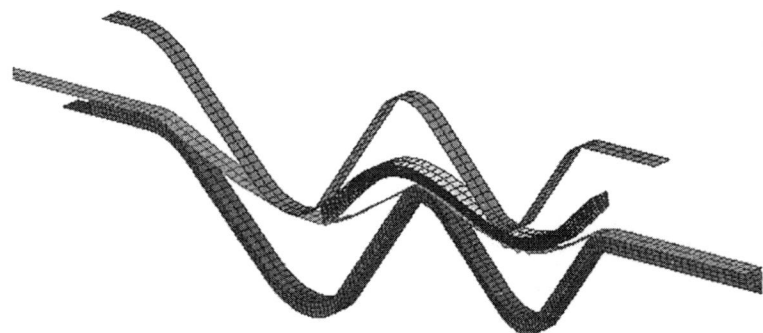

Figure 2. *Forming of a Z reinforcement. Forming stage*

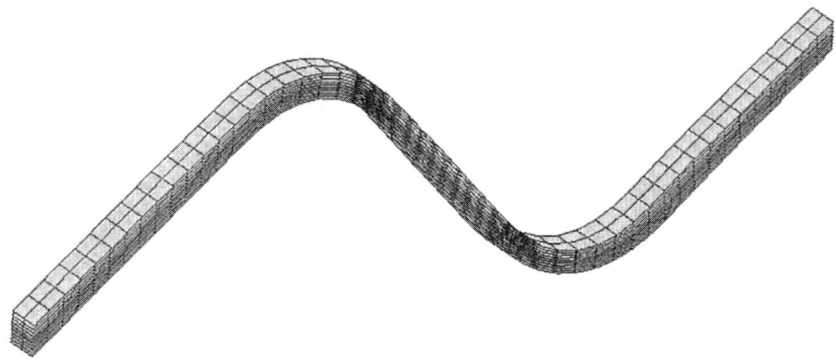

Figure 3. *Final shape of the composite*

Figure 4. *Interply sliding during the forming stage*

3. The re-consolidation stage

3.1. *Experiments*

An important aspect for the quality of the final part is the absence of porosities in the thickness of the composite. These porosities are a source of possible fracture in the service life of the composite and they must be avoided especially for aeronautical applications. That is the reason for the re-consolidation stage 3 of the forming process (Figure 1). In order to investigate the apparition of these voids and their resorption during forming and compaction stages, experiments have been performed at CCR EADS Suresnes on the Z shape reinforcement. The forming process has been stopped at different level of manufacture (Figure 5) and micrographies have been carried out in order to measure the porosities. The measures and micrographies have been carried out at different places in the specimen (Figure 6) especially in a flat and in a curved part. H is the height before the end of forming.

Figure 5. *Parts obtained at different stage of the process*

100   Prediction of Defects in Material Processing

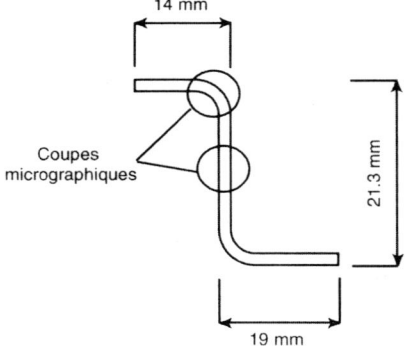

Figure 6. *Positions of the micrographies, on the Z profil*

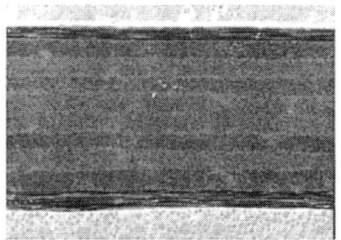   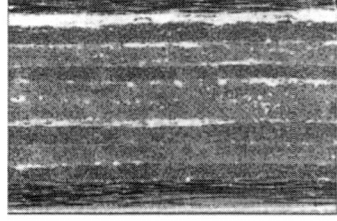

Figure 7a. *Material before heating*   **Figure 7b.** *Material after heating*

Figure 7a shows the state of the composite before forming. This composite plate has been pre-consolidated in an autoclave and no void is present. Figure 7b is obtained after one minute's heating. Many voids and porosities can be seen in the part. Figure 7c corresponds to a position at the beginning of the forming process (H = 11.3 mm). Porosities are still present both in flat and curved parts. Figure 7d corresponds to H = 5.3 mm. There are still voids in flat parts but there is no porosity in curved parts of the reinforcement. For H = 0, Figure 7e, all the voids have disappeared. This last stage corresponds to the end of the re-consolidation phase where a pressure is applied on the tools. It can be seen that this stage is essential to the final quality of the part. In the curved zones of the part, the re-consolidation is done during the forming, accounting for the geometrical effect that leads to transverse stresses during forming.

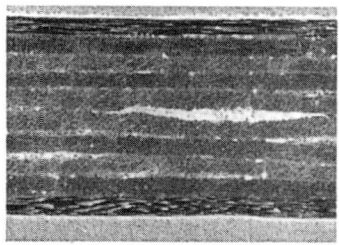 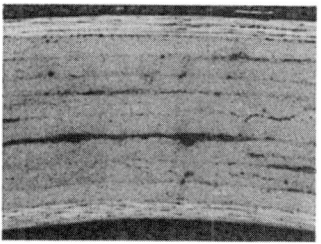

Figure 7c. *Material state in flat and curved parts for H = 11.3 mm*

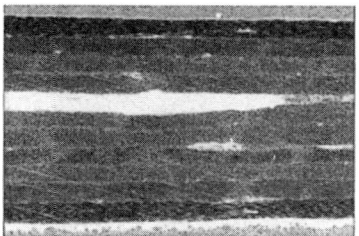 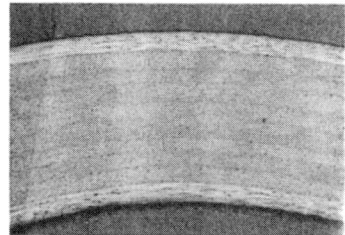

Figure 7d. *Material state in flat and curved parts for H = 5.3 mm*

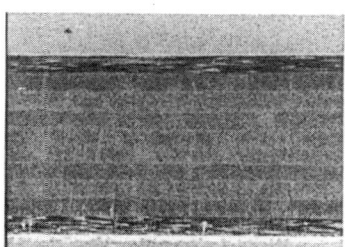 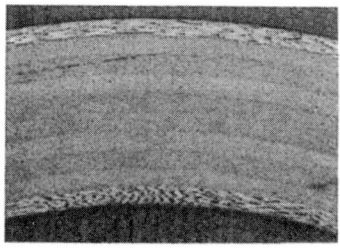

Figure 7e. *Material state at the end of the process H = 0*

3.2. Shell element with pinching

The previous experiments show clearly that porosities appear during the heating and forming stages and that the reconsolidation is essential for the final quality of the part. Consequently the numerical simulations have to describe this stage.

The re-consolidation have been studied in (Lee *et al.*, 1987) and some models have been proposed for the local consolidation. These studies have shown that the re-consolidation depends on the stress state in the laminate and mainly on the normal stress in the consolidation stage (Figure 1). This stress is not present in classical shell theory. Some finite elements with stress/strain through-the-thickness have been proposed (Simo et al., 1990; Butcher et al., 1994; Bischoff and Ramm, 1997; Bletzinger et al., 2000). In the present work a new shell element is used where a through-the-thickness strain degree of freedom is introduced (Coquery, 1999). The thickness stress is given by the constitutive law. The position of a point (Figure 7) in the shell is classically expressed by:

$$x = \bar{x} + \tilde{z}\hat{X} \qquad [7]$$

with $\bar{x}$ the projection of this point on the midsurface and $\tilde{z}$ the position of this point on the normal $\hat{X}$.

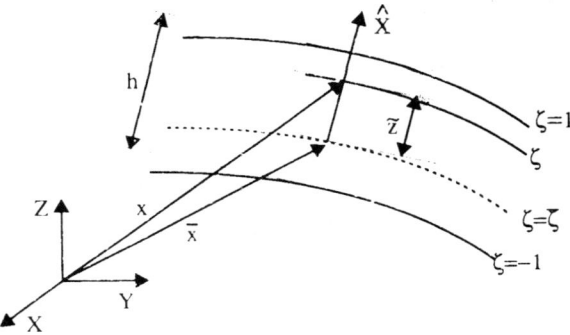

Figure 8. *Kinematics of the shell*

The displacement is obtained by subtracting the initial position from the current position.

$$u = (\bar{x}^1 - \bar{x}^0) + \tilde{z}^1\hat{X}^1 - \tilde{z}^0\hat{X}^0 \qquad [8]$$

Let $\Delta h = \tilde{z}^1 - \tilde{z}^0$ the displacement of the point in the thickness and β the through-the-thickness strain, $\beta = \dfrac{\Delta h}{\tilde{z}^0}$

In the case of small rotations between two states the displacement expression becomes:

$$u = \bar{u} - \tilde{z}^0\hat{X}^1 \times \theta + z\beta\hat{X}^1 \qquad [9]$$

If $\beta = 0$ equation [9] leads to the classical shell kinematics without pinching. We consider β = cte through-the-thickness. β is a additional pinching degree of freedom.

The strain tensor $\varepsilon(u) = \frac{1}{2}\left[\nabla(u) + \nabla^T(u)\right]$ can be derived from to the displacement [9]. In an orthogonal frame ($\hat{e}_1, \hat{e}_2, \hat{e}_3 = \hat{X}$) the membrane bending stain components are:

$$\begin{Bmatrix} \varepsilon_{11} \\ \varepsilon_{22} \\ 2\varepsilon_{12} \end{Bmatrix} = \begin{Bmatrix} u^m_{x,1} \\ u^m_{y,2} \\ u^m_{y,1} + u^m_{x,2} \end{Bmatrix} + z \begin{Bmatrix} \theta_{y,1} \\ -\theta_{x,2} \\ -\theta_{x,2} + \theta_{y,1} \end{Bmatrix} \qquad [10]$$

The transverse shears are:

$$\begin{Bmatrix} 2\varepsilon_{13} \\ 2\varepsilon_{23} \end{Bmatrix} = \begin{Bmatrix} u^m_{z,1} + \theta_y \\ u^m_{z,2} - \theta_x \end{Bmatrix} + z \begin{Bmatrix} \beta_{,1} \\ \beta_{,2} \end{Bmatrix} \qquad [11]$$

and the through-the-thickness strain (pinching) is:

$$\{\varepsilon_{33}\} = \beta \qquad [12]$$

The transverse shear strains are modified by the pinching. In contrast to the classical shell, the normal stress through-the-thickness is not zero. It is deduced from $\varepsilon_{33} = \beta$ using the compaction behaviour law.

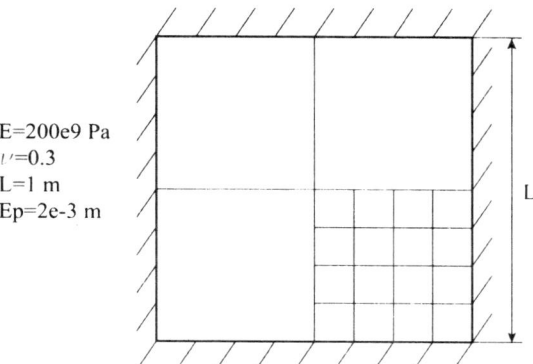

Figure 9. *Clamped square plate under pressure*

It has been shown (Coquery, 1999; Cheruet, 2001) that this element exhibits a "pinching" locking. To avoid this locking it is necessary to modify the constitutive relation in order to remove the coupling between pinching and bending (Coquery, 1999; Cheruet, 2001). The example presented in Figures 9 and 10 (clamped square plate under pressure) shows the pinching locking with is obtained with a complete behaviour law and the accurate result if pinching and bending are not coupled.

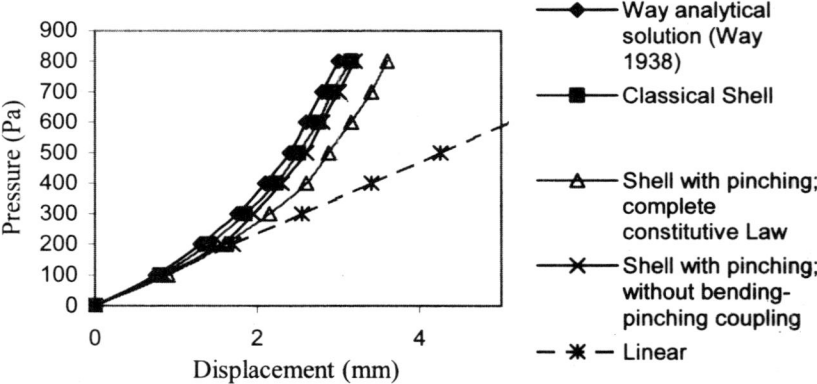

Figure 10. *Displacement at the center of a plate under pressure*

3.3. Simulation of forming and re-consolidation stage of a Z profile

Using the shell element described in 3.2, the through-the-thickness stress is computed in the case of an initially flat ply which is formed in a Z shape (Figure 2) and then compacted.

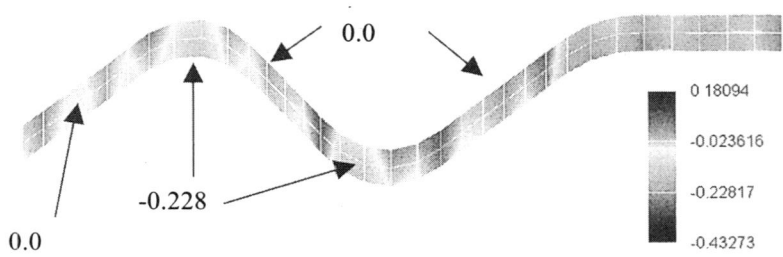

Figure 11. *Stress component through-the-thickness at the end of forming*

During the forming and at the end of the forming stage (Figure 11), the through-the-thickness stress is equal to zero in the most part of the part except in the radius of the tools. That confirms the observations on the micrographies that have shown that the consolidation happens only near the radius in the forming stage.

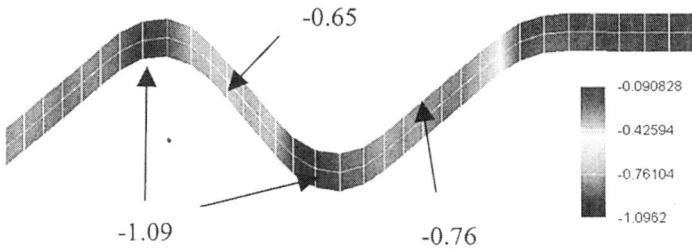

Figure 12. *Stress component through-the-thickness after the re-consolidation stage*

After the compaction phase, all the part is in compression (Figure 12). The larger value of the stress is obtained in the radius, but the full part is in compression. The value of the thickness stress depends of the angle of the curved part of the Z profile. In the present example, the through-the-thickness constitutive law has been supposed linear. That is not a correct assumption, because the porosities lead to very non-linear compaction laws (Gutowski, 1985; Boaxing *et al.*, 1999). Determination and identification of the good compaction law is one of the main points of further work on this subject.

4. Conclusion

The experimental analysis of the forming of a thermoplastic composite has shown the existence of porosities induced by the process and the importance of the re-consolidation phase. To simulate this compaction stage a shell element with pinching is presented. This element has a degree of freedom in the thickness and give the stress and strain through-the-thickness. The computation gives compressive through-the-thickness stresses in the zones where the experiments have shown that the laminate is consolidated; for instance, in the curved parts during the forming stage. The practical efficiency of the approach depends on the compaction constitutive relation that will have now to be investigated.

5. References

Baoxing C., Chou T.W., "Compaction of woven-fabric preforms in liquid composite molding processes: single layer deformation", *Composite Science and Tech.*, 1999, 59, pp. 1519–1526.

Belytschko T., "An overview of semidiscretisation and time integration procedures", *Computation methods for transient analysis*, Ed. T. Belytschko and T.J.R. Hughes, Elsevier Science, 1983, pp. 1–65.

Bletzinger K.U., Bischoff M., Ramm E., "A unified approach for shear-locking – free triangular shell finite elements", *Computer and structures*, Vol. 75, pp. 321–334, 2000

Bischoff M., Ramm E., "Shear deformable shell elements for large strains and rotations", *International Journal for Numerical Methods in Engineering*, Vol. 40, pp. 4427–4449, 1997.

Boisse P, Daniel J.L., Hivet G., D. Soulat, "A simplified explicit approach for simulations of fibre fabric deformation during manufacturing preforms for R.T.M. process", *International Journal of Forming Processes*, Vol. 3, No. 3–4, pp. 331–353, 2001

Butcher N., Ramm E., Roehl D., "Three dimensional extension of non-linear shell formulation based on the enhanced assumed strain concept", *Int. J. for Num. In Engng.*, 1994, Vol. 37, pp. 2551–2568.

Cheruet A., Analyse et simulation de la mise en forme des composites thermoplastiques, Thèse de doctorat, Université d'Orléans, 2001.

Cheruet A., Soulat D., Boisse P., Soccard E., Maison-Le Poec S, "Analysis of the reconsolidation of thermoplastic composites forming", *Proc. of the Fifth International ESAFORM conference on Material Forming*, pp. 291–294, 2002

Coquery M.H., Modélisation d'un joint de culasse multifeuille, Thèse de doctorat, ENSAM Paris, 1999.

De Luca, P., Pickett, A.K., "Industrial examples of forming non-metallic parts using PAM-FORM^{TM}", *Proceedings of PAM'98*, PSI/ESI Group, Tours, France, 1998, pp. 1–18.

Gutowski TG., "A resin flow/fibre deformation model for composites", SAMPE Quart., 1985, 16 (4), pp. 58–64

Lee W., Springer G., "A model of the manufacturing process of thermoplastic matrix composites", *Journal of Composite Materials*, Vol. 21, pp. 1017–1055, 1987.

Maison, S., Thibout C., Garrigues, C.,. Garcin, J.L, Payen H., Sibois, H., Coiffer, C., Vautey P., "Technical developments in thermoplastic composites fuselages", *SAMPE journal*, Vol. 34, No. 5, 1998, pp. 33–39.

Pora J., "Composite Materials in the Airbus A380 - From History to Future", *Proceedings of ICCM13*, CD-ROM, 2001

Simo J.C., Fox D.D., Rifai M.S, "On a stress resultant geometrically exact shell model. Part 4: Variable thickness shells with through-the-thickness stretching", *Comp. Meth. in App. Mech. and Engng*, 1990, Vol. 79, pp. 21–70.

Way S., "Uniformaly loaded, clamped, rectangular plates with large deformation", *Proc 5th International Congress of Applied Mechanics*, Cambridge, MA, 1938.

Chapter 10

Predicting Material Defects in Reactive Polymeric Flows

Francisco Chinesta
Laboratoire de Mécanique des Systèmes et des Procédés, UMR CNRS, ENSAM-ESEM, Paris, France

1. Introduction

Reactive polymeric systems are widely used in foam blowing or in resin transfer molding (RTM) processes. Numerical modelling of reactive processes is very difficult because the flow kinematics is coupled with the chemical reacting kinetics, which depends mainly on the reaction time (time elapsed since the reactants' mixing). The kinematics – chemical kinetics coupling – is a key point in foam blowing, being less important for example in usual RTM processes, where the chemical formulation can be adjusted in order to finish the mold filling before a significant rise in the resin viscosity.

The polymerization reaction increases the material viscosity and the expansion induces the material growth, generating a porous structure. When the polymerization takes place much faster than the foam expansion, the material becomes too rigid before the conclusion of the expansion. In this case the cellular walls can be broken with direct consequences for the mechanical properties of the conformed pieces. On the other hand, when the expansion finishes before the complete polymerization, the low material consistency is not enough to preserve the final geometry, and the structure collapses when the expansion is finished.

The numerical modelling of forming processes involving the flow of chemically-reacting polymeric foams requires taking into account the different problem scales. Thus, in industrial applications a macroscopic approach is suitable, whereas the macroscopic flow parameters depend on the cellular structure: porosity, size shape and orientation of the cells, cellular walls properties, etc. Moreover, the shape and orientation of the cells are induced by the flow during the foam expansion (induced anisotropy). We have analyzed the influence of process conditions on the cellular shape. Elongated cells were obtained by applying a traction or a shear in the material during its expansion (Chirivella *et al.*, 2000).

This microscopic information can be introduced in a macroscopic model by means of a homogenization technique. Thus, at each point in the material domain we can associate a characteristic volume containing some cells, whose size, shape and orientation depend on the point considered. Then, from the matrix and gas viscosities, and the cellular structure, we can compute an equivalent viscosity tensor. However, some difficulties appear in the homogenization procedure: (1) sometimes surface tension cannot be neglected, which requires a very accurate description of the inclusion geometry as well as consideration of its effect on the homogenized model; (2) the behaviour of the matrix fluid becomes quickly non-Newtonian, and its non linear character makes difficult the application of usual homogenization techniques; (3) the reaction kinetics remain uncertain for usual industrial processes; (4) the numerical tracking of a cell (microscopic level) which grows and deforms into the material flow is today an open problem, which in spite of some promising works (Sussman *et al.*, 1994, 1999; Sussman and Fatemi, 1999; Sussman and Puckett, 2000), important difficulties persist (Paredes, 2001); and (5) the physical mechanism of gas diffusion, the cells interaction and fusion, and the modelling of the final stage of the process, when the material is concentrated on the cell walls, are, among many others, important and unsolved difficulties found in the global modelling of foam growing.

In this way a macroscopic modelling where the reaction kinetics and the viscosity evolution are obtained experimentally is suitable from the point of view of its industrial application. Thus, such a numerical modelling could provide a visualization of mold filling (foam volume evolution and flow front position during the expansion reaction) and will be useful to optimize the mold and process design. From that simulation we can determine the number, size and optimal position of the evacuation orifices to avoid internal overpressures (reducing the cleaning necessities as well as the material losses). Moreover, an accurate flow front tracking allows us to predict the interactions between different flow fronts when the curing reaction is too advanced (welding lines are fragile) as well as to avoid filling defects due to the big air bubbles retained between the flow front and the mold walls.

2. Simplified flow model

In order to obtain a macroscopic simplified flow model, able to carry out fast simulations of reacting foam flows, we consider the usual balance equations:

– The equilibrium equation neglecting the inertia and mass terms gives

$$Div\underline{\underline{\sigma}} = \underline{0} \qquad [1]$$

where $\underline{\underline{\sigma}}$ is the stress tensor.

– The mass conservation

$$\frac{\partial \rho}{\partial t} + Div(\rho \underline{v}) = 0 \qquad [2]$$

where $\underline{v}$ is the velocity field and ρ the foam density. The previous equation can be rewritten as

$$\frac{\partial \rho}{\partial t} + \rho\, Div\underline{v} + \underline{v}\cdot Grad\rho = 0 \qquad [3]$$

When the foam expansion is homogeneous $\rho = \rho(t)$ and $Grad\rho = \underline{0}$. In this case equation [3] becomes

$$Div\underline{v} = -\frac{\partial \rho}{\partial t}\frac{1}{\rho} = \beta(t) \qquad [4]$$

This function can be defined from the chemical reaction kinetics or from a simple experimental test. The first procedure, as previously discussed, remains very inaccurate because it is very difficult to know all the chemical reactions and their kinetics, in order to compute accurately the evolution of the density. Thus, we prefer, as proposed in Chinesta et al. (Chinesta et al., 2000), to proceed from a direct determination of the density evolution.

— The constitutive equation establishes the relation between the stress tensor $\underline{\underline{\sigma}}$ and the strain rate tensor $\underline{\underline{D}}$

$$\underline{\underline{\sigma}} = \underline{\underline{\sigma}}(\underline{\underline{D}}, t) \qquad [5]$$

where $2\underline{\underline{D}} = Grad\,\underline{v} + (Grad\,\underline{v})^T$. For compressible flows $Tr(\underline{\underline{D}}) = Div\,\underline{v} \neq 0$.

Equation [5] indicates also that the foam rheology depends on the reaction time. The simplest constitutive relation consists of

$$\underline{\underline{\sigma}} = -p\underline{\underline{I}} + \left(\lambda - \frac{2}{3}\eta\right) Tr(\underline{\underline{D}})\,\underline{\underline{I}} + 2\eta\underline{\underline{D}} \qquad [6]$$

where λ denotes the volumetric viscosity, p the pressure field, $\underline{\underline{I}}$ the unit tensor and η the shear viscosity. If the volumetric viscosity contribution is neglected

$$\underline{\underline{\sigma}} = -p\underline{\underline{I}} + 2\eta\underline{\underline{D}} \qquad [7]$$

where $\underline{\underline{D}}'$ denotes the deviatoric strain rate tensor. The viscosity evolution of the matrix fluid can be measured using usual techniques (Richter and Macosko, 1980). The most difficult task is the determination of the foam viscosity evolution (when the foam is considered as a pseudo-homogeneous fluid). The rheological characterization is difficult on account of the impossibility to proceed with standard rheometers which are not adapted to multiphase flows where the cell size is of the same order as the diameter or the gap of capillary or cone-plate rheometers respectively. Thus, some authors propose the use a of viscosity law in the form of a mixing rule that includes as control variables the concentration in reactants, the shear rate, the temperature and the gas volume fraction (Lefebvre and Keunings, 1993). Other authors prefer to describe the foam rheology by means of a simple power law model (Deshpande and Barigou, 2001) which must be identified using a non-standard device. The flow consistency and the behaviour indices of the power law equation could depend on time to describe the material polymerization.

— We can notice that thermal effects are not taken into account in the previous model in order to minimize the parameters which must be identified experimentally, although it is well known that the kinetics depends significantly on the temperature field (usually the determination of the reaction kinetics is made from the temperature evolution measured during the material curing).

3. Numerical modelling

When the foam expansion is heterogeneous, the density depends on the spatial coordinates, and the mass conservation equation is expressed by equation [3]. In this case the variational formulation of equations [1], [3] and [6] results in

$$\int_{\Omega(t)} \left(\left(-p + \left(\lambda - \frac{2}{3}\eta\right) Tr(\underline{\underline{D}}) \right) Tr(\underline{\underline{D}}^*) + 2\eta \, \underline{\underline{D}} : \underline{\underline{D}}^* \right) d\Omega = 0 \qquad [8]$$

and

$$\int_{\Omega(t)} -p^* \left(Div\underline{v} - \beta + \frac{Grad\rho}{\rho} \cdot \underline{v} \right) d\Omega = 0 \qquad [9]$$

The numerical discretisation of the flow model for an homogeneous foam expansion was discussed in a former work (Chinesta et al., 2000) From the kinematics point of view the discretisation of the flow equations governing the heterogeneous foam expansion, equations [8] and [9], do not introduce particular difficulties, if we can assign to each point in the foam domain and for each time the corresponding values of β, λ, η and ρ. Usually, the volumetric viscous term can be neglected, β derives directly from the density ρ, whose time evolution is obtained following the procedure described in the previous section. Finally, in some simple rheological models, the foam shear viscosity depends on the equivalent strain rate as well as on the time, due to the polymerization and blowing kinetics.

In this form, for each point in the fluid domain at any time, we need to know, for the fluid located at that point, the time elapsed from the beginning of the chemical reactions (reaction time). Knowing that time we can assign the density and viscosity values.

In the case of homogeneous expansions we assume that the reactants are dumped in the mould simultaneously, and in consequence the reaction time is the same for any point, which corresponds to the usual Eulerian time when we consider that the reaction starts at $t = 0$.

However, the manufacture of big pieces requires a great amount of reactants that must be distributed in large areas. In this case, we can not neglect the time elapsed in the reactants' dumping, so that, sometimes, the blowing starts before the complete reactant dumping, giving rise to a heterogeneous foam expansion. In that case, we need to label each point of the reactant fluid volume with its dumping time T_d, and transport this field during the foam expansion. Thus, we must solve the following linear advection equation

$$\begin{cases} \dfrac{dT}{dt} = \dfrac{\partial T}{\partial t} + \underline{v}\, GradT = 0 \\ T(\underline{x}, t=0) = T_d(\underline{x}) \end{cases} \qquad [10]$$

For example, if we suppose that reactants are dumped over a plate of one meter, from the left $(x = 0)$ to the right $(x = 1)$, in one second, the time label T_d could be for example $T_d = x - 1$. If we take the origin of time ($t = 0$) when the total amount of reactants is dumped, the reaction time t_r will be given by $t_r = t - T$.

When an Eulerian technique, using a fixed mesh, is adopted to solve equation [10] some difficulties appear, related to the field updating in elements whose filling stage is just starting. To illustrate this fact we consider an empty finite element at time t, which starts its filling process. In this case, as we are using a fixed mesh strategy, the field T updating in an element Ω^e, i.e. the computation of T^e at time $t + \Delta t$, makes use of its value at time t. Because this element starts its filling process at time t, the field $T(t)$ is not defined. If the simulation is carried out without paying attention to the different field initializations, great deviations appear, mainly in the neighborhood of the flow front, whose incidence cannot be neglected. For example, in the welding of several flow fronts we require active chemical kinetics to assure the material cohesion. In order to predict this kind of material defect, we need carry out simulations taking into account the hyperbolic character of advection equation, keeping good accuracy in the flow front treatment.

The main original contribution of this work is the establishment of a new general strategy able to carry out the field transport accurately even through the flow front. This technique minimizes the numerical diffusion and implicitly transports all fields with the flow front movement. The proposed technique can be also applied to general models involving advection or advection-diffusion equations.

4. Transport problems

As discussed in the last section, a transport problem appears associated with the advection of the field T, which is governed by the following linear advection equation [10]. If we consider a first order discontinuous finite element discretisation then

$$T^e(t + \Delta t) = T^e - T^e \frac{\Omega^+}{|\Omega^e|} + T^{e^-} \frac{\Omega^-}{|\Omega^e|} + T^e \beta \Delta t \qquad [11]$$

where $T^e(t+\Delta t)$ is the value of T in the element Ω^e at time $t+\Delta t$, Ω^- the fluid volume coming to the element Ω^e from the upstream element Ω^{e^-} (where the value of T is given by T^{e^-}), Ω^+ is the fluid volume leaving the element Ω^e and $|\Omega^e|$ is the volume of Ω^e.

A first problem appears suddenly if we consider the time t for which an element Ω^e starts its filling process from its upstream element Ω^{e^-}. To illustrate this limitation we consider the simplest situation of an incompressible fluid, i.e. $\beta = 0$. In this case, with $\Omega^+ = 0$, we obtain

$$T^e(t+\Delta t) = T^e + T^{e^-}\frac{\Omega^-}{|\Omega^e|} \qquad [12]$$

which establishes that solution at $t+\Delta t$ is dominated by the solution existing in the element at time t, at which this element is empty, and consequently the property T is not defined. The question is: what is the value to assign to T before the beginning of its filling process?

An accurate and consistent discrete form of the conservation equation can be obtained by considering a discontinuous finite element discretisation of the equation which results from the addition of equation [10] multiplied by the fluid fraction I, whose evolution is governed by the advection equation (Chinesta et al. 2000)

$$\frac{dI}{dt} = \frac{\partial I}{\partial t} + \underline{v}\,Grad\,I = 0 \qquad [13]$$

with equation [13] multiplied by the field T, as proposed in Chinesta et al. (Chinesta et al., 2002)

$$\frac{d(IT)}{dt} = \frac{\partial(IT)}{\partial t} + \underline{v}\,Grad(IT) = 0 \qquad [14]$$

which results in

$$T^e(t+\Delta t) = \sum T^{e_i^-}\frac{\Omega_i^-}{I^e(t+\Delta t)|\Omega^e|} +$$
$$+ T^e\left(\frac{I^e(t)}{I^e(t+\Delta t)} - \frac{\Omega^+}{I^e(t+\Delta t)|\Omega^e|}\right) + \frac{\beta T^e I^e(t)\Delta t}{I^e(t+\Delta t)} \qquad [15]$$

where Ω_i^- is the volume of fluid coming to Ω^e from the upstream element $\Omega^{e_i^-}$; with

$$I^e(t+\Delta t) = I^e + \sum_i I^{e_i^-} \frac{\Omega_i^-}{|\Omega^e|} - I^e \frac{\Omega^+}{|\Omega^e|} + \beta I^e \Delta t =$$

$$= I^e + \sum_i \frac{\Omega_i^-}{|\Omega^e|} - \frac{\Omega^+}{|\Omega^e|} + \beta I^e \Delta t$$

[16]

5. Numerical example

We consider a numerical example involving a welding line development from two reacting flows with different reaction times. Figure 1 shows the foam presence function and the velocity field at three steps in the mold filling. Figure 2 depicts the transport of the dumping time. The actualization of the field T is carried out by using equations [15] and [16].

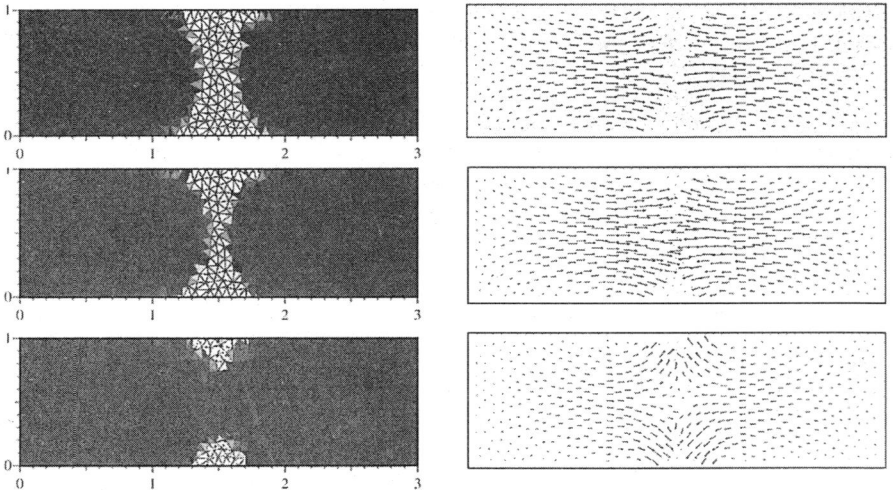

Figure 1. *Foam presence and velocity field during the expansion and welding*

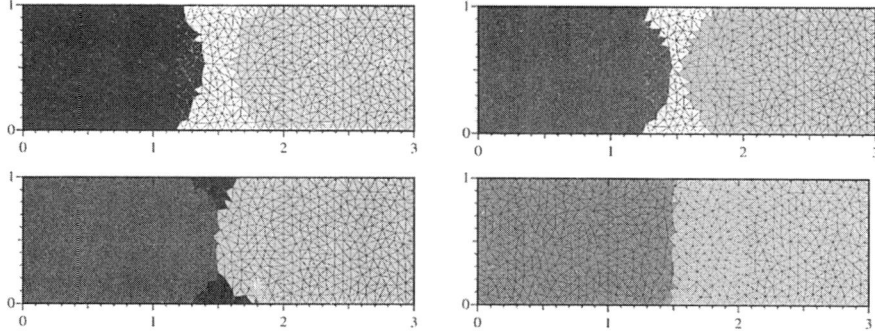

Figure 2. *Transport of the dumping time and welding line development*

6. Conclusions

From these results we can draw conclusions about the accuracy and low numerical diffusion of the numerical strategy proposed in the discretisation of equation [10]. The results are significantly improved, in relation to standard discontinuous finite element discretisations, mainly in the neighborhood of the flow front. In this way the proposed strategy can be used to compute accurately the reaction times in welding areas in 3D foam expansion process simulation, to predict defects in the material welding.

7. Bibliography

Chirivella P., Chinesta F., Godet M., Experimental analysis of the induced anisotropy in foam forming processes, Internal report LRTMM, CNAM Paris, 2000.

Chinesta F., Chaidron G., Godet M., Bermudez A., "Evaluation d'un modèle numérique simplifié de l'expansion de mousses", *Les Cahiers de Rhéologie,* Vol. 17, No. 1, 2000, pp. 403–412.

Chinesta F., Mabrouki T., Ramon A., "Some difficulties in the flow front treatment in fixed mesh simulations of composites forming processes", 5th *International Esaform Conference*, Krakow, 2002.

Deshpande N.S., Barigou M., "Foam phenomena in sudden expansion and contractions", *International Jour. of Multiphase Flow*, Vol. 27, 2001, pp. 1463–1477.

Lefebvre L., Keunings R., "Mathematical simulation of the flow of chemically-reacting polymeric foams", *Int. Conf. on Mathematical Modelling for Material Processing*, Clarendon Press, 1993, Oxford, pp. 399–417.

Paredes J., Numerical simulation of the bubbles movement inside a flow (*spanish*), Internal rapport, Departamento de Matematica Aplicada, Universidad de Santiago de Compostela, Spain, 2001.

Richter E.B., Macosko C.W., "Viscosity changes during the isothermal and adiabatic network polymerization", *Polymer Eng. and Sci.*, Vol. 20, No. 14, 1980, pp. 921–924.

Sussman M., Smereka P., Osher S., "A level set approach for computing solutions to incompressible two-phase flows", *Journal Comput. Physics*, Vol. 114, 1994, pp. 146–159.

Sussman M., Almgren A., Bell J., Colella P., Howell L., Welcome M., "An adaptative level set approach for incompressible two-phase flows", *Journal Comput. Physics*, Vol. 148, 1999, pp. 81–124.

Sussman M., Fatemi E., "An efficient interface preserving level set redistancing algorithm and its applications to interfacial incompressible fluid flows", *SIAM J. Sci. Comput.* Vol. 20, No. 4, 1999, pp. 1165–1191.

Sussman M., Puckett E., "A coupled level set and volume-of-fluid method for computing 3D and axysymmetric incompressible two-phase flows", *Journal Comput. Physics*, Vol. 162, 2000, pp. 301–337.

Chapter 11
Anisotropic Damage Model for Aluminium

Patrick Croix, Franck Lauro and Jérôme Oudin
Laboratoire LAMIH, Université de Valenciennes, France

1. Introduction

The finite element method is nowadays widely applied to crashworthiness or sheet metal forming. In order to accurately predict the damage and failure evolution occurring in such simulations, a realistic material description is required. For this purpose, this paper presents a damage model for elasto-viscoplastic materials, taking into account both the material and the damage anisotropy. On the microscale, the material is composed of a pore-free matrix, inclusions and second phase particles and microvoids randomly distributed. During the straining process, the volume of microvoids growths under hydrostatic stress and the number of microvoids increases by nucleation due to the fracture of inclusions and to the decohesion of second phase-matrix interfaces. The evolution of the geometrical form and the volume of the microvoids may lead to the coalescence of some cavities. Therefore, ductile rupture may occur when the stress-carrying capacity has vanished.

To describe the previous damage process, a coupling mechanical-damage model is used. The description of the porous material is based on the introduction of an internal variable representing the microvoid volume fraction and defined as the ratio of the microvoid volume and the material volume. The microvoid growth, nucleation and coalescence are described. The growth of existing microvoids is determined from the plastic incompressibility relation. The microvoid nucleation is related to the plastic strain and depends on the distribution of the inclusions and second phase particles. An additional term is used to ensure some damage in the case of shear loading and it corresponds to a microvoid nucleation part controlled by shear strain. The microvoid coalescence occurs when the space between neighbouring microvoids is almost equal to their length. The ductile rupture is predicted at the end of the previous damage process for a determined level of microvoid volume fraction. Damage anisotropy is introduced by means of the shape of the microvoid. The microvoid, usually initially elliptical, changes in shape and in direction as a function of the loading. The plastic flow of the porous material is predicted by the Gologanu microvoided material potential modified to take into account the material anisotropy.

A radial return algorithm is proposed for the implementation of a new constitutive law for elasto-viscoplastic porous materials for convected shell elements into an explicit finite element framework.

Accurate prediction of the damage evolution using the previous damage model implies well defined material parameters of this model. These damage material parameters are essentially the initial porosity, the geometrical shape of the microvoids, the inclusion volume fraction and the mean nucleation strain for tensile and shear parts, and the value of the microvoid volume fraction at coalescence onset. Due to their microscale, the identification of these parameters by a direct method needs expensive measurement machinings and complex measurement techniques which challenge the interest of numerical simulations.

To determine the damage material parameters of the proposed damage model, an original identification technique is proposed. This identification technique is based on the inverse method. The variation due to straining of an external variable strongly dependent on the material parameter values is first experimentally measured. An optimisation technique is then used to determine the values of material parameters which correlate the experimental variation of the external variable and its numerical prediction by the finite element method.

The damage material parameters are identified by using a tensile test of a notched specimen and by measuring the variation of the inner radius of the specimen during straining as experimental setup (Lauro et al., 2001).

Due to the anisotropy aspect of the damage parameter, three macroscopic measurements in the three orthotropic directions L, L-T and T are considered. The optimisation of the damage material parameters is performed by interfacing the optimiser and the finite element software. The damage parameters for the shear nucleation part are deduced from an Arcan test which combined tensile and shear loadings.

In this paper, the damage model formulations and its numerical implementation are first described. The identification by inverse method is then presented and applied to an anisotropic aluminium part. The Arcan test is presented and used to identify damage parameters as well as to validate the identification procedure.

2. Anisotropic damage model

This model is based on the studies of a material unit cell formed by two confocal spheroids. The GLD model (Gologanu et al., 1994) extends the well-known GTN model (Gurson, 1977) to take the microvoid shape effect into account. Indeed, the GLD model considers cavities of ellipsoidal form, whose form volume and the orientation can evolve. The Gologanu analysis of ellipsoidal voids is based on the velocity field which describes the expansion of these voids. This velocity field satisfies conditions of homogenous boundary strain rate, as did the Gurson analysis, but applied to ellipsoidal voids. The GLD model is therefore interesting as it becomes the GTN potential for spherical microvoids. This model depends on three internal parameters: the porosity f, the shape of the microvoid S $\{S = \ln(a_1/b_1)$, where a_1 and b_1 represent the minor and major semi-axes of the axisymmetric ellipsoidal cavities$\}$, and the vector e_x collinear to the axes of revolution of the microvoid.

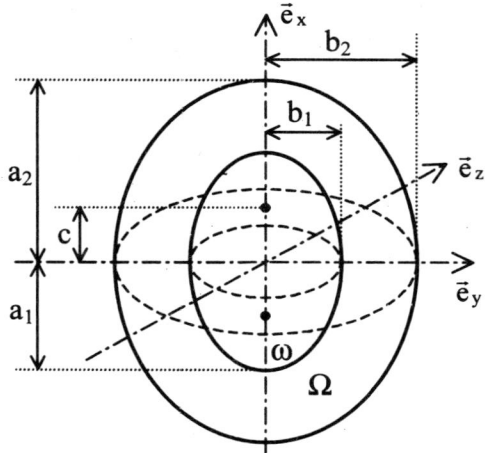

Figure 1. *The microstructure of a prolate cavity*

The plastic potential is a quadratic formulation which can also be used as a yield function as follows:

$$\phi_{evp} = C \frac{\|\sigma' + \eta \, \sigma_H \, X\|^2}{\sigma_M^2} - \varphi = 0 \qquad [1]$$

where

$$\varphi = 1 + (q_1 f^*)^2 - 2 q_1 f^* \cosh(\upsilon), \quad \upsilon = \frac{\kappa \sigma_H}{\sigma_M}, \sigma_H = (1 - 2\alpha_2)\sigma_{11} + \alpha_1 \sigma_{22} \qquad [2]$$

The parameters κ, η, C and X are given by:

$$\kappa^{-1} = \frac{1}{\sqrt{3}} + \frac{1}{\ln(f)} \left[(\sqrt{3} - 2) \ln\left(\frac{e_1}{e_2}\right) - \frac{1}{\sqrt{3}} \ln\left(\frac{3 + e_1^2 + 2\sqrt{3 + e_1^4}}{3 + e_2^2 + 2\sqrt{3 + e_2^4}}\right) + \ln\left(\frac{\sqrt{3} + \sqrt{3 + e_1^4}}{\sqrt{3} + \sqrt{3 + e_2^4}}\right) \right] \qquad [3]$$

$$\eta = -\frac{\kappa(1-f^2)\,sh}{1+f^2+f(\kappa H\,sh - 2\,ch)}$$

$$C = -\frac{\kappa f\,sh}{(1-f+\eta H)\eta}, \quad sh \equiv \sinh(\kappa H), \quad ch \equiv \cosh(\kappa H), \quad H = 2(\alpha_1 - \alpha_2) \tag{4}$$

$X = \frac{1}{3}(2\,e_x \otimes e_x - e_y \otimes e_y - e_z \otimes e_z)$, with e_x, e_y, e_z the axis of the frame of the void and where the eccentricities e_1, e_2 and the parameter α_2 are deduced from the void shape parameter S and the void volume fraction f by the following equations:

$$e_1 = \sqrt{1 - e^{-2|S|}} \tag{5}$$

$$\frac{e_2^3}{1-e_2^2} = f\,\frac{e_1^3}{1-e_1^2} \tag{6}$$

$$\alpha_2 = \frac{1+e_2^2}{3+e_2^4} \tag{7}$$

The notation $\|\bullet\|$ is used in the original GLD potential to express the calculation of the von Mises norm $\left(\|T_{ij}\| = [1.5\,T_{ij}T_{ij}]^2 \right)$. In order to consider the anisotropy of the material the von Mises norm is replaced by the Hill norm (Hill, 1948; Croix et al., 2001). It is supposed that the cavity follows the rolling direction, consequently in this work the axis e_x of the microcavity corresponds initially to the rolling direction, and both the constraint and strain tensors are defined in the base e_x, e_y, e_z.

This model is completed by the equation of the evolution of the internal shape parameter S given by:

$$\dot{S} = \frac{3}{2}\left(1+\frac{9}{2}h_T(T,\zeta)(1-\sqrt{f})^2 \frac{\alpha_1 - \alpha_1^G}{1-3\alpha_1}\right)\left(\dot{\varepsilon}_{11} - \frac{\dot{\varepsilon}_{kk}}{3}\right) + \left(\frac{1-3\alpha_1}{f} + 3\alpha_2 - 1\right)\dot{\varepsilon}_{kk} \tag{8}$$

with $h_T(T,\zeta)$ a function, dependent on the triaxiality $T \equiv \sigma_{kk}/(3\sigma_{eq})$ according to the sign of $\zeta = \sigma_{kk}\sigma'_{ii}$, given by:

$$\begin{cases} h_T(T,\zeta) = 1 - T^2 & \text{for } \zeta > 0 \\ h_T(T,\zeta) = 1 - \dfrac{T^2}{2} & \text{for } \zeta < 0 \end{cases} \tag{9}$$

and α_1 and α_1^G are obtained by:

$$\alpha_1 = \frac{1}{2e_1^2} - \frac{1-e_1^2}{2e_1^3}\tanh^{-1}(e_1),\qquad [10]$$

$$\alpha_1^G = \frac{1}{3-e_1^2} \qquad [11]$$

The evolution of the micro structural damage is represented by the current void volume fraction f, defined by $f = 1 - V_M/V_A$, where V_A, V_M are respectively the elementary apparent volume of the material and the corresponding volume of the matrix. f^* is a function of the void volume fraction f (Tvergaard and Needleman, 1984):

$$\begin{cases} f & \text{if } f \leq f_c \\ f_c + \dfrac{1/q_1 - f_c}{f_F - f_c}(f - f_c) & \text{if } f > f_c \end{cases} \qquad [12]$$

where f_c is the critical microvoid volume fraction at coalescence onset, f_F is the microvoid volume fraction when ductile fracture occurs. This specific function f^* inside the microvoid material potential describes the rapid loss of the stress-carrying capacity due to the coalescence of the neighbouring microvoids, when f reaches the limit $1/q_1$.

The microvoid volume fraction evolution has two main phases: the nucleation of the new voids and the growth of existing voids. The microvoid volume fraction rate is expressed by:

$$\dot{f} = \dot{f}_{growth} + \dot{f}_{nucleation} \qquad [13]$$

The constitution from void nucleation is controlled by plastic strain, and takes the form (Chu and Needleman, 1980):

$$\dot{f}_{nucleation} = \frac{f_N}{S_N\sqrt{2\pi}} e^{-\frac{1}{2}\left(\frac{\varepsilon_M - \varepsilon_N}{S_N}\right)^2} \dot{\varepsilon}_M = A_1\dot{\varepsilon}_M \qquad [14]$$

where f_N is the nucleated microvoid volume fraction, S_N is the Gaussian standard deviation, ε_N is the mean effective plastic strain for nucleation and ε_M is the effective plastic strain and $\dot{\varepsilon}_M$ is define by the equation [20]. $\dot{f}_{nucleation}$ value is equal to zero for negative or nil hydrostatic stress and then particularly for pure shear loading.

The growth of existing voids is given by:

$$\dot{f}_{growth} = (1-f)\dot{\varepsilon}_{kk} \qquad [15]$$

Numerically, in pure shear loading, there is no damage evolution and so no rupture. To consider the damage due to shearing, the damage evolution law is modified to become the sum of the classical law [13] and a new one for shear law, *i.e.*:

$$\dot{f} = \dot{f}_{growth} + \dot{f}_{nucleation} + \dot{f}_{shear} \qquad [16]$$

In pure shear, it is commonly accepted that the voids experience a rotation without any change in growth and it appears that nucleation can be generated. Consequently, the damage evolution law due to shearing takes the form of a statistical law, like the nucleation evolution law but considering the shearing strain and the shearing strain rate, and is so defined by:

$$\dot{f}_{shear} = \frac{f_S}{S_S\sqrt{2\pi}} e^{-\frac{1}{2}\left(\frac{\varepsilon_{xy}-\varepsilon_S}{S_S}\right)^2} \dot{\varepsilon}_{xy} \qquad [17]$$

3. Numerical implementation

The previous constitutive damage model is implemented in the finite element code PAM-SOLID^{TM} (Figure 2).

Assuming the strain rate tensor $^{n-1/2}\dot{\varepsilon}_{ij}$ computed at the previous time increment is elastic, the corresponding Cauchy stress tensor is updated by the central finite difference:

$$^n\sigma_{ij} = {^{n-1}\sigma_{ij}} + ({^nt} - {^{n-1}t})\, ^{n-1/2}C^e \dot{\varepsilon}_{ij} \qquad [18]$$

in which $^n\sigma_{ij}$ depends on $^n\sigma_{xx}$, $^n\sigma_{yy}$, $^n\sigma_{xy}$ in the case of shell elements, where t is the time and n the increment.

The corresponding potential is obtained by $\phi_{evp} = \sqrt{C}\,^n\sigma^{eq} - {^{n-1}\sigma_M}\sqrt{\varphi}$; then all the derivation comes from this equation.

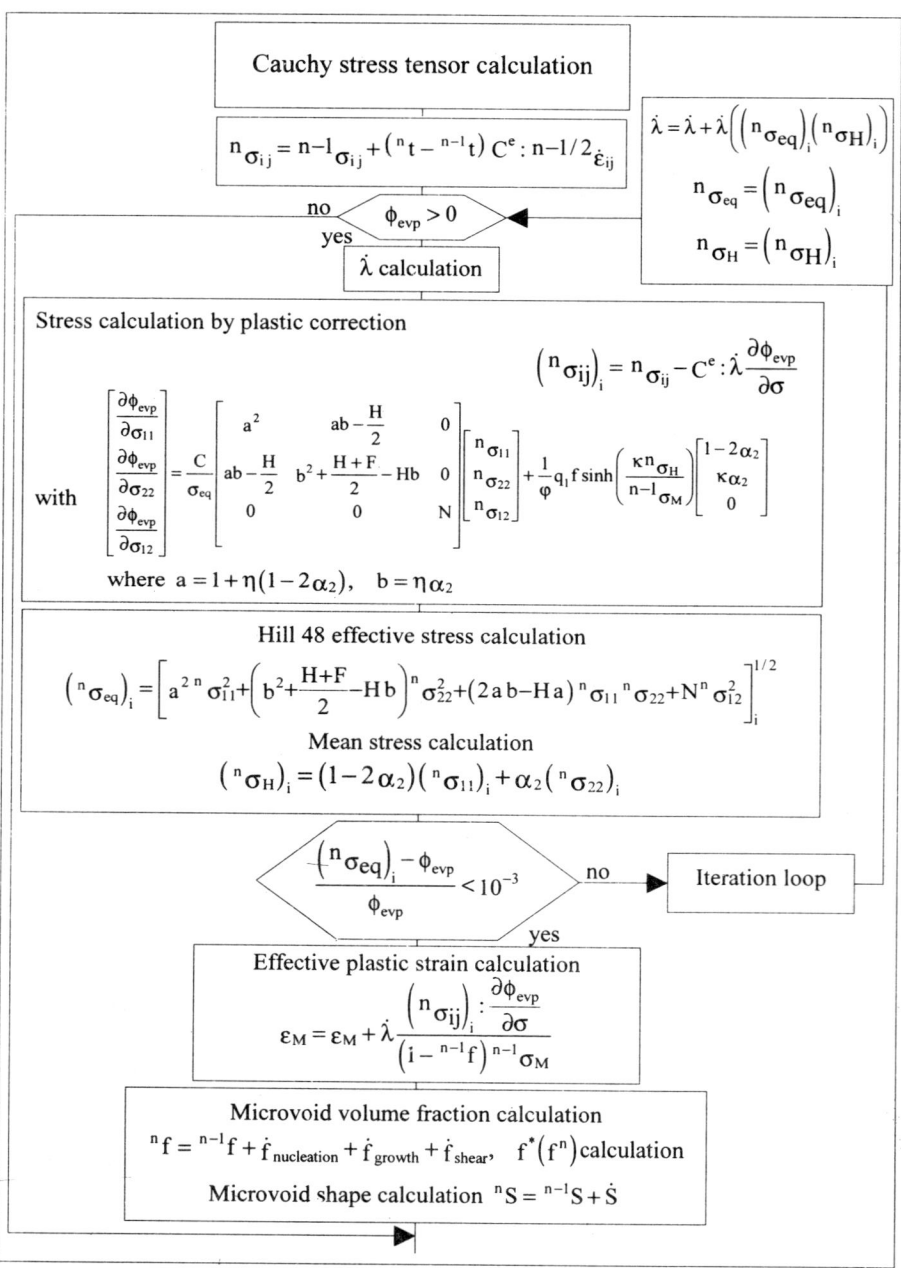

Figure 2. *The modified stress elastic prediction and plastic correction flowchart*

The plastic multiplier $\dot{\lambda}$ is deduced from the consistency condition $\phi_{evp} = 0$ and $\dot{\phi}_{evp} = 0$ leading to:

$$\phi_{evp} = \dot{\phi}_{evp} = \frac{\partial \phi_{evp}}{\partial \sigma} : \dot{\sigma} + \frac{\partial \phi_{evp}}{\partial \sigma_M} : \dot{\sigma}_M + \frac{\partial \phi_{evp}}{\partial f} : \dot{f} = 0 \qquad [19]$$

The plastic multiplier is finally expressed by

$$\dot{\lambda} = \frac{\phi_{evp}}{\frac{\partial \phi_{evp}}{\partial \sigma} : C^e : \frac{\partial \phi_{evp}}{\partial \sigma} - \frac{\partial \phi_{evp}}{\partial \sigma_M} \frac{\partial \sigma_M}{\partial \varepsilon_M} A_2 - \frac{\partial \phi_{evp}}{\partial f} \left[(1-f) \frac{\partial \phi_{evp}}{\partial \sigma} : I + A_3 \right]} \qquad [20]$$

$$\text{with } A_2 = \frac{\sigma : \frac{\partial \phi_{evp}}{\partial \sigma}}{(1-f)\sigma_M} \qquad [21]$$

$$\text{and } A_3 = A_1 \cdot A_2 \qquad [22]$$

where C^e is the isotropic material tensor and I is the second order identity tensor.

4. Identification

The direct identification of the damage parameters is complex and expensive. Thus an identification method based on an inverse technique is used. This method consists of the identification of the damage parameters by correlating an experimental and numerical macroscopic measurement strongly dependent on the parameters. The correlation is obtained by minimising a cost function which is defined by the least square approximation as

$$Q(\alpha) = \sum_{i=1}^{nb\_point} \frac{\left[Z_i^{sim}(\alpha) - Z_i^{exp} \right]^2}{\left[Z_i^{exp} \right]^2} \qquad [23]$$

where α are the material parameters, Z_i^{sim} and Z_i^{exp} are the simulated and experimental macroscopic responses and nb_point is the number of experimental points of the experimental response. An optimiser is used to find the material parameters minimising this cost function.

An iterative method which uses the cost function value as criterion, and a convergence method which takes information about gradients and second derivatives (Hessien) of the cost function into account, are the most efficient.

The optimiser is linked with the finite element code to automate the identification of the material parameters (Figure 3).

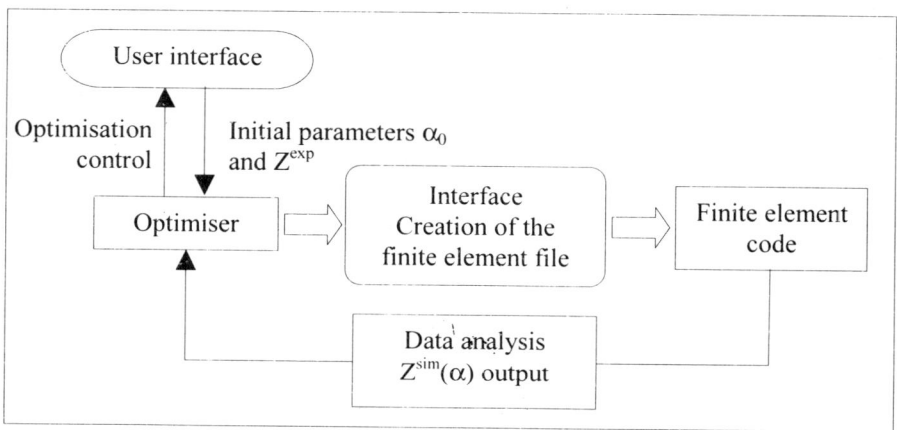

Figure 3. *Flowchart of the identification by inverse method*

The macroscopic measurement must come from a mechanical test which presents significant damage and a ductile rupture. Moreover, it must be easy to obtain. So, the tensile test of a notched specimen is used because the damage is perfectly localised in the bottom of the notch. The macroscopic measurement is the variation of the width in the bottom of the notch according to the elongation of the tensile specimen. This test presents the advantage of being simple to set up and requires few specific materials. Indeed, the testing apparatus is composed only of a tensile testing machine and two extensometers for the measurement of the macroscopic size.

In order to identify the damage parameters and particularly the shape parameter which controls the anisotropy of the damage, this procedure is carried out for three tensile tests according to 0°, 45° and 90° of the rolling direction. Then, the identification consists of the minimization of the cost function.

These tensile tests are not able to identify the damage parameters controlled by the shearing strain. These parameters are determined by Arcan type tests which present some shearing. The coalescence parameters f_C and f_F are not obtained by the inverse method because the experimental measurements are not able to seize the propagation of the crack. They are simply determined by comparing the value of the damage obtained at the instant of rupture numerically and experimentally.

The behaviour law used is obtained according to the rolling direction for an aluminium sheet. With this consideration, the determination of Hill's anisotropic parameters are made by means of three tensile tests of smooth tensile specimens taken in material according to 0°, 45° and 90° of the rolling direction.

So, the flow stress used for the numerical simulation, is described by successive tangent moduli:

E_1 = 647 MPa $\qquad \alpha_1$ = 206 MPa

E_2 = 600 MPa $\qquad \alpha_2$ = 218 Mpa

E_3 = 450 MPa $\qquad \alpha_3$ = 227 Mpa

E_4 = 300.5 MPa $\qquad \alpha_4$ = 518 Mpa

The initial yield stress, σ_y = 120 MPa, Poisson's ratio, υ = 0.3212 and Young's modulus, E = 65000 MPa, for the elastic part. The three orthotropic coefficient values are G = 0.958; F = 1.191 and N = 2.834.

The determination of the damage parameters, with tensile tests of notched specimens, is carried out by the inverse method. Due to the geometrical symmetries of the notched tensile specimen (Figure 4a), a quarter of the tensile specimen is numerically modelled (Figure 4b).

The initial damage parameters and the identified parameters are presented in Table 1.

Table 1. *Values of initial and optimised damage parameters*

Parameters	Initial values	Identified values
q_1	1.5	1.396
S	0.001	0.000961
f	0.001	0.001
f_N	0.04	0.03858
S_N	0.1	0.1202
ε_N	0.2	0.09519

The new identified damage parameters improve the experimental-numerical correlation for the three directions considered (Figures 5, 6 and 7).

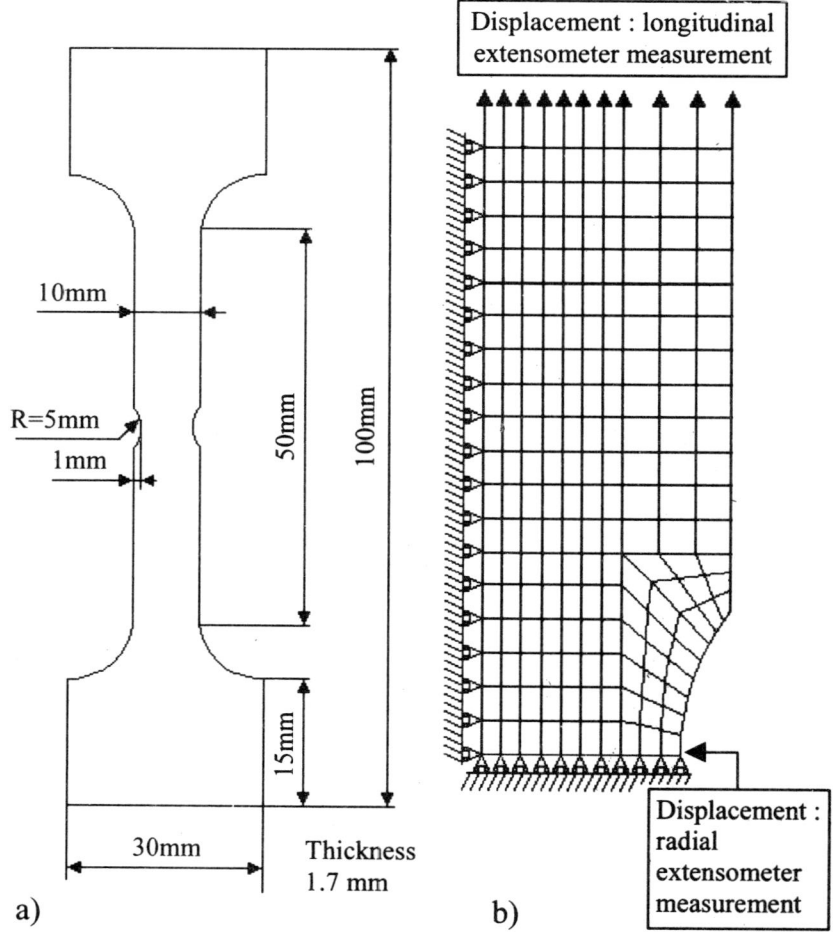

Figure 4. *a) Geometrical description of the thin notched specimen, b) Finite element modelling, boundary conditions for one quarter of the notched tensile specimen*

The identification of the damage parameters controlled by the shearing strains must be carried out. This identification can be carried out by the inverse method using an experimental test with shear loadings. The Arcan type tests are then used because these tests combine various tensile-shear loadings.

Anisotropic Damage Model for Aluminium   129

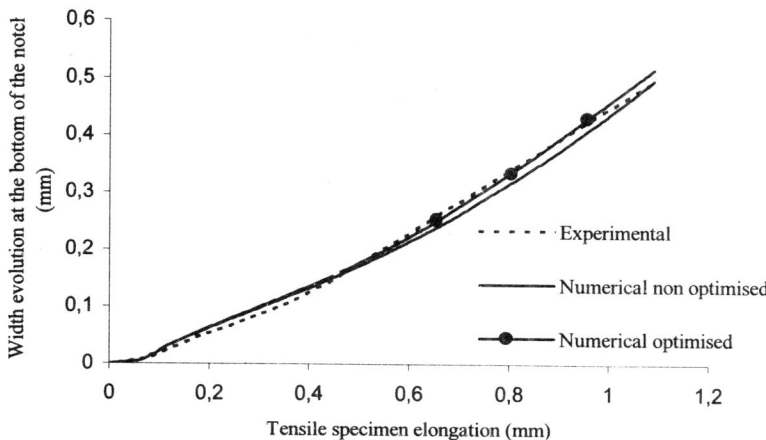

Figure 5. *Comparison of experimental-numerical width evolution at the bottom of the notch according to the tensile specimen elongation, for a specimen sample along the rolling direction*

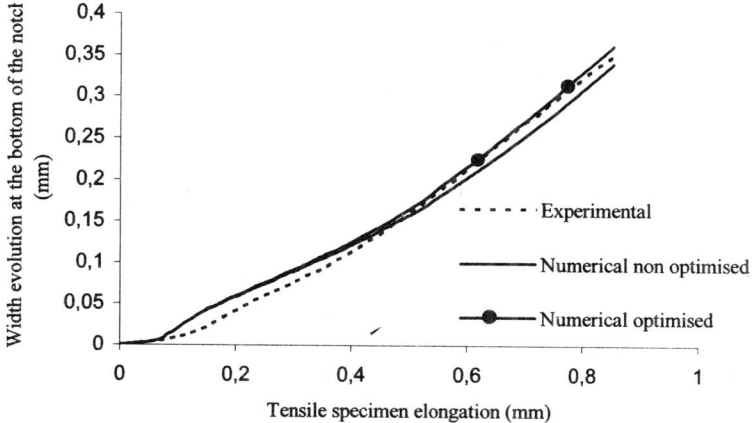

Figure 6. *Comparison of experimental-numerical width evolution at the bottom of the notch according to the tensile specimen elongation, for a specimen sample of 45° with the rolling direction*

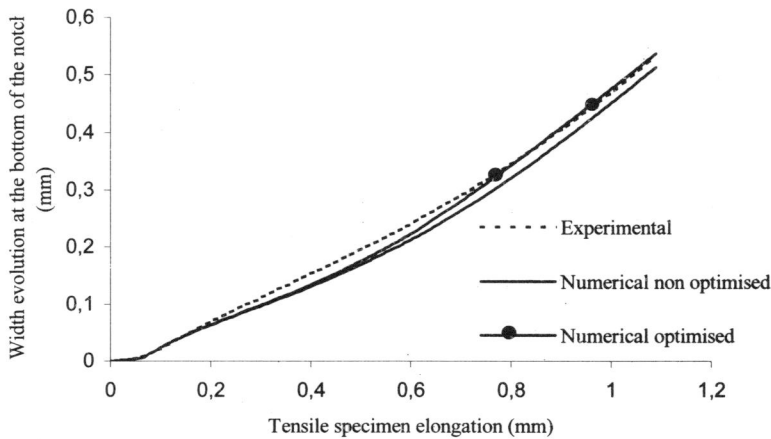

Figure 7. *Comparison of experimental-numerical width evolution at the bottom of the notch according to the tensile specimen elongation, for a specimen sample of 90° with the rolling direction*

To carry out the Arcan type tests, a specific fixture is defined for the flat test specimen. This fixture is adapted for a tensile test machine (Figure 8).

Figure 8. *Test Arcan fixture*

Various combinations of loading in traction-shearing are obtained by changing the position of the disc.

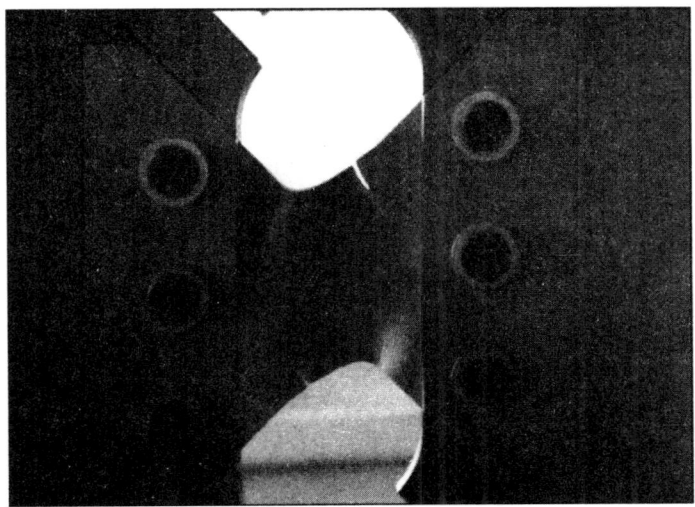

Figure 9. *Rupture occurrence for the Arcan test at 90°*

With a test at 90° (Figure 9), the rupture occurs in the tensile area. Without taking into account the evolution of damage due to shearing, the initiation of rupture is numerically incorrect. It is therefore necessary to take the damage due to shearing into account. The identification of the three damage parameters is performed by the inverse method for the Arcan test at 90°, (Table 2). For this identification the experimental measurement considered is the variation of the strength according to the displacement.

Table 2. *Values of initial and optimised damage parameters.*

Parameters	Initial values	Identified values
f_S	0.04	0.01
S_S	0.1	0.2
ε_S	0.2	0.1

The crack propagations obtained numerically and in experiments are similar (Figure 10). The damage parameters obtained give a better prediction of the instant of rupture, approximatively at 5.2 mm of displacement for numerical simulation and experimental test. Then, the loss of carrying capacity is relatively in good correlation with experiment due to the elimination of rough element (creation of step on the end of the curve, Figure 11). A finer mesh will probably lead to a better representation of the loss of the carrying capacity.

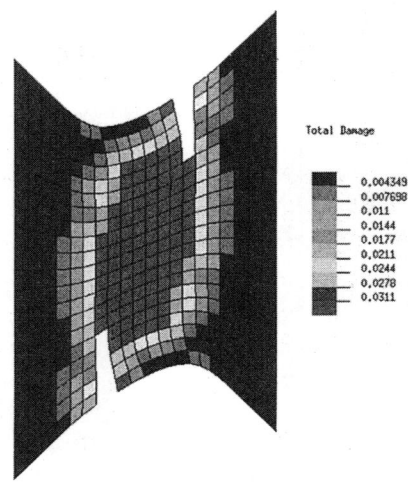

Figure 10. *Damage result of the numerical simulation of the Arcan test at 90°*

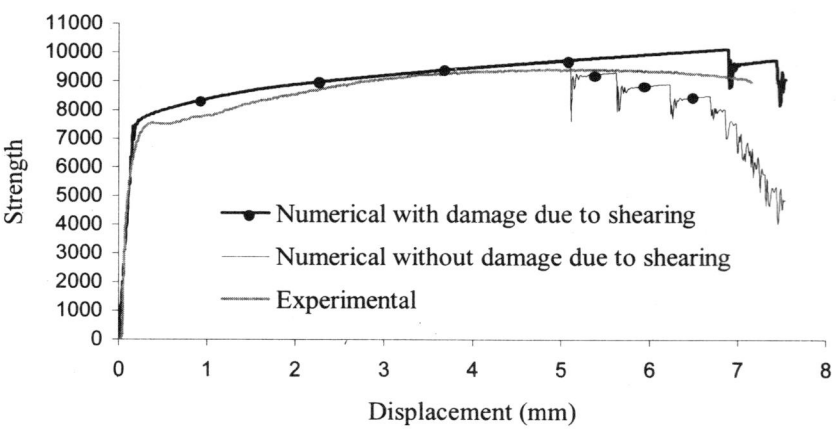

Figure 11. *Curved effort displacement for the Arcan test at 90°, numerical result with identified and non identified damage parameters due to shearing, and experimental results*

5. Conclusion

This paper presents a coupled damage model which takes into account the anisotropy of both the damage and the material behaviour. This model corresponds to the Gologanu-Leblond-Devaux model which takes the voids shape effect into account and in which the von Mises yield stress is replaced by the anisotropic Hill 1948 yield stress. The void nucleation, growth and coalescence are modelled. In order to consider the evolution of damage due to shearing, an original formulation for the rate of damage evolution is proposed. The evolution of the microvoid shape and orientation due to the deformation is considered. The ductile fracture is predicted at the complete loss of the stress-carrying capacity.

The anisotropic damage material parameters are identified by using an inverse method. This method uses an optimiser which minimises the gap between experimental and numerical simulations. The identification procedure uses two inverse methods. At first, the identification of the damage parameters by an inverse method using notched tensile specimens which come from 0°, 45° and 90° with the rolling direction. This identification gives the reference parameters (ε_N, f_N, S_N, S, f, q_1), then the damage parameters of coalescence are deduced by comparison with the initiation of rupture. Finally, the experimental results of Arcan type tests are considered in order to obtain the non-optimised damage parameters, *i.e.*: ε_S, f_S, S_S.

6. Bibliography

Lauro F., Bennani B., Croix P., Oudin J., "Identification of the damage parameters for anisotropic materials by inverse technique: Application to an aluminium", *Journal of Materials Processing Technology*, Vol. 118, No. 1–3, 2001, pp. 472–477.

Gologanu M., Leblond J.B., Devaux J., "Numerical and theoretical Study of coalescence of cavities in periodically Voided solids", *Computational Material Modelling*, Vol. 42, 1994, pp. 223–244.

Gurson A.L., "Continuum theory of ductile rupture by void nucleation and growth: Part I – Yield criteria and flow rules for porous ductile media", *Engineering Material Technology*, Vol. 99, 1977, pp. 2–15.

Hill R., "A theory of the yielding and plastic flow of anisotropic metals", *The hydrodynamics of non-Newtonian fluids*, 1948, pp. 281–297.

Croix P., Lauro. F., Oudin J., Christlein J., "Anisotropic damage applied to numerical ductile rupture", *Revue européenne des éléments finis*, Vol. 10, 2001, pp. 311–326.

Tvergaard V., Needleman A., "Analysis of the cup-cone fracture in around tensile bar", *Acta Metallurgica*, Vol. 32, 1984, pp. 157–169.

Chu C.C., Needleman A., "Void nucleation effects in biaxially stretched sheets", *Engineering Material Technology*, Vol.. 102, 1980, pp. 249–256.

Chapter 12

Shape Defects Measurement in 3D Sheet Metal Stamping Processes

Leonardo D'Acquisto and Livan Fratini
University of Palermo, viale delle Scienze, Palermo, Italy

1. Introduction

Three dimensional stamping operations are often affected by the insurgence of defects such as shape defects and ductile fractures. So far as the former are regarded, they are usually associated with the insurgence of wrinkles or puckers both on the flange and on the sheet metal between the punch and the die. It should be observed that elastic springback has to be considered as a shape defect occurring during sheet metal stamping processes since it introduces deviations from the desired final shape; consequently, the stamped sheet does not conform to the design specifications and could result in being unsuitable for the application (Makinouchi, 1996).

Springback takes place at the end of the stamping operation after removing the forming tools: actually almost all the sheet forming processes are characterised by a significant amount of deformation introduced by bending mechanics, the distribution of strains along the sheet thickness being strongly inhomogeneous. Such a distribution, together with the elastic-plastic behaviour of the workpiece determines the occurrence of springback after the removal of the forming tools (Mickalich *et al.*, 1988; Lange, 1985).

In modern automotive industries the tendency to reduce the weight of the stamped parts has led to the use of new materials such as high strength steels and aluminum alloys. It is well known from tensile tests that the elastic amount of the total strain, which is recovered if the load is released, is equal to the ratio of the stress before unloading to the Young's modulus. Thus the tendency to elastic springback increases with increasing the strain hardening coefficient and by decreasing the elastic stiffness; consequently springback plays a very important role for the above mentioned materials (Forcellese *et al.*, 1998).

In the last two decades some mathematical models were proposed to predict springback effects for very simple geometries and process mechanics such as bending. Actually these models were characterised by strong geometrical assumptions and for this reason they cannot be used to predict springback in complex three-dimensional industrial processes. More recently powerful numerical techniques based on the finite element method have been proposed to simulate sheet stamping processes. Nevertheless, the application of numerical techniques to the prediction of the springback effect requires a rather high skill level on the part of the analyst since large scattering in the results is obtained by varying of some numerical parameters (Forcellese *et al.*, 1998).

What is more, for the measurement of out-of-plane surfaces several techniques based both on contact and no contact procedures present useful characteristics, showing different sensitivity levels and, utilizing different set-ups which are very complex and troublesome to implement within an industrial environment. In such techniques, the use of traditional CMMs is widely utilized; such equipment is quite effective, but it is time consuming since it performs a single-point measurement, thus requiring a great number of points to describe completely a surface with a low level of uncertainty (Kobayashi, 1991).

In the present paper the authors presents the application of an innovative optical technique to measure the springback effect on fully 3D stamped parts. Such technique, already developed by the authors, is aimed to digitise and measure 3D surfaces: in particular, moving from the classical shadow Moiré technique, new features were implemented enabling one to overcome certain shortcomings, intrinsic in the use of the original Moiré method (Post et al.,1994; D'Acquisto et al., 2001).

The proposed measurement technique was utilized to measure the final shape of a set of aluminum AA 6061 deep drawn square boxes. In particular the springback measurement resulted from the geometrical comparison of the measured final shape of the drawn part with the ideal one evaluated through the CAD geometry of the utilized forming dies. Furthermore, the influence of the level of the blankholder force on the elastic springback phenomenon was investigated.

2. The deep drawing of square boxes

The elastic springback which occurs after the deep drawing of square boxes was considered (Figure 1). Such a process shows basic deformation mechanics which is determined by the restraining actions applied to the drawing part as it is pushed inside the die by the punch. Such actions start from the combined effect of two operative parameters, namely the blankholder force and the lubricating conditions. A too heavy combination of the two effects would determine excessive action on the blank with subsequent necking phenomena and tearings; on the other hand a too low combination of blankholder force and lubricating condition would lead to the insurgence of wrinkles. So far as the process mechanics is regarded, box drawing is characterized by the variation of metal flow rates along the straight walls and around the corners of the deforming flange; in this way an uneven material distribution around the box walls arises (Ahmetoglu et al., 1995; Fratini et al. 1995; Doege et al., 1987; Doege et al., 1994).

It should be observed that the process mechanics occurring at the four box corners is quite different from what happens in the four straight sides. Actually, so far as the box corners are regarded, typical circumferencial compressive stresses are observed characterizing the process mechanics of the axial symmetric deep drawing operations. In this way, at the corners, blankholding is needed to avoid defects such as wrinkling due to the described stress state. At the same time, if the tensile load in the corner wall is too severe due to the blank-holder pressure acting on a large surface, the metal may tear and this defect can progress into the flat wall. The latter consideration leads to the conclusion that blank-holder force to be applied in the corners has to be carefully checked (Yossifon et al., 1995).

138   Prediction of Defects in Material Processing

On the other hand, as the flat wall is regarded, no hoop stresses occur and consequently no compression on the flange would be required to control metal flow toward the die radius; in this way the process cannot properly be referred to as drawing. The process mechanics which occurs at the straight box walls is typically a bending and straightening of the sheet metal: consequently the process mechanics is less sensitive to the blank-holder force.

The above considerations show that in the process design a compromising level of the blank-holder force has to be found out, able to supply an acceptable value of the restraining action for each region of the flange avoiding the insurgence both of wrinkles and of tearings.

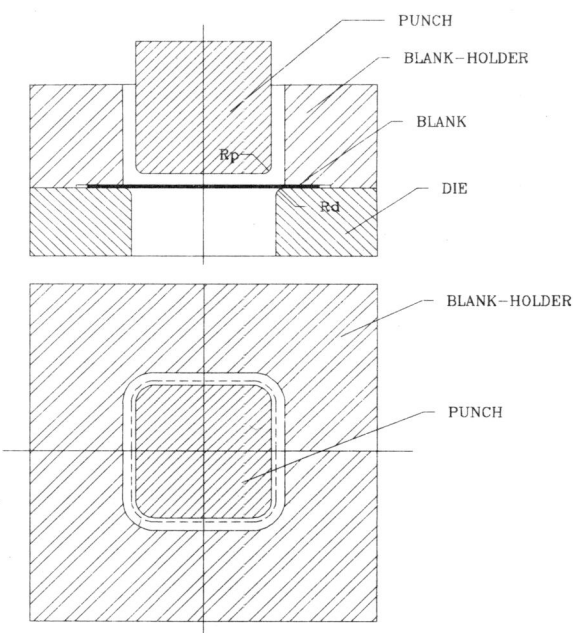

Figure 1. *The process scratch*

What is more, the levels of restraining forces locally determine the stretching mechanics induced in the sheet metal and in this way quantitatively determine the elastic springback. Such a phenomenon determines an opening of the cup walls which at the end of the process presents a greater tapering with respect to the one given by the die design and in particular by the punch lateral surfaces tapering and by the clearance between the punch and the die. Actually the four corners represent a geometrical discontinuity in the stamped part and they determine the stiffness of the square box. Nearby, at the four corners the highest levels of accumulated plastic strain are observed and, as a consequence, a different elastic deformation after the dies opening is expected.

The deep drawing process considered (Figure 1) is characterized by a drawing ratio, defined as the ratio between the diameter of the undeformed blank and the punch diameter, R = 90/45 = 2.0. The drawn parts obtained showed an average height equal to 25 mm. The punch and the die radii (Rp and Rd respectively) are equal to 3 mm, furthermore, in order to favor the ejection of the stamped parts the punch lateral surfaces have got a tapering with a ratio equal to 4:1000. An aluminum alloy AA 6061 was utilized for the tests developed with an initial thickness equal to 0.5 mm. The deep drawing operations were developed utilizing a blankholder activated by a pneumo-oleodynamic system and monitored through a personal computer; a screw mechanical press was utilized with a ram speed equal to 2 mm/s. Different levels of blankhoder force have been taken into account referring to the initial blankholder force on the blank expressed as percentage of the material yield stress (180 N/mm^2). Finally the lubrication between the blank and the dies has been obtained utilizing an extreme pressure grease.

3. The measurement technique

The characterization of the amount of springback occurring after the stamped part is relieved from the dies is performed in terms of the change that has occurred in the outer lateral surface of the stamped part.

A measurement technique suitable to measure 3D object surfaces has therefore to be employed to measure the actual shape of the stamped part after it has been extracted from the dies. It will be then compared to the ideal shape of the stamped part in order to characterize and calculate the amount of the elastic springback occurring. A number of techniques are currently suitable to measure 3D object surfaces. They are based both on contact and no contact procedures and present different sensitivities.

An optical technique to measure and digitize 3D surfaces based on the shadow moirè method and on the Fourier transform method (D'Acquisto *et al.*, 2002) is employed in the present paper to measure the outer lateral surface of the stamped part after springback has occurred. In previous papers the authors have compared the results of the proposed measurement technique with those obtained from a CMM (ZEISS Prismo Vast 7 HTG), showing a good agreement and highlighting the effectiveness of the proposed technique (D'Acquisto *et al.*, 2001; D'Acquisto *et al.*, 2002).

The employed measurement technique, moving from the classical shadow moirè requires a few set-up changes in order to introduce a carrier fringe pattern to significantly improve spatial resolution, overcoming some inherent shortcomings in the conventional method. The optical signal acquired is processed by the Fourier transform method (Takeda *et al.*, 1982; Sciammarella *et al.*, 1986). The measurement results obtained show adequate accuracy to resolve out-of-plane elevations of some micrometers over a depth range of a few millimeters. Such characteristics enable the developed technique to be very suitable for the measurement of stamped parts, usually characterized by large flat areas, assuring proper resolution and a low level of uncertainty. Furthermore, this non-contact measurement technique guarantees the absence of any interaction (instrument-measurand) which could alter the measurand itself.

With reference to Figure 2, the proposed method then relates distance z between the measurement surface on the prismatic bar and the points on the specimen profile to the phase of light intensity signal visualized on the specimen caused by the mechanical interference between the moiré grid and its shadow projected from the light source.

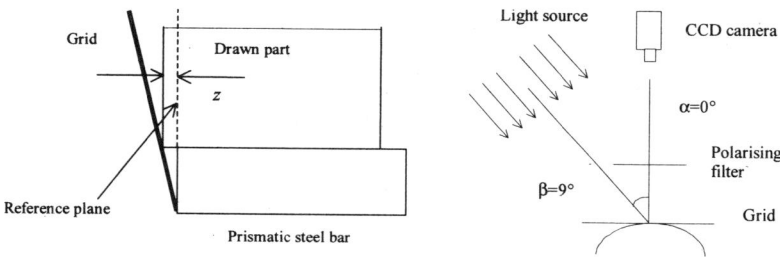

Figure 2. *a) specimen grid lay-out*   *b) image acquisition scheme*

The following parameters, with reference to the Figure 2, have been chosen in order to acquire the images utilized in the measurement procedure:
- vision angle $\alpha = 0°$; lighting angle $\beta = 9°$;
- distance between the grid plane and the camera 1700 mm; stop $f = 32$

The experimental set up utilized consists of the following items: a 35 mW He-Ne laser light source; a Moiré grid on glass plate (Graticules mod. SAG4) with rectilinear parallel fringes with pitch $p = 0,127$ mm; a prismatic block to hold the specimen (painted white matt to visualize the carrier optical signal on the front surface); a 10 bit CCD B/W digital camera (SVS-Vistek CA085A10) (1280x1024 pixel); a camera lens Nikkor AF Micro 70–210mm, 1:4 D; a Pentium III class PC.

The real profile to measure and the use of a fringe carrier signal give rise in the presence of noise sources to a light intensity signal expressed by the following equation, in which, for reason of simplicity, a column ordered data processing has been indicated:

$$I(x) = I_0 + I_1 \cos[\phi(x) + \omega_c x] + I_n$$

when $\omega_c \cdot x \gg \phi(x)$, being $I_0(x)$ the background intensity of the signal, $I_1(x)$ the intensity of the modulation term, $I_n(x)$ the intensity of the high frequency added noise and $\omega_c = 2 \cdot \pi \cdot f_c$ the carrier frequency (rad/mm). The unknown phase $\phi(x)$ contains the required information related to the geometry of the investigated lateral surfaces. To carry out the evaluation of $\phi(x)$ it is also necessary to remove the terms $I_0(x)$ and $I_n(x)$, by filtering (Sciammarella et al., 1967), to obtain the *in-phase* signal $I_f = I_1 \cos(\phi(x) + \omega_c^* x)$.

Introducing the use of a *quadrature* signal I_q, the wrapped phase $\phi'(x)$, variable in the range $+/-\pi$ is described by the following relation, when the information on the sign of $I_q(x)$ and $I_f(x)$ are both retained:

$$\phi'(x) = tan^{-1}(I_q(x)/I_f(x))$$

The continuous phase $\phi(x)$ is obtained by the unwrapping process described by equation: $\phi(x) = \phi'(x) + 2 \cdot k \cdot \pi$, with k integer number. This process requires the apriori knowledge of the sign of k (Sciammarella *et al.*, 1986).

The Fourier transform method enables one to obtain in a semi-automated manner the wrapped phase along a profile or a surface. In the unwrapping procedure, when an initial fringe pattern with constant spatial frequency is introduced by the use of a carrier signal, the sign of each 2π jump is obtained by evaluating the phase and frequency change in the final fringe pattern with respect to those of the carrier fringe. In the procedure utilised, it is given by subtracting $\phi_c(x) = \omega x$ from the phase obtained by the application of the Fourier transform method to the experimental fringes signal.

The proposed measurement technique allows maximum levels of combined standard uncertainty (ISO, 1993) $u_c(z)$ of about 10 μm to be obtained. Assuming a coverage factor $k = 2$, the extended uncertainty is then equal to: $U = k \cdot u_c(z) = 21.4$ μm.

A proper rebuilding procedure is realised to merge the single digitised surfaces into the whole lateral surface of the investigated stamped part.

4. The results obtained

The measurement technique described above has been applied to the quantification of elastic springback occurring after a deep drawing operation. Eight different images for each specimen were acquired in order to measure the actual lateral surface of the stamped box. A procedure aimed at assembling the different part surfaces was set up in order to obtain the complete lateral surface of the fully three-dimensional stamped part (D'Acquisto *et al.*, 2002). In particular a rebuilding procedure based on a least square fit technique was used, until a perfect fitting of each couple of contiguous images was obtained. In other words, assuming the reference system of the first acquired image as the global one, all other local reference systems were roto-traslated to coincide with the global one.

The eight images acquired for the evaluation of the total lateral surface of a drawn part manufactured with an initial blankholder pressure equal to 3% of the AA 6061 yield stress (σ_0) are reported in Figure 3. In Figure 4 the final rebuilding of the lateral surface of the specimen is shown.

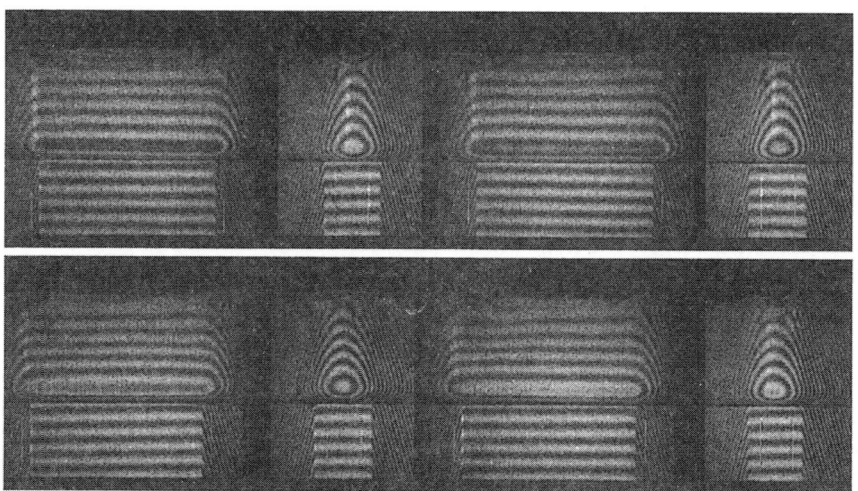

Figure 3. *Single acquired modulated fringe patterns*

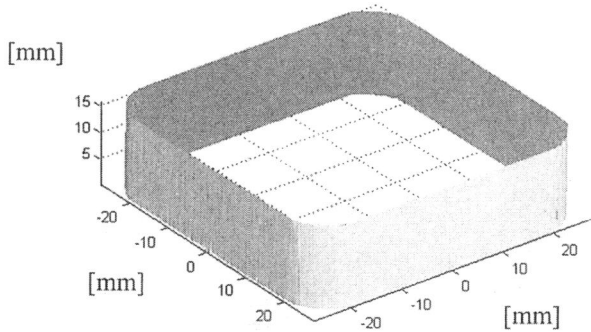

Figure 4. *Rebuilt total lateral surface of the stamped part ($BHP_i = 3\% \ \sigma_0$)*

In the present paper in order to quantify the elastic springback of the drawn part, two numerical parameters are proposed, namely Re_1 and Re_2, which allow one to quickly describe the different shape of the measured surfaces with respect to the ideal ones. The first parameter (Re_1) represents a percentage volume error between the volume obtained considering the lateral surface of the measured specimen (V) and the ideal volume evaluated on the basis of the ideal CAD specimen geometry (V_id):

$$Re_1 = [(V - V\_id)/V\_id]*100$$

Such a parameter provides quantitative information regarding the global discrepancy between the two specimen volumes, *i.e.* it gives an average quantitative indication of the springback effect on the specimen shape. Furthermore, in order to complete the information given by the parameter Re_1 a second parameter (Re_2) is introduced with the aim of quantifying the geometrical difference between the measured section area (A) and the ideal one (A_id) at a fixed specimen height z:

$$Re_2 = \{[A(z)-A\_id(z)]/A\_id(z)\}*100$$

The proposed measurement technique has been then utilized to investigate the influence of the blankholder force level on the elastic springback phenomenon. In particular, constant levels of such a variable, corresponding to an initial blankholder pressure (BHP_i) on the blank equal to 0.5%, 1%, 2%, 3% and 4% of the material yield stress respectively, have been tested. The BHP_i values were chosen in such a way to avoid the insurgence of both wrinkles and ductile fractures in the stamped part.

The results obtained are shown in Table 1; it should be observed that for a fixed lubricating condition, as the blankholder force level increases, the restraining forces increase, producing a sort of stretching in the drawn part walls which reduces the elastic springback. Actually, for increasing levels of the blankholder force the obtained volumes tend to become closer to the ideal one.

Table 1. *The Re_1 values*

BHP_i	Re_1
0.5%	3,56%
1.0%	2,70%
2.0%	2,19%
3.0%	1,87%
4.0%0	1,30%

Actually, index Re_1 gives a quite global result and does not allow a deep investigation of the elastic deformation to occur in the drawn part with the elastic springback phenomenon. Such deformation is unevenly distributed in the stamped part on the basis of the residual stress state at the end of the forming process. In Figure 5 the Re_2 index is plotted vs. the number of pixels of the acquired images (1 pixel = 0.51 mm) for a deep drawn box with a BHP_i equal to 3% of the aluminum yield stress. The figure shows that the sectional distortion is not regular all along the specimen height since both the residual stress state after forming and the part stiffness are not uniform in the drawn part; in particular a maximum percentage difference of about 0.85% is observed.

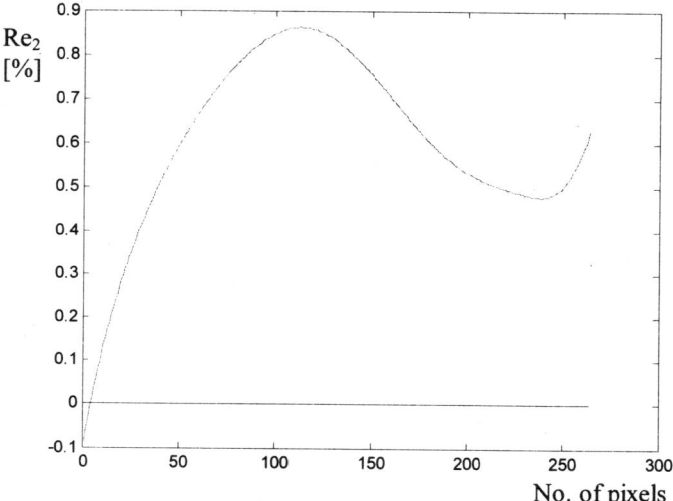

Figure 5. *Re_2 vs. No. of pixels in a drawn box (BHP_i = 3% σ_0)*

Another interesting result is obtained splitting the above reported Re_2 result for the corners ($Re_{2\,s}$) from the flat walls ($Re_{2\,p}$) of the specimen. In Figure 6 such Re_2 functions are reported.

In particular, so far as the four corners are concerned, a monotonic increasing trend of the Re_2 parameter is observed with the number of pixels, *i.e.* with a z coordinate taken on the measured specimen. On the other hand a quite complex trend is observed for the flat walls, similar to the behaviour of the global Re_2 function. Such different trends of the Re_2 parameters have to be carefully investigated through a FEM analysis and correlated to the different residual stress states after forming, occurring in the four corners and in the flat walls of the drawn box.

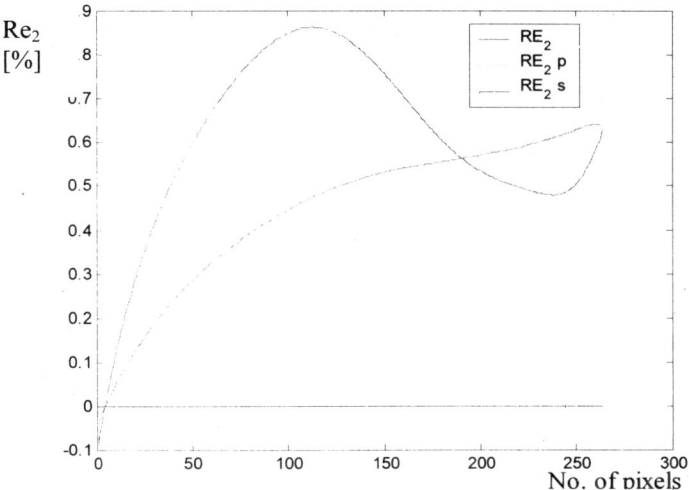

Figure 6. Re_2 vs. No. of pixels for corners ($Re_{2\ s}$) and for flat walls ($Re_{2\ p}$) ($BHP_i = 3\%\ \sigma_0$)

5. Conclusions

In this paper a previously developed measuring technique for out-of-plane surfaces was applied to the determination of the elastic springback occurring in the deep drawing of square boxes. In particular, an aluminum alloy was taken into account and several levels of the blankholder force were investigated with the aim of highlighting the influence of such a process parameters on the investigated shape defect. What is more, a proper coefficient was introduced in order to quantify the investigated phenomenon.

Acknowledgements

This work has been made using MURST (Italian Ministry for University and Scientific Research) funds.

6. Bibliography

Ahmetoglu M, Broek T. R., Kinzel G., Altan T., "Control of blankholder force to eliminate wrinkling and fracture in deep-drawing rectangular parts", *Annals of CIRP*, Vol. 44/1, 1995, pp. 247–250.

D'Acquisto L., Fratini L., "An optical technique for springback measurement in axisymmetrical deep drawing operations", *Journal of Manufacturing Processes*, 2001, Vol. 3, No. 1, pp. 29–37.

D'Acquisto L., Fratini L., Siddiolo A., "A modified moire technique for 3D surface", *Measurement science and technology*, 2002, Vol. 13, pp. 613–622.

Doege E., Boinski F., "Computer-aided design and calculation of deep drawn components", *Journal of Materials Processing Technology*, Vol. 46, 1994, pp. 321–331.

Doege E., Sommer N., "Blank-holder pressure and blank-holder layout in deep drawing of thin sheet metal", *Advanced Technology of Plasticity*, 1987, pp. 1305–1314.

Forcellese A., Fratini L., Gabrielli F., Micari F., The evaluation of springback in 3D stamping and coining processes, *J. of Materials Proc. Technology*, Vol. 80–81, 1998, pp. 108–112.

Fratini L., Lo Casto S., Micari F., "Deep drawing of square boxes: analysis of the influence of the geometrical parameters by numerical simulations and experimental tests", *Proc. of Numiform '95*, 1995, pp. 705–709.

ISO – Guide to the Expression of Uncertainty in Measurement, 1993, Switzerland, first ed.

Kobayashi A., "New optical measurements and their applications in industry", *Measurement*, Vol. 9, No. 2, 1991, pp. 88–96.

Lange, K., *Handbook of metal forming*, eds. McGraw Hill, 1985.

Mickalich M.K., Wenner M.L., Calculation of springback and its variation in channel forming operations, GMR- 6108, *General motors research publication*, General Motors research lab, 1988, Warren, MI.

Makinouchi A., Sheet forming simulation in industry, *J. of Materials Proc. Technology*, Vol. 60, 1996, pp. 19–26.

Post D., Han B., Ifjiu P., *High sensitivity Moiré*, Springer Verlag, New York, 1994.

Sciammarella C.A., Ahnadshahi M.A., "Detection of fringe pattern information using a computer based method", *Proc. of the VIIIth Int. Conf. on Experimental Stress Analysis*, 1986, pp. 359–368.

Sciammarella C.A., Sturgeon, D.L., "Digital filtering techniques applied to the interpolation of moirè-fringe data", *Experimental Mechanics*, Vol. 7, No. 11, 1967, SEM, pp. 468–475.

Siegert K., Dannenmann E., Wagner S., Galaiko A., "Closed-loop control system for blankholder forces in deep drawing", *Annals of CIRP*, Vol. 44/1, 1995, pp. 251–254.

Takeda M., Ina H., Kobayashi S., "Fourier transform method of fringe pattern analysis for computer based topography and interferometry", *J. Opt. Soc. of Am.*, Vol.72, 1982, pp. 156–160.

Yossifon S., Sweeney K., Ahmetoglu M, Altan T., "On the acceptable blank-holder force range in the deep drawing process", *Journal of Materials Processing Technology*, Vol. 33, 1995, pp. 175–194.

Chapter 13

Cavity Defects and Failure Study of Ceramic Components

Ioannis Doltsinis
Faculty of Aerospace Engineering, University of Stuttgart, Germany

1. Introduction

The present study deals with aspects of brittle and quasi-brittle failure of brittle materials on the microscopic and on the continuum level. It refers particularly to ceramics. The significance of microstructural parameters is investigated by a numerical model that progressively accounts for separation of grain interfaces. The microcracking model [DOL 98] requests specification of the material structure which can be subject to variations in conjunction with synthetic Monte Carlo sampling. Beyond the evolution of damage, the strength of the material specimen is estimated by this computational approach which applies fracture mechanics to a microstructure subject to statistical variability as suggested in the literature [DAV 80].

Under tensile actions, the brittle mode of failure by separation of the specimen is predominant, but any structural disorder promotes damage by distributed microcracking prior to ultimate failure. Compressive loading may close microcracks, separation of grain interfaces is mainly by sliding which may be opposed by friction. The quasi-brittle mode of failure prevails, synergies between interacting cracks while damage progresses form localized patterns and lead to ultimate failure. Structural disorder can be introduced by the nature and size of grain phases, thermal eigenstrains stemming from manufacturing, and pores. The presence of pores was found to favour distributed damage as well, but if the loading is by internal pressure in the pores the process of microcracking is such that pores are bridged and failure by separation of the material specimen is quite straightforward [DOD 00]. Apart from the dependence on overall structural parameters like porosity, pore size and -shape, random sampling demonstrates the sensitivity of the material strength to the topology of the microstructure. Statistical analysis explores the variability of strength in the sample, and the significance of varying dimensions of the material specimen. The frequency distribution of the strength suits with Weibull statistics, and indicates an affinity with the size effect of the weakest link hypothesis [WEI 51]. The scatter of observations related to failure of brittle materials must be considered a feature which cannot be separated from physical aspects [FRE 68]. In this context the frequency or probability of appearance of a specified value is of interest rather than mean values of properties.

For an assessment of the reliability, that is the probability of successful employment of structural components, experimental laboratory tests on ceramics are discussed and interpreted. The appertaining statistics is utilized in order to deduce the variability of the material strength within the specific component under investigation [DDS 01]. In a discretized representation of the structural part by finite elements, the size effect from the Weibull approach is employed for the definition of the strength of individual elements. An element is assumed to fail when the maximum principal stress attains the strength allocated to the element volume. At the same time it is examined whether the size of an hypothetical critical crack would be contained in the element. In such a case local failure could propagate rapidly through the component. Implementation is based on a stationary stress field for each level of the loading, while the strength distribution within the finite element mesh is repeatedly sampled. Thus, the frequency of locations prone to incipient failure is obtained at the respective load level. The approach is applied to a ceramic filter support and confirms the probability of failure initiation at fluid pressures comparable with those actually observed. Ultimate failure is commonly assessed by the transition from the statistics of uniaxial testing to the dimensions of the structural component and the stress distribution in it. This essentially requires evaluation of a stress volume integral subsequent to finite element analysis. An interesting issue is that of mesh improvement, which is proposed to steer by the element contribution to the error in the stress volume integral [DON 91]. This adaptation criterion observes the statistical properties of the material, and appears well-suited for the present purpose.

2. Micromechanical analysis of progressive damage and failure

2.1. *Computational approach*

The origin of our micromechanical approach to brittle microcracking of solids considers the impact of applied actions on the material structure [DOL 99]. For this purpose, two-dimensional (mosaic) models are created on the computer starting with a regular hexagonal lattice associated with the finest grain phase, then distorting and introducing at random additional phases and pores, Figure 1. Overall characteristics from laboratory analysis like volume resp. area fraction of phases and specific perimeter are observed. The procedure provides us with an artificial material specimen with defined structure. The grain interfaces constitute a network of potential cracks, which presumes interfacial fracturing.

Elementary stress solutions for cavities of various shapes (circular, elliptic, triangular) compiled from the literature [DOD 00] supplement the stress fields emanating from interacting cracks [KAC 87] in order to estimate the straining of grain interfaces by the applied loads. The microcracking algorithm that operates on the artificial structure of the material specimen requires the course of applied stresses as an input. At each incremental step during the loading process, the tractions along grain interfaces are computed while the kinematic activity of cracks (crack closure, frictional sliding of interfaces) is considered [DOD 99], [DAT 02].

Material damage progresses by separation of grain interfaces. The Griffith energy criterion examines possible transition of elastic strain energy to crack

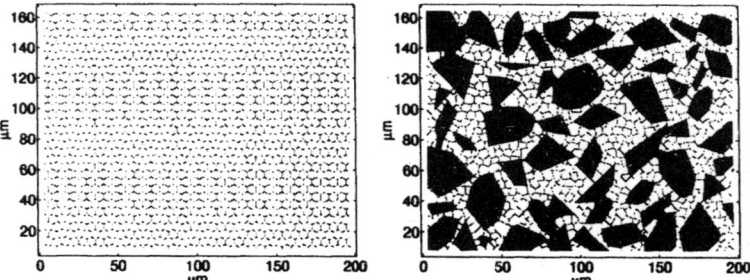

Figure 1. *Stages in modelling a material structure: 1) Undistorted regular hexagonal lattice 2) Distortion and implantation of a second grain phase 3) Pores to be introduced as a third phase*

surface energy. It is assumed that the jth grain interface fails completely and forms a crack of the same length l_j if

$$\int_{l_j} G \mathrm{d}l \geq (2\gamma + \mathcal{E})_j \, l_j. \tag{1}$$

The elastic energy G per unit crack length is released during separation of the grain interface at constant stress. Evaluation of the energy release integral is simple. Under boundary loading it can be based on the average tractions with magnitude t_j actually computed at the considered state. From familiar relationships [DOL 98],

$$\int_{l_j} G \mathrm{d}l = \int_{l_j} \frac{\pi l t^2}{2E} \mathrm{d}l = \frac{\pi l_j^2 t_j^2}{4E}. \tag{2}$$

The elastic modulus E refers to the embedding isotropic continuum in plane stress. For the plane strain condition assumed in the present case studies, $E \Leftarrow E/(1 - \nu^2)$. If the material is stressed by internal pressure, stress intensity factors are used instead as outlined later below. The specific surface energy γ may vary between grain interfaces. It is modified by the release of stored energy $\mathcal{E}$ per unit crack length. In ceramics cooling-down from manufacturing temperature induces misfit strains in individual grains (eigenstrains) which store elastic energy in the material locally varying in dependence of the material structure [BUR 85]. When the energy release from the eigenstrains counterbalances the specific surface energy ($\mathcal{E} = -2\gamma$) separation of the grain interface can occur spontaneously at zero applied stress.

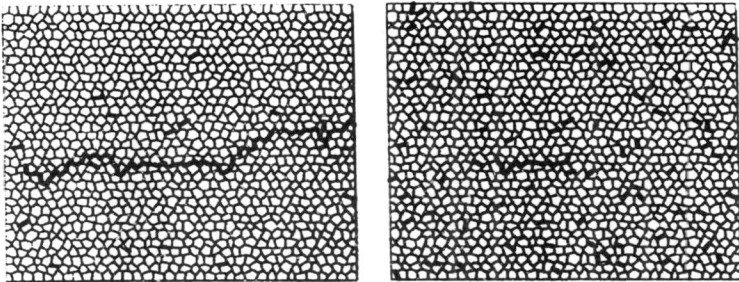

Figure 2. *Patterns of 100 microcracks under vertical loading and varying magnitude of eigenstrain*

2.2. Damage under boundary loads

Figure 2 displays a two-dimensional grain structure in which the energy of eigenstrains is randomly distributed. Microcracking is induced by the application of tensile stress along the vertical direction. The different patterns formed by 100 microcracks depend strongly on the participation of the eigenstrains. At low eigenstrains (left) the specimen fails by separation due to the tension. Increasing the intensity of the random eigenstrains by a factor of three is seen to introduce distributed damage in the rather regular structure (right). Cracks localize and lead to separation at higher tensile stress. As a rule, prevalence of tension implies brittle behaviour; failure occurs by separation essentially perpendicular to the applied stress.

Under compressive loading damage by distributed microcracking precedes failure. The specimen fails when microcracks localize to a band across the specimen. Localization is a consequence of synergy effects between interacting cracks as demonstrated in Figure 3 for vertical compression. In the upper row, initial cracks appearing in the middle part of the specimen attract subsequent nucleations, such that a pronounced oblique band is formed. The displayed material structure represents a single realisation out of a statistical sample comprising five hundred units that differ as for the topology of the two grain phases while overall characteristics are kept constant. The second row in the figure shows that the microcracks are rather randomly distributed when interactions are suppressed, and the same crack density is now obtained at higher compressive stresses. The scatter in the appertaining stress values is reduced, which implies stabilization of the synthetic sampling after two hundred random realizations of the material structure. The nature of compressive loading hinders crack opening. The separation of grain interfaces is rather due to sliding motion where friction plays a significant role and usually delays damage.

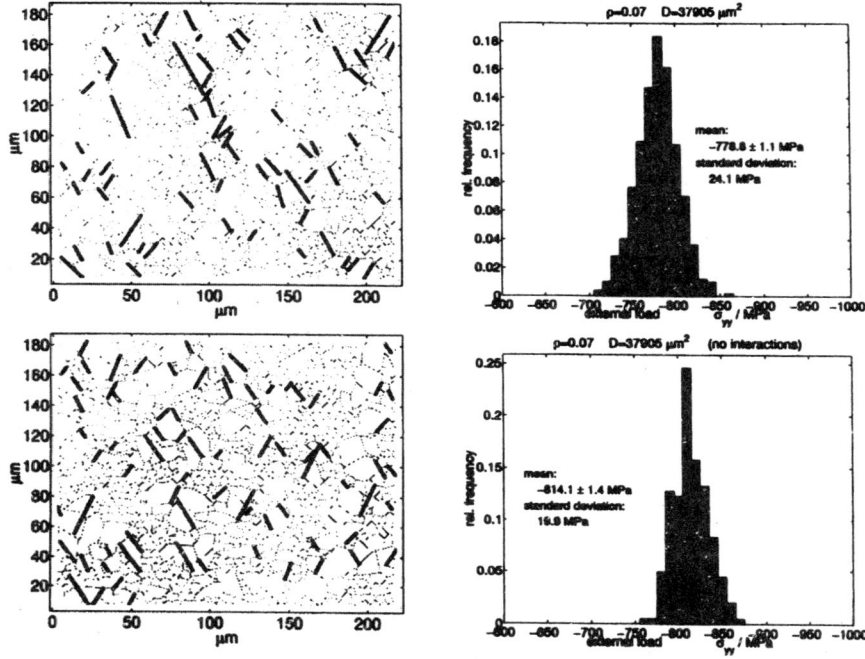

Figure 3. *Microcrack patterns forming under vertical compression. Upper: interacting cracks, lower: no interaction*

Further to the dependence of the material response on the nature of the applied stress, any disorder in the material structure prolongs the phase of quasi-brittle behaviour, that is progressive damage prior to ultimate failure. Disorder can be introduced by the size and shape of grains, different material phases and pores as well as by local eigenstrains.

2.3. Failure under internal pressure

Pressure in the pores is the predominant loading mode in certain applications of brittle materials as for ceramic elements in filtration equipments. Elementary analytical solutions for isolated cavities may be composed to an approximate stress distribution which indicates locations in the microstructure prone to failure. Cracks along grain interfaces emanate as a rule from the boundaries of pressurized pores. They tend to bridge pores and lead to failure by separation of the material specimen (percolation). Instead of complete stress distributions, we base here the fracture criterion on the stress intensity factor K for pore-crack configurations available in the literature [BER 66]. With reference to the straight crack emanating from an elliptical pore in Figure 4 there is

$$\left(\frac{K}{\pi t}\right)^2 = \left(a + \frac{l}{2}\right) \frac{(1+l/a)^2 - 1}{(1+l/a)^2 - m} \frac{(1+l/a)+1}{(1+l/a)+m}, \qquad m = \frac{a-b}{a+b} \qquad (3)$$

where t denotes biaxial tension resp. the pressure acting in the pore. A graphical representation of the above relationship is included in Figure 4. Since K^2 is proportional to the energy release rate in crack extension, the fracture criterion of eqn (1) for the jth grain interface is evaluated as

$$\int_{l_j} G \, dl = \int_{l_j} \frac{K^2}{E} \, dl, \qquad (4)$$

and E is the elastic modulus of the surrounding isotropic continuum in plane stress or in plane strain depending on the case under investigation. The local nature of stress intensity factors prohibits the consideration of interactions, but these are not that significant here as for interior cracks since internal pressure is the essential crack driving force. Also less significant in the present case are effects introduced by the boundaries of the material specimen.

The analytical expression for the stress intensity factor has to be extended as for arbitrary crack directions [DOD 00]. Furthermore, flow networks (agglomerates of pores and cracks) forming while damage progresses are represented by equivalent longer cavities, which allow application of the stress intensity factor from eqn (3) and the approximation for alternative crack directions. Figure 5

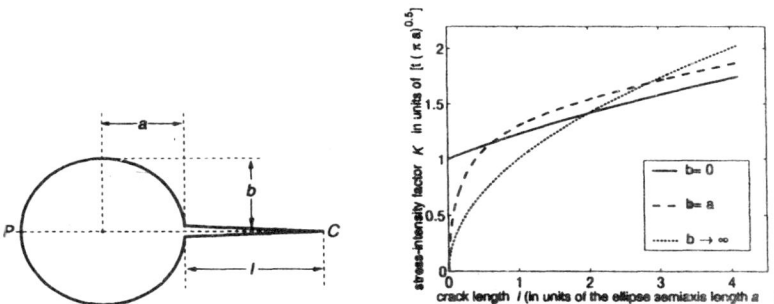

Figure 4. *Crack emanating from an elliptical pore in the direction of a semi-axis. Stress intensity factor*

demonstrates two different modes of failure under internal pressure depending on the topology of the pores. Quasi-brittle damage shown in the upper row is characterized by distributed microcracks, brittle failure occurs by separation of the specimen as in the lower row. The respective behaviour is reflected in the strain response.

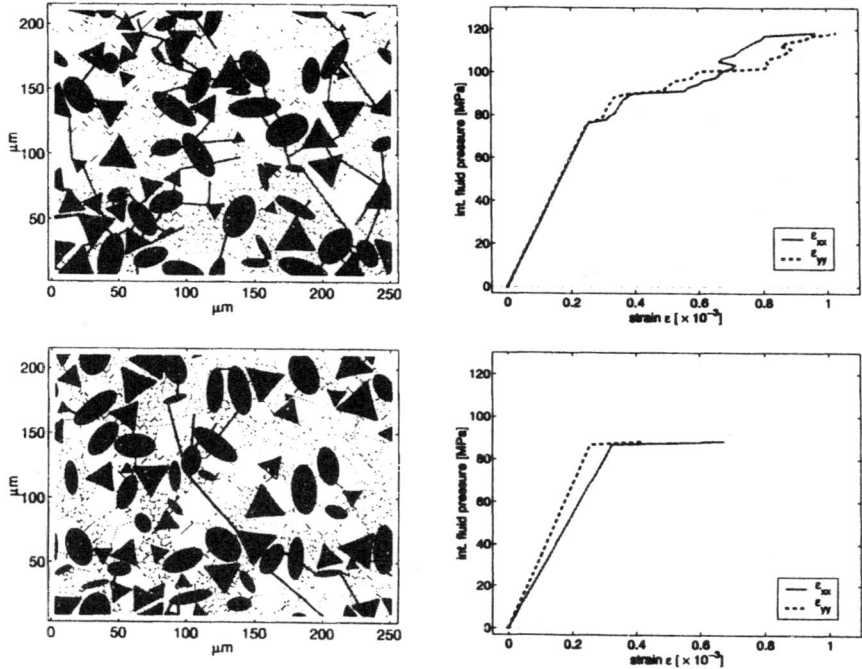

Figure 5. *Quasi-brittle (above) and brittle mode of failure (below) under internal pressure*

Figure 6 collects results from percolation studies on statistical samples of porous microstructures. Distinct realizations have been obtained by randomly positioning the pores within the grain network, all other characteristics kept constant; the circular pores are of the same size. The histograms refer to the intensity of the internal pressure when the specimen separates. Synthetic sampling has been continued until the mean value of the strength stabilized. Comparison between the first and the second row indicates the effect of an increasing specimen size. The ultimate pressure is lowered as suggested by the weakest link assumption while the sensitivity to the topology of the pores diminishes, since the specimen becomes more homogeneous. Thereby, the statistical sample could be reduced from one hundred realizations to fifty. At fixed specimen size the frequency distribution of the strength suits the better with Weibull statistics (see eqn (5) in the next section) the larger the dimensions. At the same time the exponent m increases signifying enhanced homogeneity, and the characteristic strength decreases. Comparison between the second and the third row reveals the influence of the pore size quantified by the specific perimeter of the porous phase. At constant porosity, the higher specific perimeter (smaller pores) displaces in the third row the ultimate pressure to higher values.

Failure Study of Ceramic Components 155

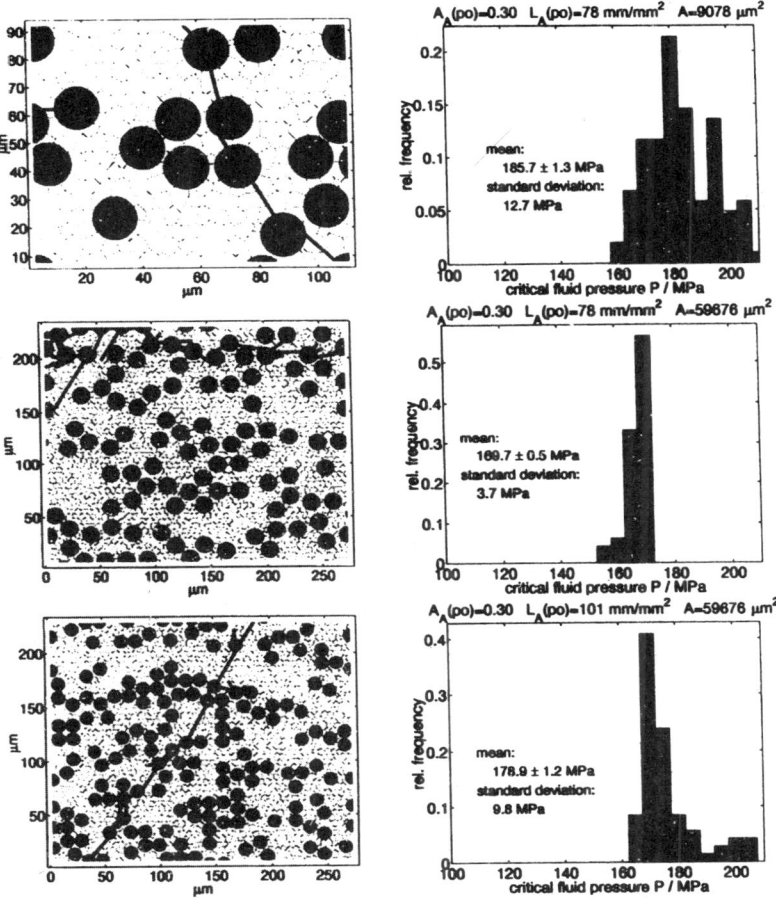

Figure 6. *Above and middle: effect of specimen size. Increasing specimen area A; specific perimeter of porous phase is $L_A = 78$ mm/mm^2, porosity $A_A = 0.30$. The diameter of the pores is 15.5 µm, the average grain size 5.4 µm. Middle and below: the specific perimeter of the porous phase increases to $L_A = 101$ mm/mm^2*

Figure 7 examines the influence of the pore shape. For the same porosity and specific perimeter of the porous phase, the material with elliptical cavities sustains lower pressure than the circular. The scatter of the strength when varying the microstructure is increased since both position and orientation of the pores are now involved.

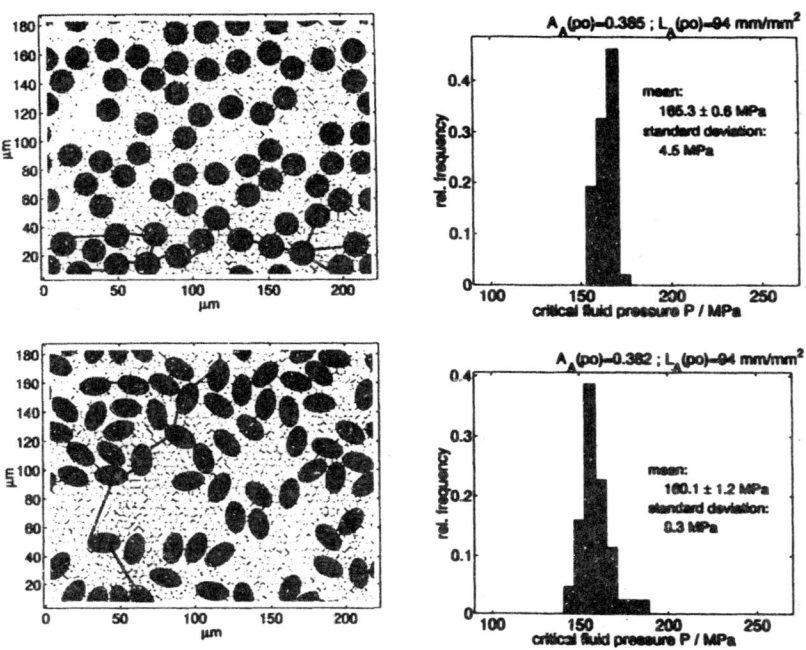

Figure 7. *Influence of pore shape. Percolation studies for equal porosity* $A_A = 0.38$ *and specific perimeter of the porous phase* $L_A = 94\ mm/mm^2$

3. Strength of structural components

3.1. *Statistical description of material*

In a structural component locations become critical where the stress to be sustained equals or exceeds the brittle material strength. The stress distribution in the loaded component can be determined by elastic stress analysis, but the strength of brittle materials appears to vary rather randomly within the part. Each part of a series will exhibit a different spatial distribution of material strength, which may have an impact on the identification of critical locations. Since the material strength constitutes a random field, statistical description is appropriate. When the statistical characteristics are available, the distribution of strength can be sampled in the component following a Monte Carlo technique and the frequency is registered at which single locations may become critical under the applied loading [DDS 01].

The proposed concept is exemplified for the ceramic filter support shown in the perspective view of Figure 8. Under operational conditions the filter support is subject to an overpressure by the fluid in the channels. The porosity of the

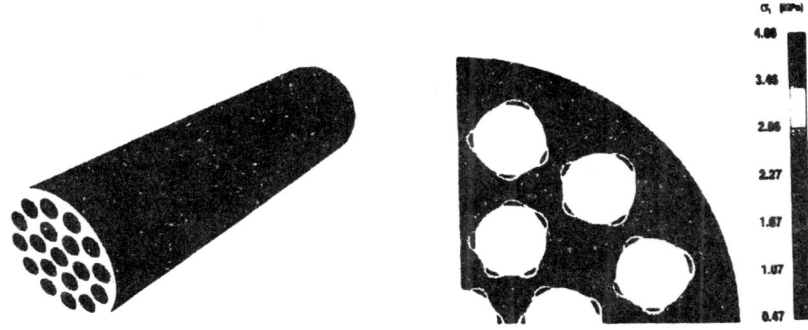

Figure 8. *Ceramic filter support. Variation of maximum principal stress σ_1 for $\Delta p = 5.0$ MPa between channels and outer surface*

material degrades the overpressure from the channels towards the ambient value at the outer surface. A finite element analysis of the elastic system supplies the stress distribution. For brittle failure the maximum principal stress σ_1 is considered decisive, and is plotted in Figure 8 as a numerical result for plane strain conditions.

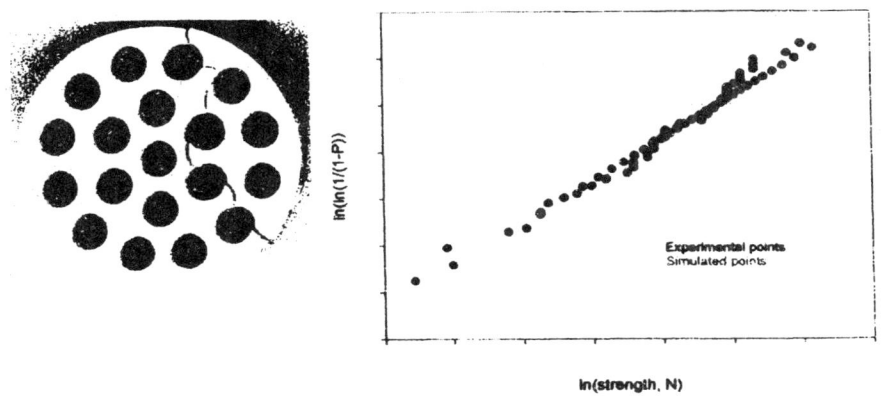

Figure 9. *Brazilian test. Frequency distribution of load at failure (Ismra-Lermat)*

For the random variation of strength we utilize statistical data from the diametrical compression test on discs cut along the axis of the filter support [OSV 99]. The observed frequency of failure up to a value of the compressive force

F, interpreted as probability of failure P_f, is plotted in Figure 9. It fits reasonably with the following distribution [WEI 39],

$$P_f = 1 - \exp\left[-\left(\frac{F}{F_0}\right)^m\right]. \tag{5}$$

The exponent m and the characteristic force F_0 are constants, and therefore eqn (5) is represented by a straight line in the $\ln[-\ln(1 - P_f)] - \ln F$ diagram. The quantity $1 - P_f = P_s$ is the survival probability of the test specimen.

The counterpart of eqn (5) in terms of a uniaxial stress σ reads

$$P_f = 1 - \exp\left[-\left(\frac{\sigma}{\sigma_0}\right)^m\right]. \tag{6}$$

The characteristic strength σ_0 cannot be obtained as a direct result of the experiment, but has to be deduced from the force applied by the testing machine. For this purpose the multiaxial probabilistic approach outlined in the subsequent section (see eqn (8)) is applied in conjunction with a finite element stress analysis of the Brazilian test for the support, which provides us with an interpretation of the experiment in terms of the characteristic strength σ_0.

Figure 10. *Realizations of random strength distribution in finite element mesh*

3.2. Critical locations

The input σ_0, m specified from experimental data defines the random distribution of the material strength within the component. In the discretized representation, distribution of strength among the finite elements includes weighting by the element volume in order to account for the effect of size on the strength. A finite element is assumed to fail when the maximum principal stress (σ_1) resulting from the pressure loading attains the strength allocated to the element volume. At the same time, it is examined whether the critical length of an hypothetical crack placed in the local stress field is confined within the dimensions of the element. In such a case local failure might be considered to grow rapidly through the element and the location becomes critical. Repeated sampling of the strength distribution in the finite element mesh (Figure 10) and comparison with the applied stress field indicates the frequency of positions prone to incipient failure in the support and the associated level of fluid pressure. Figure 11 refers to 1000 observations in the strength distribution and underlines that the vicinity of the flow channels and the ligaments are most critical to the initiation of failure at pressure values comparable with actual operation conditions. It thus can be concluded that the distribution of applied stress overwhelms the scatter in material strength. In the figure, the results from the entire cross section have been projected on a 30° section, and have been mirrored in order to increase visibility.

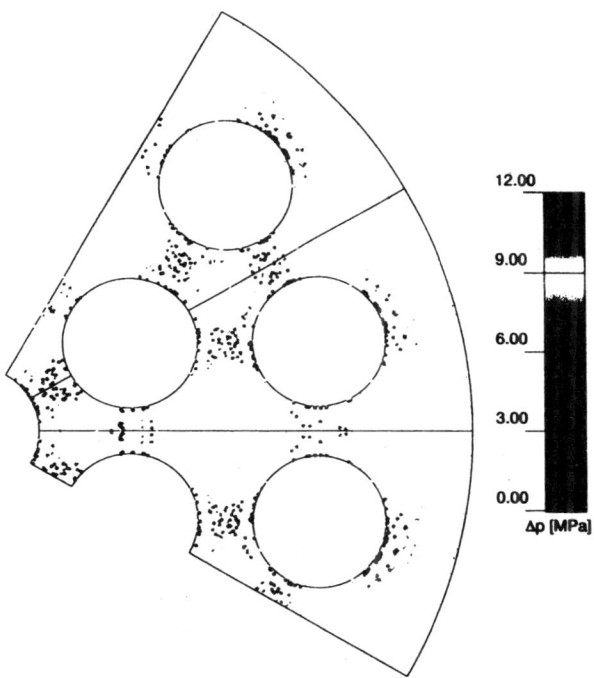

Figure 11. *Critical locations and pressure level from 1000 observations*

3.3. Probability of failure

The exponential description of the failure probability of uniaxial specimens by eqn (6) facilitates incorporation of the size effect [WEI 51]. The condition for the survival of a specimen consisting of a sequence of n members is that none of the parts will fail. Assuming statistical independence, the probability of survival for the assembly equals the product of the survival probabilities of the parts

$$P_{sn} = 1 - P_{fn} = \exp\left[-n\left(\frac{\sigma}{\sigma_0}\right)^m\right] \tag{7}$$

Equation (7) presumes a unique distribution function for all members, and $n = l/l_0 = V/V_0$ is the scale factor of the projected length l to the test specimen length l_0 generalized also for the volume V.

The probability of failure for specimens of volume V can be obtained from data for specimens of volume V_0 as by

$$\begin{aligned} P_{fV} &= 1 - \exp\left[-\frac{V}{V_0}\left(\frac{\sigma}{\sigma_0}\right)^m\right], \\ &= 1 - \exp\left[-\frac{1}{V_0}\int_V \left(\frac{\sigma}{\sigma_0}\right)^m dV\right], \\ &= 1 - \exp\left[-\frac{1}{V_0}\int_V \left(\frac{\sigma_1}{\sigma_0}\right)^m dV\right]. \end{aligned} \tag{8}$$

Equation (8) interprets eqn (7) for the volume effect, and introduces additional extensions. The second line refers to a stress σ varying along the uniaxial specimen, while the third expression extends the failure probability to multidimensional stress states in the considered volume assuming dominance of the maximum principal stress σ_1. The above is useful for the transition from uniaxial test data to the failure probability of structural components subject to multiaxial stresses. An essential assumption is that the material is statistically homogeneous, which implies independence of its strength distribution function of the position in the material space.

A more sophisticated approach to the failure probability of structural components goes back to the work in [SFS 73]. Accordingly,

$$P_{fV} = 1 - \exp\left[-\frac{V}{V_0}\left(\frac{\sigma_N}{\sigma_0}\right)^m \Sigma\right] \tag{9}$$

The parameters σ_0, m are from uniaxial tests referring to the specimen volume V_0. The volume of the component under investigation is V, and σ_N denotes a reference stress that characterizes the magnitude of the applied loading. The quantity

$$\Sigma = \int_V \left[\left(\frac{\sigma_1}{\sigma_N H(\sigma_1)} \right)^m + \left(\frac{\sigma_2}{\sigma_N H(\sigma_2)} \right)^m + \left(\frac{\sigma_3}{\sigma_N H(\sigma_3)} \right)^m \right] \frac{dV}{V}, \quad (10)$$

is known as the *stress volume integral* (SVI). It accounts for the distribution of the principal stresses $\sigma_1, \sigma_2, \sigma_3$ in the component, weighted by the Heaviside function $H(\sigma)$. For tensile stress $H(\sigma) = 1$, while $H(\sigma) = -\alpha$ for compressive stress and α measures the ratio of average strength under tension or compression. Divided by the reference stress σ_N the integral is independent of the magnitude of the loading, and is made dimensionless by the volume V of the component. Different assumptions as for the participation of the stress components in eqn (10) have been investigated in [DON 91].

In the discretized representation of the component the stress volume integral Σ is computed as the sum of volume integrals over all finite elements constituting the numerical model for the elastic analysis. In this connection we discuss modification of the finite element mesh adapted to the computation of the stress volume integral. Mesh modification by relocating nodal point coordinates relies on the determination of new coordinates such that the distribution of a certain quantity taken as the quality criterion is balanced within the discretization model. The objective quantity is used as a weighting factor w for the computation of new nodal point coordinates by the instruction

$$\mathbf{x}_i = \left(\sum_{q=1}^{ie} w_q \right)^{-1} \sum_{q=1}^{ie} w_q \bar{\mathbf{x}}_q \quad (i = 1, \cdots, N). \quad (11)$$

Modification by eqn (11) extends over all N mesh nodal points. The coordinates $\mathbf{x}_i$ of the ith nodal point are computed as a function of the coordinates $\bar{\mathbf{x}}_q$ of the centre of gravity of all ie elements attached to the nodal point weighted by the respective w_q (Figure 12). The procedure tends to attract nodal points where the weight w is higher. Ideally, mesh modification specifies new nodal point positions such that $w_q = \text{const.}$ $(q = 1, \cdots, nel)$ in all nel elements in the mesh. Also, comparing different discretizations, that with a lower $\sum_{q=1}^{nel} w_q$ is preferred.

Proposed measures of quality of numerical finite element solutions are based on the difference between σ, the immediate usually discontinuous result of the computation, and σ^* obtained therefrom after smoothing over the domain of the problem [ZIZ 92]. The difference may be used to define an energy norm of

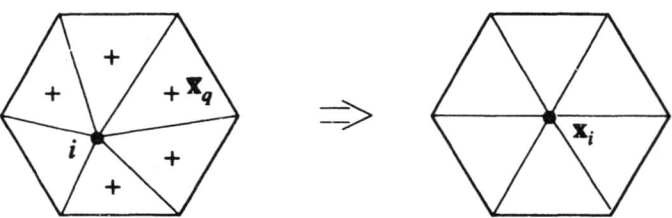

Figure 12. *Modified nodal point position from weighted element coordinates*

the error in the solution

$$\|e\|^2 = \sum_{q=1}^{nel} \int_{V_q} [\sigma - \sigma^*]^t [\epsilon - \epsilon^*] \, dV_q = \sum_{q=1}^{nel} \|e\|_q^2, \qquad (12)$$

which is employed as the weighting factor in eqn (11): $w_q \Leftarrow \|e\|_q$. For the stress volume integral we found it suitable to define instead the error criterion by two evaluations as

$$e = |\Sigma(\sigma) - \Sigma(\sigma^*)| = \sum_{q=1}^{nel} |\Sigma(\sigma) - \Sigma(\sigma^*)|_q \qquad (13)$$

and $w_q \Leftarrow |\Sigma(\sigma) - \Sigma(\sigma^*)|_q$ enters eqn (11) for the computation of modified nodal point coordinates. Numerical investigations suggest a superiority of the second, specific criterion.

4. Concluding remarks

The random variability of strength inherent to brittle materials like ceramics poses the task of reliability for a structural component, that is the probability of successful employment. The presented micromechanical model proves helpful when studying the significance of the material structure for failure on this scale. On the continuum level, random sampling of the strength variation within the component under investigation provides us with information on locations which may become critical under the actual stress distribution. In a different combination of finite element stress analysis with statistical test results, an extended Weibull approach forms the basis for the reliability assessment of structural components. In this connection, adaptive mesh improvement is proposed with reference to the error in the stress volume integral rather than to the customary energy norm.

5. References

[BER 66] Berezhnitskii L.T., Propagation of cracks terminating at the edge of a curvilinear hole in a plate, *Soviet Materials Science* 2, 1966, 16 – 23.

[BUR 85] Buresch F.E., Relations between the damage in and microstructure of ceramics, *Materials Science and Engineering* 71, 1985, 187 – 194.

[DAT 02] Dattke R., *Modelling the microstructure and simulation of progressive fracturing in brittle porous ceramics*, Doctoral Dissertation, Faculty of Aerospace Engineering, University of Stuttgart (submitted), 2002.

[DAV 80] Davidge R.W., Combination of fracture mechanics, probability, and micromechanical models of crack growth in ceramic systems, *Metal Science* 14, 1980, 459 – 462.

[DOL 98] Doltsinis I., Issues in modelling distributed fracturing in brittle solids with microstructure, in A. Idelsohn et al. (eds), *Computational Mechanics – New Trends and Applications: Fourth World Congress on Computational Mechanics*, Buenos Aires, CD-Rom, CIMNE, Barcelona, 1998.

[DOD 99] Doltsinis I. and Dattke R., Studies on porous fracturing solids, in J. Bento et al. (eds.) *Computational Methods in Engineering and Science*, Elsevier, Amsterdam, 1999.

[DOD 00] Doltsinis I. and Dattke R., Modelling the damage of ceramics under pore pressure, in ECCOMAS 2000 CD-Rom Proceedings, CIMNE, Barcelona, 2000. Published in *Comput. Meths Appl. Mech. Engng* 191, 2001, 29 – 46.

[DDS 01] Doltsinis I., Dattke R., Schimmler M., Statistical aspects of failure of brittle ceramic materials, in S. Valliapan and N. Khalili (eds.), *Computational Mechanics – New Frontiers for the New Millenium*, Elsevier, Amsterdam, 2001.

[DON 91] Doltsinis I. and Nötzel G., Investigations on the failure of a ceramic rotor, Workshop: *FE-Modelling of the Mechanical Behaviour of Materials*, Oct. 22, 1991, Max-Planck-Institut für Metallforschung, Stuttgart, FRG.

[FRE 68] Freudenthal A.M., Statistical approach to brittle fracture, Chapter 6, in H. Liebowitz (ed.), *Fracture*, Volume II, Academic Press, New York and London, 1968.

[KAC 87] Kachanov M., Elastic Solids with many cracks: a single method of analysis, *Internat. J. Solids and Structures* 23, 1987, 23 – 43.

[OSV 99] Osterstock F. and Vansse O., Results of Brazilian tests, *Private Communication*, ISMRA-LERMAT, Caen, France, 1999.

[SFS 73] Stanley P., Fessler H. and Sivill A.D., An engineer's approach to the prediction of failure probability of brittle components, Proceed. *British Ceramic Society* 22, 1973.

[WEI 39] Weibull W., A statistical theory of strength of materials, *Proc. Ing. Vetensk. Akad.* 151, 1939.

[WEI 51] Weibull W., A statistical distribution function of wide applicability, *ASME, Journal of Applied Mechanics*, 1951, 293 – 297.

[ZIZ 92] Zienkiewicz O.C. and Zhu J.Z., Automatic adaptive analysis – The new look of finite elements, in P. Ladevèze and O.C. Zienkiewicz (eds), *New Advances in Computational Structural Mechanics*, Elsevier, Amsterdam, 1992.

Chapter 14

Prediction of Necking Initiation during the Bending of Metal-rubber Profiles by FEM Simulations of the Forming Process

Monique Gaspérini and Cristian Teodosiu
*Laboratoire des Propriétés Mécaniques et Thermodynamiques des Matériaux,
CNRS/Université Paris, France*

David Boscher
*Laboratoire des Propriétés Mécaniques et Thermodynamiques des Matériaux,
CNRS/Université Paris, France, and Metzeler Automotive Profil Systems, Nanterre,
France*

Eric Hoferlin
OCAS N.V., Zelzate, Belgium

1. Introduction

Despite its importance for the optimisation of forming processes and of material choice, the capability of FEM simulations to predict the occurrence of necking in complex products is limited by the difficulty taking into account simultaneously (i) an accurate description of the material behaviour, (ii) the precise real boundary conditions and the different stages of an industrial process (iii) relevant necking criteria, since the classical models for necking have been developed for simple homogeneous stress states. However, necking criteria applied to post-processing results in an FEM simulation of a real multi-stage process and is expected to allow location of the zone of risk in the piece and its sensitivity to the material and process parameters. The present paper illustrates this approach for a glass-run channel for automotive made of aluminium alloy covered with rubber. The formability of this product is limited by the appearance of necking on the aluminium alloy during the final bending, as illustrated in Figure 1, which is the last stage of a complex thermo-mechanical loading from a flat sheet: sheet forming in rolls, extrusion and vulcanisation of the EPDM rubber, forming of this bi-material straight profile, then arching and bending to the desired shape. Following preliminary work (Boscher et al., 2000), this study focuses on the FEM simulation of the process and on necking prediction using a Marciniak-Kuczynski (Marciniak et al., 1967) – hereafter named MK – approach permitting one to take into account arbitrary strain paths (Hoferlin et al., 2000).

2. Experimental characterisation

2.1. Material behaviour

For the strain levels involved in the zone of risk, the rubber may be considered as hyperelastic, and a Rivlin potential has been used: $W = 0.9503(I_1- 3)+0.0933(I_2-3)+0.0081(I_1-3)^2$. The aluminium alloy (Al-Mn 1%) used in the study is, at the beginning of the process, a partly recrystallised sheet with a weak crystallographic texture. As a first approximation, standard plasticity with a von Mises yield surface and isotropic hardening is used in the FEM simulations. A Swift law: $Y = C(\varepsilon_0 + \overline{\varepsilon}^p)^n$, $C = 207.99$ MPa, $\varepsilon_0 = 0.00341$, $n = 0.2157$, describes the hardening behaviour in tensile test of the initial sheet after the heat treatment due to the vulcanisation of the rubber, and is compatible with the non saturating behaviour for larger strains observed through shear tests.

2.2. Strain measurements

Sections of the profile normal to the initial rolling direction were cut after successive stages of the process in order to compare their shapes with the FEM simulations. Moreover, strain distribution on the available part of the rubber in the bent zone close to the zone of risk was obtained by grid measurements, using a serigraphic mesh with 2 mm step size.

2.3. Necking features

By imposing different rotation angles and pressure conditions on the holding tools, conditions for necking occurrence were analysed on an industrial prototype of bending. Friction of rubber inside the holding tools, far from the bent part of the profile, controls the tensile effect superimposed on bending. Due to the small bend die radius (15mm), either wrinkles or necking were observed, whatever the pressure in the holding tools. In the case considered for the FEM simulation, the pressure conditions led to a relative displacement of the profile in the tools equal to 28 mm. Localized necking on the bent profile was visible with the naked eye after a rotation larger than 55°, failure being observed for 64°, as shown in Figure 1.

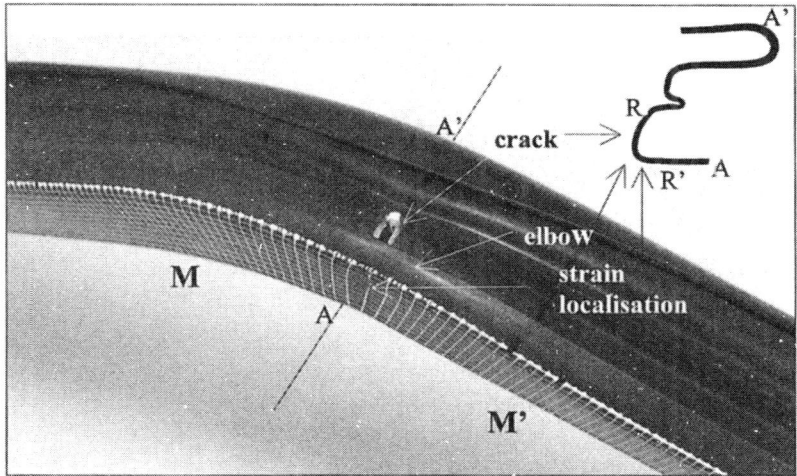

Figure 1. *Failure of the bent profile after 64° rotation of the bending tools. The RR' curve delimitates the zone of risk on the section AA' of the sheet. MM' fibre is referred to in Figure 5*

It is worth noting that while strain localisation was evidenced by grids deposited on the rubber on the lower flat part of the profile, below the "elbow", open cracks were already active inside the wall above it, suggesting that the necking could progress from above the elbow towards below the elbow.

3. FEM simulation of the bending process

3.1. *Numerical procedure*

Analysis of the whole multi-stage process from the initial flat metallic sheet to the final bent bi-material profile has been made by use of FEM simulations with the MSC-Marc code, involving four main sequences: (i) pre-forming of the metal sheet, (ii) forming of the straight metal-rubber profile, (iii) setting the profile in the bending tools (iv) bending the profile. Stages (i) to (iii) were simulated under the assumption of plane strain, whereas stage (iv) was performed by 3D simulations using solid elements. 869 quadrilateral plane strain elements were used in the 2D simulations (510 with 4 Gauss points for metal, 359 with 5 Gauss points for rubber), and 2261 hexahedral bilinear elements (646 for the metal and 1615 for the rubber) in the 3D simulations. The total Lagrangian scheme was used for the rubber elements, and updated Lagrangian scheme for the metal elements. Adhesion between metal and rubber was considered as perfect, consistently with experimental observations.

3.2. *2D simulations of the forming of the straight metal-rubber profile*

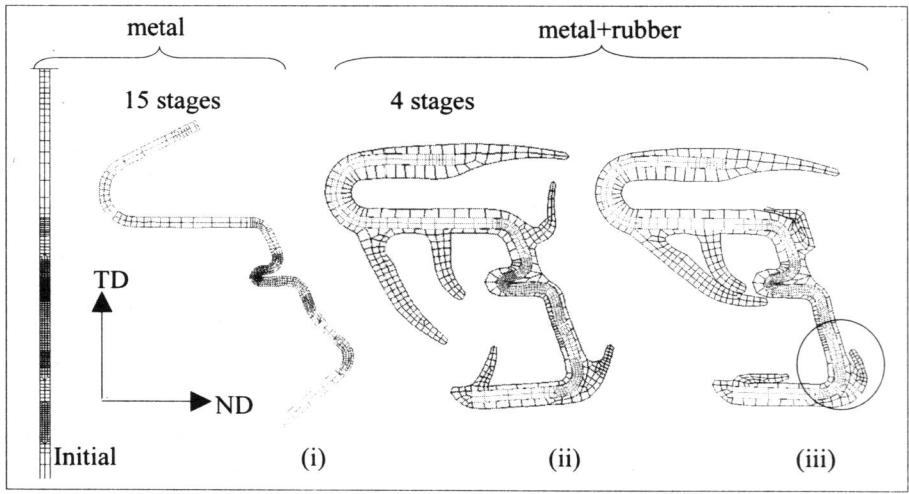

Figure 2. *Evolution of the 2D FEM meshes for the stages before bending*

Considering the reference frame of the initial sheet (RD, TD, ND), stages (i) to (iii) lead mainly to shape the section of the profile in the plane (TD,ND). The 15 successive pairs of rolls used in stage (i) were described by the successive displacement and withdraw of 15 pairs of rigid tools stamping the sheet, considering a Coulomb friction coefficient equal to 0.1 between the tools and the sheet, and taking into account the springback of the profile between the successive passes. In the same way, the 4 pairs of tools used in stage (ii) were described, with a friction coefficient equal to 1 between the tools and the rubber. Stage (iii) consists mainly of compression on rubber only, with low potential effect on the zone of risk. Figure 2 illustrates the different stages.

The link between the different stages was made by transferring the deformed geometry and the equivalent plastic strain at each integration point.

3.3. *3D simulation of bending*

During stage (iv), the profile so-formed had to be bent around the bending die axis parallel to TD. To reduce the CPU time, the section normal to RD obtained at the end of the previous stages was rezoned with a coarser mesh, then expanded in the RD. In order to minimize the length of the profile to be meshed without introducing artificial end effects, special attention was paid to the boundary conditions. As one end of the profile was blocked in the tools, the corresponding nodes were locked. At the other end, the complex effect of friction of the rubber in the holding tools, far from the bent zone, was taken into account by applying the resulting relative displacement in the tools measured on the prototype as a kinematical boundary condition, coupled with the incremental rotation of the wiper die. The geometry before bending is shown in Figure 3.

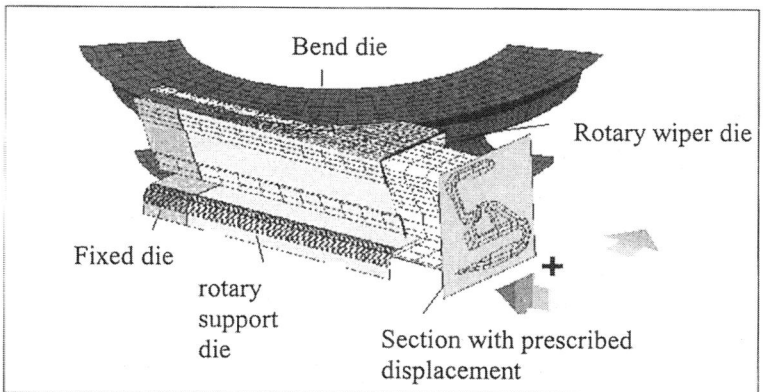

Figure 3. *Tools and profile geometry before bending*

In the present case, rotation of the tools up to 65° was imposed, which corresponded to the attainment of a saturating strain field progressing without change from section to section in the bent zone, consistent with the strain measurements. The simulation of bending presented here was made without taking into account the initial hardening produced by the previous equivalent plastic strain accumulated during the previous stage of forming of the straight profile.

4. Analysis of FEM results

The FEM simulations permit one to estimate the strain distribution throughout the profile at the different stages of the process, leading to accumulated equivalent plastic strain from around 0 to 2 in the zones of high curvature, with values up to 0.5 in the zone of risk before bending. During the pre-forming stage (i) and the forming stage (ii), each pair of stamping tools acts successively on different parts of the section, and, at a given point, the strain-path is approximately linear. In the zone of risk, during the formation of the "elbow" by plane bending, the inner fibre of the section in a TD,ND plane undergoes compressive stresses whereas the outer fibre undergoes tensile stresses. Figure 4 shows the evolution of the total equivalent plastic strain along the outer fibre. Stage (iii) provokes only elastic strains; the rubber is highly compressed against the tools, before being relaxed in the zone of risk during the bending stage (iv).

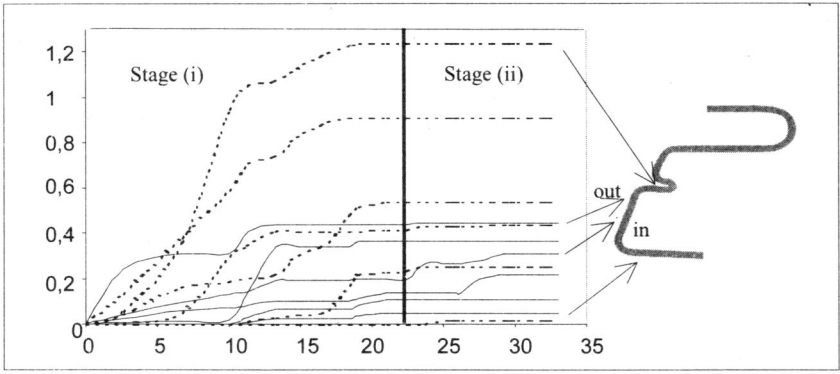

Figure 4. *Equivalent plastic strain history during stages (i) and (ii) for the outer fibre. The x axis is arbitrary numerical time. Solid lines refer to the zone of risk*

Bending corresponds to significant strain-path changes for the sheet, since the main extension strains are then evidenced in the material direction initially parallel to RD, which was the zero strain direction during the previous stages. The comparison with strain measurement on rubber shows that before necking, qualitative agreement both on the strain level and on the shift of the strain maximum is obtained despite a wider angular spreading for the simulated strain distribution, as shown in Figure 5.

Prediction of Necking Initiation   171

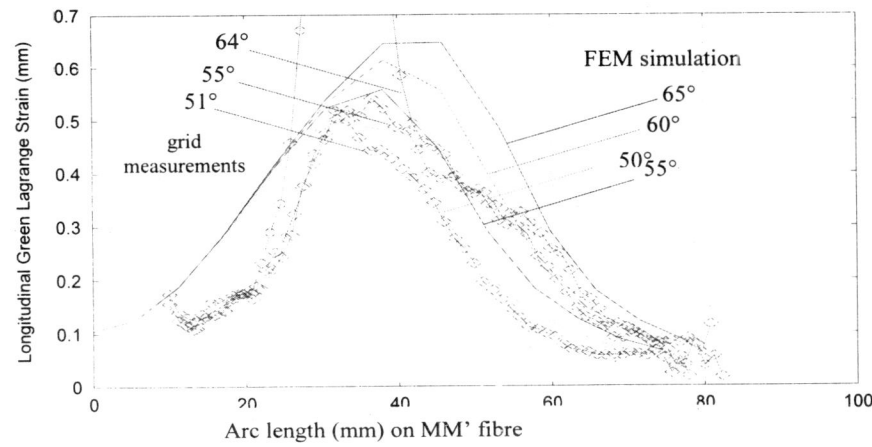

Figure 5. *Extensional strain distribution during bending along MM' (see Figure 1)*

In the zone of risk, extension is rather homogeneous through the thickness, but strain gradients up to 10% are obtained on the two negative principal strains, which could be due to the differences in rubber-tools contact conditions.

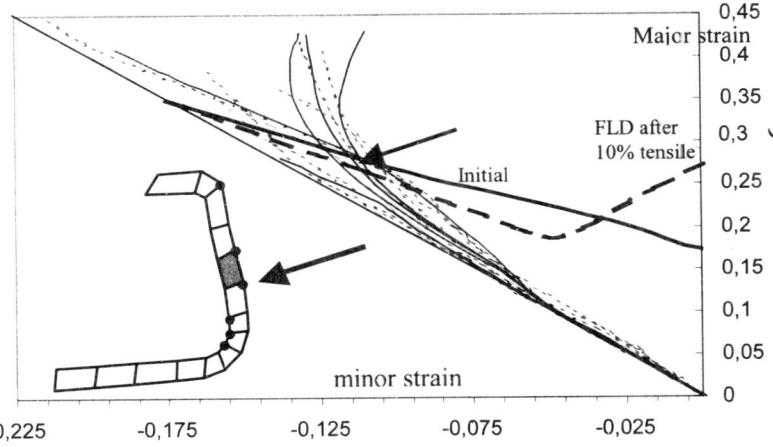

Figure 6. *Predicted strain-path on the most bent section and comparison with predicted FLDs. Solid (resp. dashed) lines concern outer (resp. inner) fibre*

Both on external and internal fibres, the strain paths correspond first to uniaxial tension, then some points evolve more or less rapidly towards plane strain in the RD,ND plane, as shown in Figure 6 for the most deformed section of the bent profile. In this case sheet thinning increases, which may favour early necking initiation compared with the surroundings. Whereas simple considerations on pure bending of beams involve only uniaxial tensile deformation, the FEM simulations permits exhibition of more complex strain-paths. The consequences of the MK-type approach is now discussed.

5. Prediction of necking initiation thanks to MK-analysis

For a given homogeneous strain state on a piece of sheet (matrix) containing a small groove, the classical MK-analysis consists of computing until a critical value the strain ratio between the matrix and the groove using the constitutive equations and the force equilibrium relations at the groove/matrix interface. MK-analysis has been widely used to predict forming limit diagrams (FLD) for linear strain-paths and for some simple non-linear strain-paths involving mainly uniaxial and/or biaxial tensions (Barata da Rocha et al., 1984). An extension to MK-analysis for arbitrary strain-paths, which permits to use strain paths resulting from FEM (Hoferlin et al., 2000), has been applied here to bending, after validation on FLD.

5.1. *Experimental and predicted FLD for linear and non-linear paths*

Experimental FLD have been determined at the IFU in Stuttgart for the initial sheet and after 10% plane strain obtained by uniaxial testing along TD of wide samples, approaching the strain-path change encountered at the beginning of bending by the external skin of the sheet in the zone of risk.

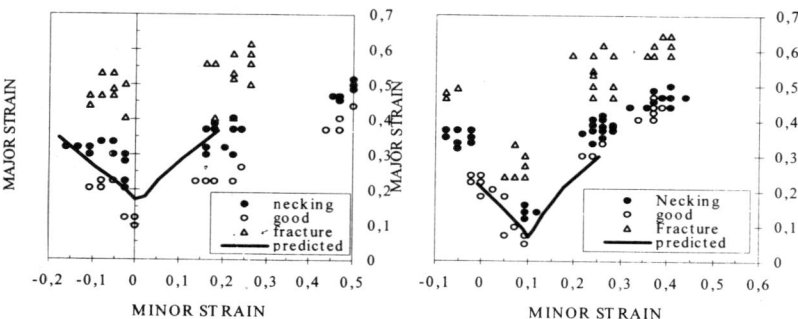

Figure 7. *Experimental and predicted FLD for the initial sheet (left) and after 10% plane strain (right)*

Figure 7 shows the qualitative agreement obtained between the predicted and measured FLD, despite the inherent dispersion of the experimental data, which are consistent with other results on the same alloy (Kohara, 1993).The considered prestrain lowers the FLD, the prediction being rather more severe than the experiment. Predicted FLD after 10% tensile strain is also below the initial FLD. The first point where the FEM strain paths intersects the predicted FLD is indicated by an arrow in Figure 6, which corresponds to 48.4° rotation of the bending tools.

5.2. MK-analysis on FEM results

MK-analysis was conducted for a selected set of the zone of risk. As a first step, only the bending stage (iv) was considered. For each element, a mean strain history was determined from the nodes coordinate history obtained by FEM simulation, then the MK criterion was tested at each time increment until it was achieved. The lowest time increment defines then the critical element for necking.

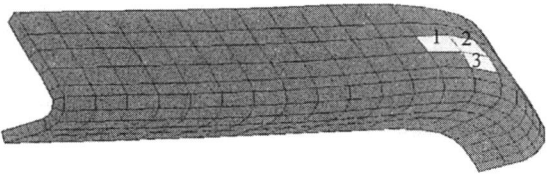

Figure 8. *First elements to localized during bending according the MK-analysis*

Among the elements of the most bent section, necking would begin after 40.7° rotation for the grey element in Figure 6. But when considering the whole bent part, necking initiation is predicted earlier, in the same area of the profile section, but for only 23.1° rotation as illustrated by Figure 8. The well formed neck seen with the naked eye after 55° is expected later than the very beginning of necking initiation, however the prediction is here too severe to be directly used for selection of the bending process parameters.

This too conservative prediction could have two different reasons. First, both strain gradients through the sheet thickness, due to bending, and rubber effects were neglected in the MK-analysis. Both could act to delay incipient necking. Second, the assumption of isotropic behaviour with isotropic hardening, permitting reasonable computing times, is only a rough approximation of the real behaviour of the aluminium alloy, which affects strain distribution and MK-analysis. Work in progress shows recovery effects and orthogonal strain-paths effects during laboratory testing, leading to a better fit by the microstructural model proposed by *Teodosiu et al.* (Teodosiu *et al.*, 1998).

6. Conclusion

The present work has shown that the complex multi-stage bending of a metal-rubber profile could be successfully simulated by FEM calculations. Thanks to the precise description of the geometry and of the tools, the strain distribution permits qualitative understanding of strain-path evolution and the risk of necking. Encouraging results of MK-analysis in the final bending stage has been obtained, which is expected to be improved by a better material description.

Acknowledgements

The authors thank Metzeler for its financial support. They are grateful to OCAS for permitting scientific collaboration for this study.

7. References

Barata da Rocha A., Barlat F., Jalinier J. M., "Prediction of the Forming Limit Diagrams of Anisotropic Sheets in linear and non-linear loading", *Materials Science and Engineering*, Vol. 68, 1984–1985, pp. 151–164.

Boscher D., Gaspérini M., "Simulation of the bending process of an aluminium-rubber profile", *Proceedings of the 7th International Conference on Aluminium Alloys, (ICAA7)*, Charlottesville, Virginia, 9–14 April 2000, ed. E.A. Starke, T.H. Sanders, W.A. Cassada, pp. 1793–1798.

Hoferlin E., Van Bael A., Van Houtte P., Brasseur E., Moriau O., "Influence of texture and strain-path change on the forming limit depth of a squared box with reversed drawing", *Proceedings of J.J. Jonas Symposium on Thermomechanical Processing of Steel*, Montreal, Canada, 20–23 August 2000, ed. S. Yue, E. Essadiqi, pp. 81–95.

Kohara S., "Forming-limit curves of aluminum and aluminum alloy sheet and effects of strain path on the curves", *Journal of Materials Processing Technology*, Vol. 38, 1993, pp. 723–735.

Marciniak Z., Kuczynski K., "Limit strains in the processes of stretch-forming sheet metal", *Int. J. Mech. Sc.*, Vol. 9, 1967, pp. 609–620.

Teodosiu C., Hu Z., "Microstructure in the continuum modelling of plastic anisotropy", *Proc. 19th Riso Int. Sypm. on Materials Science*, Roskilde, Denmark, September 1998, ed. Risø National Laboratory, 1998, pp. 149–168.

Chapter 15

Application of the Variational Self-consistent Model to the Deformation Textures of Titanium

Pierre Gilormini
Laboratoire de Mécanique et Technologie, ENS de Cachan, CNRS, Cachan, France

Yi Liu and Pedro Ponte Castañeda
Department of Mechanical Engineering and Applied Mechanics, University of Pennsylvania, Philadelphia, USA

1. Introduction

Forming processes involve large finite strains that are known to induce strong anisotropy of the mechanical properties of polycrystalline metals, especially when their crystal structure is hexagonal. This is due to the reorientation of the crystalline axes induced by plastic strain, and to the highly anisotropic behavior of hexagonal crystals. The prediction of the evolution of the so-called deformation texture, and of the corresponding anisotropy, has been performed for several decades by means of various models, as reported for instance in [KOC 98].

The aim of this paper, which reports on a preliminary study (more extensive results will be presented in [LIU 02]), is to show that a method that has been proposed recently for computing the overall properties of polycrystals ([BOT 95], [NEB 01], [GIL 01]) is also able to predict deformation textures with interesting features. A comparison will be made with the results of the most popular model that is presently used ([MOL 87], [LEB 93]), and with some experimental textures measured on titanium polycrystals obtained by Balasubramanian and Anand [BAL 02].

2. Variational procedure

The variational procedure that is used in this study follows from the original paper by Ponte Castañeda [PON 91] on nonlinear composites. Applications to polycrystals with fixed textures have already been presented in several papers, including [BOT 95] for face-centered cubic crystals, [NEB 01] for other cubic structures, and [GIL 01] for hexagonal crystals. Consequently, the details of the procedure will not be repeated here, and only the essential equations will be recalled, as they apply specifically to polycrystals.

The approach applies to crystals where the behavior of the slip systems is governed by a convex potential. This includes viscoplastic crystals, for which the standard example of the potential obeys a power law:

$$\varphi_{(s)}^{(g)} = \frac{m}{1+m} \tau_{0(s)}^{(g)} \dot{\gamma}_0 \left(\frac{|\tau_{(s)}^{(g)}|}{\tau_{0(s)}^{(g)}} \right)^{\frac{1+m}{m}} \qquad [1]$$

where the slip rate sensitivity m is between 0 (rate insensitive limit) and 1 (linear behavior), $\tau_{(s)}^{(g)} = m_{(s)}^{(g)} \cdot \sigma \cdot n_{(s)}^{(g)}$ is the resolved shear stress on the system s of grain g, through the stress tensor σ ($m_{(s)}^{(g)}$ and $n_{(s)}^{(g)}$ are unit vectors parallel to the slip direction and normal to the slip plane, respectively), $\tau_{0(s)}^{(g)}$ is a reference shear stress and $\dot{\gamma}_0$ a reference slip rate (assumed the same for all systems for convenience). Such a power-law potential is assumed to hold in this work for each slip system in the polycrystal, although this is not a limitation of the approach ([GIL 01] for instance).

Moreover, m will be assumed the same for all systems, which is not required by the approach either (an example with a non-uniform m is also given in [GIL 01]).

Basically, the variational procedure introduces a linear comparison polycrystal in a systematic manner. Its definition involves the same slip systems as in the nonlinear polycrystal, but obeying quadratic potentials :

$$\widehat{\varphi}_{(s)}^{(g)} = \frac{1}{2\eta_{(s)}^{(g)}} \left(\tau_{(s)}^{(g)}\right)^2 \qquad [2]$$

A complete equivalence between the nonlinear and comparison polycrystals is obtained when the stiffness $\eta_{(s)}^{(g)}$ of each system is allowed to vary not only from system to system in any grain, and from grain to grain, but even inside each grain, provided that a maximization problem can be solved to optimize the choice of the $\eta_{(s)}^{(g)}$ field. In practical applications, $\eta_{(s)}^{(g)}$ is assumed uniform in each grain, which leads to an underestimate of the effective stress potential $\Phi(\Sigma)$ of the polycrystal:

$$\Phi(\Sigma) \geq \max_{\eta_{(s)}^{(g)} \geq 0} \left[\frac{1}{2}\Sigma : M : \Sigma - \sum_g w^{(g)} \sum_s V_{(s)}^{(g)}\right] \qquad [3]$$

where $w^{(g)}$ is the volume fraction of grains with crystallographic orientation g, and where $V_{(s)}^{(g)}$ has the following expression in the special case of a power law:

$$V_{(s)}^{(g)} = \frac{1-m}{1+m} \frac{\tau_{0(s)}^{(g)} \dot{\gamma}_0}{2} \left(\frac{\tau_{0(s)}^{(g)}}{\eta_{(s)}^{(g)} \dot{\gamma}_0}\right)^{\frac{1+m}{1-m}} \qquad [4]$$

If M is evaluated in the linear comparison polycrystal by the self-consistent model, which is widely considered as pertinent for such computations, equation [3] gives a self consistent estimate of the response of the nonlinear polycrystal. The computation of M involves the compliance of each grain in the linear comparison polycrystal, which is given by

$$M^{(g)} = \sum_s \frac{1}{\eta_{(s)}^{(g)}} \mu_{(s)}^{(g)} \otimes \mu_{(s)}^{(g)}$$
$$\text{with} \quad \mu_{(s)}^{(g)} = \frac{1}{2}\left(n_{(s)}^{(g)} \otimes m_{(s)}^{(g)} + m_{(s)}^{(g)} \otimes n_{(s)}^{(g)}\right) \qquad [5]$$

The nonlinear self-consistent estimate that will be obtained with this procedure will be in agreement with the bounds that can be defined on the nonlinear polycrystal by using lower bounds of M in equation [3].

This brief outline of the method shows that a large optimization problem has to be solved to apply the variational procedure, as shown in equation [3]: the number of optimization variables is equal to the total number of slip systems available in the polycrystal. Simple algorithms were feasible as long as transversely isotropic polycrystals were considered, without texture evolution. The Powell method was used in [NEB 01] and [GIL 01], for instance, because a small number of crystallographic orientations was sufficient. This method does not use partial derivatives and cannot be used for the very large numbers of grains (and, consequently, variables) that are involved in texture predictions. The present work makes use of a more efficient method, that is more suitable for the large scale of the optimization problem considered, namely the modification of the BFGS method proposed by Liu and Nocedal [LIU 89]. This requires the partial derivatives of the function to optimize, which were computed from analytical formulae derived from equation [3] by using symbolic programming.

After the linear comparison polycrystal has been obtained through the optimization procedure, there remains to allow the microstructure evolve before repeating the process at the next time increment. As explained in [PON 99], the texture change is computed from the slip rates obtained for each crystal orientation in the linear comparison polycrystal, which lead to the plastic spin and, finally, to the rotation rate of the crystallographic axes of the grains.

3. Material and loading conditions considered

The above procedure has been applied to the titanium polycrystal described in [BAL 02]. Twenty-four slip systems were considered, belonging to 4 families of systems: 3 $\{0001\}\langle 11\bar{2}0\rangle$ systems for basal slip, 3 $\{10\bar{1}0\}\langle 11\bar{2}0\rangle$ for prismatic slip, 12 $\{10\bar{1}1\}\langle 11\bar{2}3\rangle$ for first-order pyramidal $\langle c+a\rangle$ slip, and 6 $\{11\bar{2}2\}\langle 11\bar{2}3\rangle$ for second-order pyramidal $\langle c+a\rangle$ slip. A slip-rate sensitivity of $m = 0.16$ has been used on all systems, as well as a reference slip rate of $\dot{\gamma}_0 = 10^{-3}$ per second. Different reference shear stresses were taken into account: the initial values were $\tau_{0(s)}^{(g)} = 8.2$ MPa for basal and prismatic slip, and $\tau_{0(s)}^{(g)} = 82$ MPa for pyramidal slip, with a hardening law that increased all $\tau_{0(s)}^{(g)}$ values with the same rate

$$\dot{\tau}_{0(s)}^{(g)} = h_0 \sum_r \left[1 - \frac{\tau_{0(r)}^{(g)}}{\tau_{1(r)}} \left(\frac{\dot{\gamma}_0}{|\dot{\gamma}_{(r)}^{(g)}|} \right)^n \right] |\dot{\gamma}_{(r)}^{(g)}| \qquad [6]$$

with $h_0 = 12$ MPa, $n = 0.1$, and $\tau_{1(r)} = 18$ MPa for basal and prismatic systems and 180 MPa for pyramidal systems in the above summation. It should be noted that,

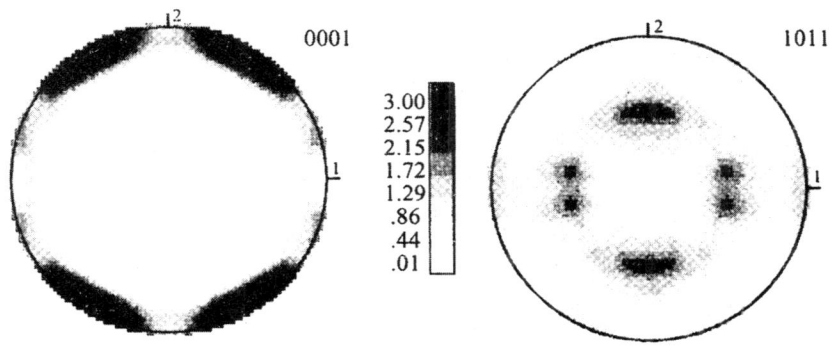

Figure 1. $\{0001\}$ and $\{10\bar{1}1\}$ pole figures of the initial texture

because of the lack of direct measurements on single crystals, these values were deduced by Balasubramanian and Anand [BAL 02] by fitting the predictions of their finite element model to experimental measurements obtained with titanium polycrystals. As a result, these data are probably not the best possible for getting good agreement between experiments and other models. They will nevertheless be used in this study as a reasonable starting set of values that allows some preliminary comparisons between models. Therefore, only qualitative comparisons with experimental results will be possible. More quantitative analyzes would require new fittings (one for each model, actually), and are presently underway.

The texture is not isotropic initially, as shown in Figure 1, where orthotropic symmetry has been applied to the same set of 729 orientations as used by Balasubramanian and Anand [BAL 02] to simulate their experimental pole figures. Note that Figure 1, like the other pole figures in this paper, use equal-area projection. The simulations presented below use the same data file and symmetrization and, consequently, $24 \times 729 = 17,496$ variables are involved in the maximization procedure of the self-consistent variational model.

Two types of loadings have been applied to the above initial texture: uniaxial and plane-strain compressions. In the former, a reduction in height $\Delta H/H_0$ of 63% (i.e., an axial strain of -1) was prescribed along axis 3 (see axes definition in Figure 1), with a constant axial strain rate $\dot H/H$ of 10^{-3} per second, keeping all lateral sides stress-free. In plane strain compression, no displacement was allowed along axis 2, a reduction in height of 59% was applied along axis 3 (i.e., a log strain of -0.9), with a constant strain rate of 10^{-3} per second, and the sides normal to axis 1 were kept stress free. These two cases correspond to some of the experiments of [BAL 02], with friction neglected in our simulations. They are also typical of the (much more complex, of course) conditions that prevail in such forming processes as forging and rolling, respectively.

180 Prediction of Defects in Material Processing

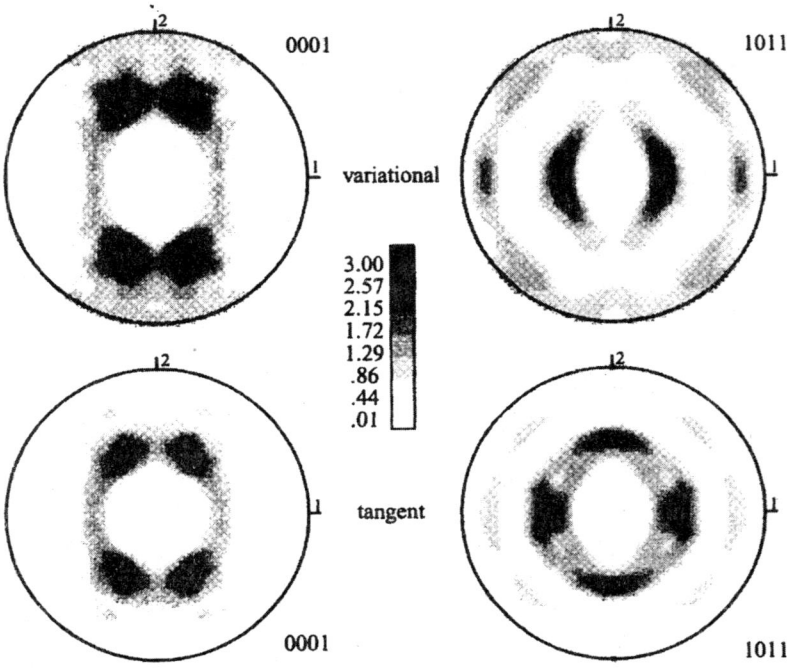

Figure 2. *Pole figures of the textures predicted by the variational procedure (above) and the tangent model (below) after uniaxial compression*

4. Results

In the present paper, which reports a preliminary study, the results obtained with the variational procedure are compared only to those given by the tangent model of [MOL 87] and [LEB 93]. This approach applies the self-consistent model to a tangent approximation of the nonlinear behavior of the slip systems, and is widely used to simulate the texture evolution in metals as well as in minerals (many examples are presented in [KOC 98], for instance). The VPSC5 program developed by R. Lebensohn and C. Tomé has been used in the simulations that are presented here. A well-known problem with the tangent model is that it tends to the uniform-stress lower bound when the rate sensitivity m decreases [LEB 93]. It is worth mentioning that the value of $m = 0.16$ used in the simulations belongs to this sensitive range, where significant differences with the variational procedure can be expected.

Figure 2 shows the pole figures predicted by the variational procedure and by the tangent model after the uniaxial compression has been applied during 100 equal time increments. It can be observed that the two models used with the same data do predict different results. The tangent model predicts a $\{10\bar{1}1\}$ pole figure with four groups of high intensity areas, including two located along the axis 2. This is absent from the predictions of the variational procedure as well as from the experimental pole figures shown in [BAL 02] (and from the finite element simulations in the same reference). A closer analysis of the differences between the two models can be performed by looking at the system activity, defined as

$$A_f = \frac{\sum_g w^{(g)} \sum_{s \in f} |\dot{\gamma}_{(s)}^{(g)}|}{\sum_g w^{(g)} \sum_s |\dot{\gamma}_{(s)}^{(g)}|}$$ [7]

for a system family f (basal, or prismatic, or pyramidal), where a system is considered as active ($\dot{\gamma}_{(s)}^{(g)} \neq 0$) for a given grain orientation g if it contributes to at least 5% of the sum of all slip rates pertaining to this orientation. Figure 3 shows that the activity in the prismatic systems, which decreases when compression proceeds, is larger in the tangent model, while that of the basal systems (which increases) is smaller than in the variational model. The latter also predicts a significant activity of the pyramidal systems, that is almost absent in the results for the tangent model. The activities predicted by the (uniform strain) Taylor model are also shown in the figure, and it may be observed that they are very different from what the two models considered in this paper suggest. The pyramidal activity, for instance, is very large in the Taylor model, which is due to a complete lack of grains interaction: the model requires all grains to deform equally and consequently pyramidal systems are activated in most grains. The self-consistent model, on the opposite, allows each grain to deform differently, with a larger contribution of the soft systems. Another quantity related to this result is the average number of (significantly) active systems: the value computed from the Taylor model was about 10, whereas the variational and tangent models used respectively about 5 and 3 systems per grain on average (recall that prismatic and basal systems amount to a total of 6).

For plane strain compression, where 90 increments were used, the differences in the predicted textures is less pronounced, as can be observed in Figure 4. The predictions of the variational model are close to the finite element results of [BAL 02]. The larger vertical scatter of the high-intensity areas in the $\{10\bar{1}1\}$ pole figure of the tangent model is more similar to what is observed in the experiments of [BAL 02], but the $\{0001\}$ pole figure, like for other models, lacks the high intensities measured at both ends of the 1 axis. The system activities obtained in plane strain are not shown: they lead to conclusions similar to what was mentioned above for simple compression. Further details and simulations will be found in [LIU 02].

5. Conclusion

It has been demonstrated that the variational procedure, combined with the self-consistent model, can be applied to the computation of the texture evolution induced by the deformation of polycrystals.

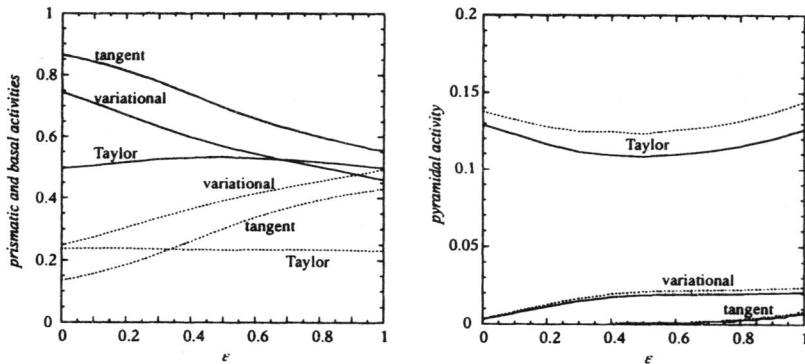

Figure 3. *System activities predicted by various models in the simulation of uniaxial compression. Prismatic (solid lines) and basal (broken lines) systems on the left. First-order (solid lines) and second-order (broken lines) pyramidal systems on the right*

For titanium polycrystals, comparison with another model, that is in wide use, has shown that some differences in the predictions are generated. Moreover, qualitative agreement with experimental pole figures has also been found. Now that the computational issues have been addressed, more quantitative comparison with experimental results is required. This is presently underway.

Acknowledgements

R. Lebensohn (IFIR, Argentina) and C. Tomé (LANL, USA) are gratefully acknowledged for making some of their programs available. We are grateful to L. Anand (MIT, USA) for kindly providing the initial texture data file and valuable discussions.

6. References

[BAL 02] BALASUBRAMANIAN S., ANAND L., "Plasticity of initially textured hexagonal polycrystals at high homologous temperatures: application to titanium", *Acta Materialia*, vol. 50, 2002, p. 133-148.

[BOT 95] DE BOTTON G., PONTE CASTAÑEDA P., "Variational estimates for the creep behavior of polycrystals", *Proceedings of the Royal Society of London A*, vol. 448, 1995, p. 121-142.

[GIL 01] GILORMINI P., NEBOZHYN M. V., PONTE CASTAÑEDA P., "Accurate estimates for the creep behavior of hexagonal polycrystals", *Acta Metallurgica*, vol. 49, 2001, p. 329-337.

Deformation Textures of Titanium 183

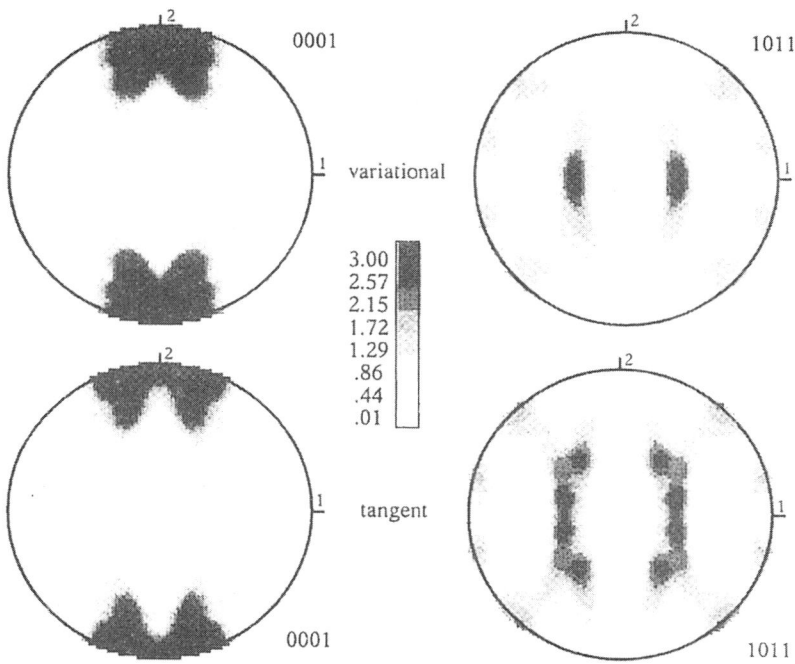

Figure 4. *Pole figures of the textures predicted by the variational procedure (above) and the tangent model (below) after plane strain compression*

[KOC 98] KOCKS U. F., TOMÉ C. N., WENK H. R., *Texture and Anisotropy*, Cambridge University Press, Cambridge, 1998.

[LEB 93] LEBENSOHN R. A., TOMÉ C. N., "A self-consistent anisotropic approach for the simulation of plastic deformation and texture development of polycrystals: Application to Zirconium alloys", *Acta Metallurgica*, vol. 41, 1993, p. 2611-2624.

[LIU 89] LIU D. C., NOCEDAL J., "On the limited memory BFGS method large scale optimization", *Mathematical Programming*, vol. 45, 1989, p. 503-528.

[LIU 02] LIU Y., GILORMINI P., PONTE CASTAÑEDA P., "Variational self-consistent estimates for texture evolution in hexagonal polycrystals", in preparation, 2002.

[MOL 87] MOLINARI A., CANOVA G. R., AHZI S., "A self-consistent approach to the large deformation polycrystal viscoelasticity", *Acta Metallurgica*, vol. 35, 1987, p. 2983-2994.

[NEB 01] NEBOZHYN M. V., GILORMINI P., PONTE CASTAÑEDA P., "Variational self-consistent estimates for cubic viscoplastic polycrystals: the effect of grain anisotropy and shape", *Journal of the Mechanics and Physics of Solids*, vol. 49, 2001, p. 313-340.

[PON 91] PONTE CASTAÑEDA P., "The effective mechanical properties of nonlinear isotropic composites", *Journal of the Mechanics and Physics of Solids*, vol. 39, 1991, p. 45-71.

[PON 99] PONTE CASTAÑEDA P., "Nonlinear polycrystals with microstructure evolution" INAN E., MARKOV K., Eds., *Continuum Models and Discrete Systems. Proceedings of the 9th International Symposium*, World Scientific Publishing Co., 1999, p. 228-235.

Chapter 16

Influence of Strain-hardening and Damage on the Solid-state Drawing of Poly(oxymethylene)

Christian G'Sell
Laboratoire de Physique des Matériaux, Ecole des Mines de Nancy, France

Ian M. Ward
IRC on Polymer Science and Technology, University of Leeds, UK

1. Introduction

It has been shown by many authors, in the 70's, that the mechanical performances of polymers (Young's modulus and strength) could be improved considerably if macromolecular chains were oriented along a particular axis rather than distributed evenly in all directions. This is because the covalent bonds along the chains are much stiffer than the van der Waals bonds (or the hydrogen bonds) between the chains. Of course, the material with oriented chains in no longer isotropic and the improvement of the properties is effective only along the principal direction parallel to the mean orientation axis.

Among the techniques developed to produce oriented polymers, die-drawing in the solid state was intensively studied by one of the authors (Coates and Ward, 1979, 1981; Gibson and Ward 1980). In this technique, a polymer billet is forced to pass through a converging conical die by means of a pulling force, the material undergoing cross-section reduction, axial stretching and unidirectional chain orientation. Several products can be obtained by this technique including filaments, rods, sheets and tubes (Richardson *et al.*, 1986). The critical parameters of this process are: i) overall reduction of cross-section, ii) cone angle, iii) die lubrication, iv) stretching speed, v) operating temperature and, v) constitutive behaviour of the polymer in the plastic range, up to large deformations (Kukureka *et al.*, 1992).

Many polymers are suitable for the die-drawing process, either amorphous (poly(vinyl chloride) and poly(ethylene terephtalate)), or semi-crystalline polyethylene, polypropylene and poly(oxymethylene)). Considerable interest has been shown to high-value technical applications like wire ropes. Substitution of steel by a polymer is motivated principally by the low density of the organic materials (about 1000 to 1400 kg/m^3). Originally, only the core of the wire ropes was made of plastic (for elevator systems after Taraiya *et al.*, 2000). New technologies, specially for off-shore oil extraction technology, are now developed to replace also the twisted steel cables by plastic filaments. In the latter application, poly(oxymethylene), POM for short, appears as a very good candidate since, although its density is somewhat higher than others, it is highly crystalline and exhibits a low friction coefficient. After being stretched at 150 °C to an elongation ratio $\lambda \approx 10$ (that is for a Hencky "natural" strain of $\varepsilon = \ln \lambda \approx 2.3$), POM acquires in the drawing direction a Young's modulus $E \approx 13.000$ MPa (while measured at room temperature), that is 4 times the standard undeformed modulus (Hope *et al.*, 1982)

Optimal control of die drawing operation can be achieved only at the cost of a comprehensive modelling of the process, for example by the means of finite-element calculation. Although the implementation of large strains and tool friction is now straightforward in modern computer codes, the determination of constitutive equations of solid polymers in the plastic stage requires several precautions. First the stress-strain-strain rate equation should be determined *intrinsically*, that is at the scale of a representative volume element (RVE) much smaller than the size of the neck which forms quite early in POM. After early works (Coates and Ward, 1978;

G'Sell and Jonas, 1979), novel methods became available for determining the plastic behaviour locally within the neck by means of video-controlled materials testing systems (G'Sell et al., 1992), so that true stress and true strain could be measured in real time. In the case of POM, such techniques enabled the evolution of strain hardening to be assessed which is directly connected to the development of crystalline texture during a stretching operation (Dahoun et al., 1994).

Several authors presented direct evidence of damage during the stretch-forming of POM, generally based on *post-mortem* microscopic observations (Hope et al., 1981), as well as from measurements of density or viscoelastic modulus (Hope et al., 1982, Dahoun, 1992). These investigations showed that in this semi-crystalline polymer, the plastic shear processes are constitutionally associated with crystallite tilting and fragmentation as soon as large deformations are attained. Although the latter phenomenon is now well identified and partially documented, it is interesting to note that its influence on the constitutive equations has never been seriously taken into account in the finite-element modelling of the die-drawing process, which is still based on a constant volume (isochoric) approximation. This is because quantitative assessment of damage during the course of mechanical tests was not easily feasible by means of the conventional mechanical testing methods and even of the early versions of the video-controlled procedures. Even more, since the microstructure and properties of stretched polymers become anisotropic, a complete treatment of damage development should specifically take in consideration the influence of cavitation on the transverse resistance of the stretched-formed products.

It is thus obvious, closing these introducing remarks, that the quantitative characterisation of damage in the 3-D modelling of plastic behaviour of POM is a very ambitious goal which will require much experimental and theoretical research effort before being completely operational for process optimisation. The objective of the present work is more modest and will be restricted to the *in-situ* determination of volume changes simultaneously to stress and strain during the deformation of POM specimens under uniaxial tensile testing at a temperature compatible with the die-drawing process. We will recall in the section below the experimental conditions of this investigation. Subsequently we will present and discuss the results obtained under these conditions.

2. Material and methods

The POM investigated in this work is a commercial grade identified as Delrin® 7031 by Du Pont de Nemours (Wilmington, Delaware, USA). This acetal resin is characterised by a number-average molecular mass equal to $\overline{M}_n = 66,000$ g/mol, with a polydispersity index of 2 (like in most POM grades). It exhibits high-viscosity in the melt and was originally developed for use in extrusion. In the present stretch-forming application, the relatively high molecular weight is favourable for deformation stability. POM is a thermoplastic homopolymer obtained by addition of polymerisation of formaldehyde. Its chains are mostly linear, with a simple monomer unit (-CH$_2$-O-) which provides it with ample molecular mobility. Its limited chemical stability at elevated temperatures necessitates short processing times in order to avoid unwanted yellowish coloration of the final product. Under normal cooling conditions, the material crystallizes in a hexagonal structure with the following cell parameters: a = b = 0.447 nm, c = 1.739 nm, $\alpha = \beta = 90$ degrees, $\gamma = 120$ degrees. The oversized $\bar{c}$ axis is controlled by the extended 2*9/5 helix structure of the crystallized chain in this allotropic form (Figure 1). With 9 monomers in the unit cell, the crystal density is equal to $\rho_c = 1491$ kg/m^3 (Wunderlich, 1973).

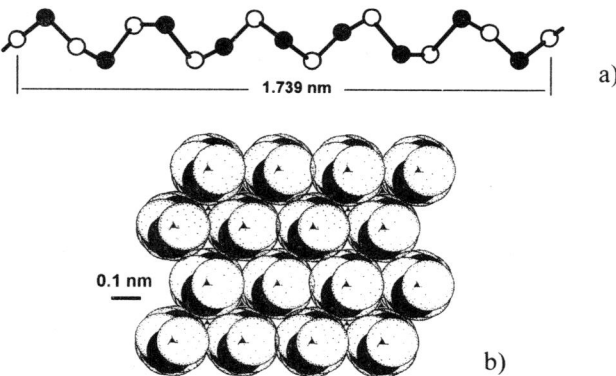

Figure 1. *Crystalline structure of POM a) 2*9/5 helix b) basal plane configuration*

The material was processed by compression moulding in the shape of 4 mm thick plates. It is characterised by its melting temperature $T_m = 178$ °C (determined by DSC according to ISO 3146C) and its high density, $\rho = 1420$ kg/m^3 (determined by ISO 1183). The latter value, compared with the densities of the crystal lattice, $\rho_c = 1491$ kg/m^3, and the density of the amorphous structure, $\rho_a = 1215$ kg/m^3 (Wunderlich, 1973), indicates a high index of crystallinity, close to 70 vol.%. Although previous authors observed spherulites with diameters larger than 100 µm in some POM specimens (*e.g.* Wunderlich, 1973), observations in microtomed slices of this grade revealed a much finer spherulite morphology (diameter lower than 10 µm), presumably because of the presence of nucleating agents.

The specimens for mechanical testing were machined out of the plates by means of a computer-controlled milling machine (Figure 2a). The calibrated length between the gripping heads is 50 mm long and 10 mm wide, with a progressive reduction of the width (minimum 9 mm) in a central zone (10 mm long) in order to localize the necking process in this predefined region. A novel technique was developed by one of the authors (G'Sell and Hiver, 1991; G'Sell *et al.*, 1992) in order to assess the plastic response of the material locally (VideoTraction® by Apollor, Vandoeuvre, France). It is based on the continuous video analysis of a set of seven ink markers printed on one main face of the specimen in the central region. Every 50 ms, the image of the marked zone (Figure 2b) is captured by the digitising board of a the fast processor PC, the seven dots are identified and the coordinates (x_1^i, x_3^i) of their centre of gravity are measured.

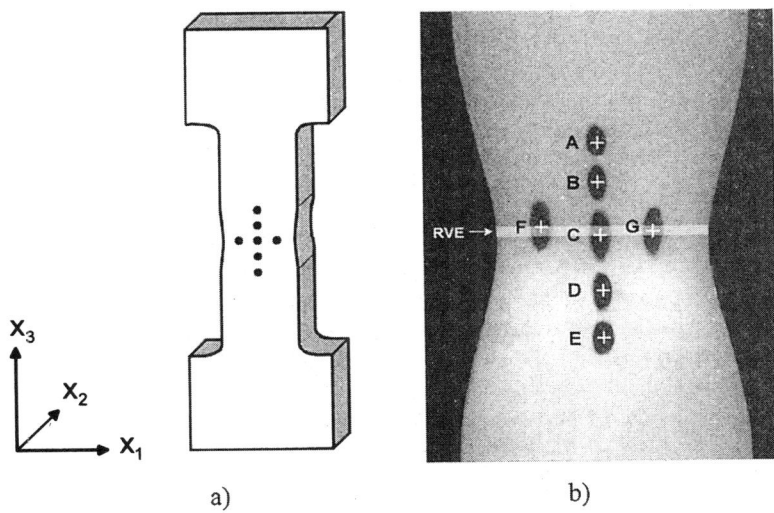

a) b)

Figure 2. *Tensile sample with dot markers for the video-controlled testing of POM a) Schematic diagram b) real video image after necking ($\varepsilon = 0.71$)*

The distribution of axial deformation is characterised from the displacement of the 5 dots (A, B, C, D, E) aligned along the tensile axis x_3. The "true axial strain" (Hencky natural strain) is calculated for each couple of markers:

$$\varepsilon_{33}^{AB} = \ln\left(\frac{AB}{A_oB_o}\right) \quad \varepsilon_{33}^{BC} = \ln\left(\frac{BC}{B_oC_o}\right)$$

$$\varepsilon_{33}^{CD} = \ln\left(\frac{CD}{C_oD_o}\right) \quad \varepsilon_{33}^{DE} = \ln\left(\frac{DE}{D_oE_o}\right)$$

where (A_o, B_o, C_o, D_o, E_o) are the initial positions of the axial markers. Each value of the above axial strains is associated with the x_3 coordinate of the mid-point between the two corresponding markers. The value of the axial true strain in the RVE at the centre of the neck (defined from the set of transversal dots (F, C, G) and represented in light grey in Figure 2b) can thus be determined by means of a non-linear polynomial interpolation. It is denoted ε_{33} or more simply ε.

For its part, the "true transverse strain", ε_{11} or simply ε_t, is determined from the displacement of the markers (F, C, G) aligned x_1. Both values of strains (negative) obtained from marker pairs (F, C) and (C, G) are first computed separately by:

$$\varepsilon_{11}^{FC} = \ln\left(\frac{FC}{F_oC_o}\right) \qquad \varepsilon_{11}^{CG} = \ln\left(\frac{CG}{C_oG_o}\right)$$

The system checks systematically that these two values are very close to one another. Discrepancies indicate that oblique shear bands propagate, instead of a symmetrical, a diffuse neck form and then the test should be rejected. If this is not the case, the true transverse strain, ε_t, is obtained by averaging the two above strains.

In the present state of the method, no way is possible to assess the second transverse strain, ε_{22}, separately. Consequently, an approximation of transverse isotropy must be made, so that ε_{22} is simply taken equal to ε_{11}. In most polymers this condition is actually verified, unless through-thickness thermal gradients would promote some preferential growth of the crystallites in the x_2 direction. Indeed, for POM samples, a small anisotropy ($\varepsilon_{22} / \varepsilon_{11} \approx 1.05$) was actually observed, which was assessed at the end of the tensile test by measuring the change of width and thickness at the level of the RVE, by means of a precision calliper. Assuming that the anisotropy ratio, $R = \varepsilon_{22} / \varepsilon_{11}$ is constant during the whole test, actual values of true strains, ($\varepsilon_{11}, \varepsilon_{22}, \varepsilon_{33}$), are thus readily accessible.

Assuming the other approximation that the reference axes, (x_1, x_2, x_3) coincide with principal axes of strain tensor in the RVE thanks to the symmetry in the centre of the neck, one obtains finally the "volume strain" in the Hencky formalism by the equation:

$$\varepsilon_v = \varepsilon_{11} + \varepsilon_{22} + \varepsilon_{33} = (1+R)\,\varepsilon_t + \varepsilon$$

By the way, the volume strain such defined is equal to $\ln(V/V_o)$, where V and V_o hold for the current and initial volume of the RVE. It should be noted that, at small strain, this definition reduces to the nominal dilatation of the material, $(V-V_o)/V_o$. Also, in the elastic range, volume strain is linked to the Poisson's ratio since, within the approximation transverse isotropy, one writes $\varepsilon_v = (1 - 2\nu)\,\varepsilon$. At large deformation, if the variation of volume strain is no longer linear with axial strain, one should introduce the notion of "tangent Poisson's ratio" defined as $\nu_T = (1-D)/2$, where $D = d\varepsilon_v / d\varepsilon$ is the slope of the $\varepsilon_v(\varepsilon)$ curve.

Finally, we define the "axial true stress" by the Cauchy stress σ_{33} in the RVE. We denote it σ for short. True stress is the load divided by unit actual cross-section. It is computed in real time during the test from the load cell signal and the video measurements through the equation

$$\sigma = F / S = (F / S_o) \exp[-(1+R)\varepsilon_t]$$

In the VidéoTraction® system, not only all parameters are calculated by the microcomputer, but also the tensile machine can be controlled dynamically in such a way that any variable of the system is regulated to vary according to a prescribed rate. In the analysis of the mechanical behaviour of polymers, it is particularly important to run tests at constant true strain rate $\dot{\varepsilon} = d\varepsilon / dt$. This condition is readily fulfilled by means of the numerical PID regulation loop build in the system, which adjusts automatically the speed of the actuator in real time. The diagram in Figure 3 represents the whole system, with all its functions (video measurement, load cell monitoring, actuator control, data display and storage).

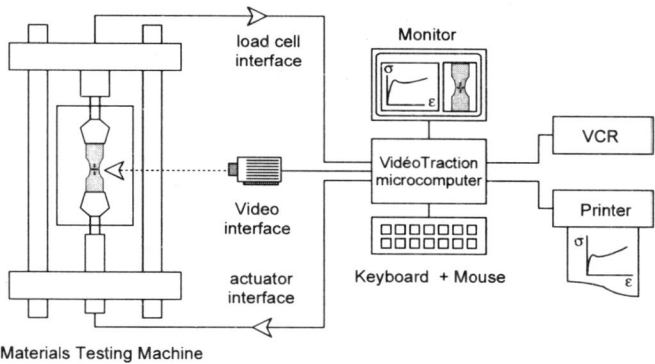

Figure 3. *Diagram of the VidéoTraction® system*

3. Experimental results and discussion

The mechanical properties of the POM grade were investigated at 160 °C, that is at the die drawing temperature for the fabrication of monofilaments. The stress-strain curves, obtained by the special tensile procedure presented above, are presented in Figure 4 for three values of the strain-rate: $\dot{\varepsilon} = 10^{-4}$ s^{-1}, 10^{-3} s^{-1} and 10^{-2} s^{-1}.

The plastic response thus recorded exhibits features common with most semi-crystalline polymers such as polypropylene at 70 °C (G'Sell *et al.*, 1997). The first stage, for $0 < \varepsilon < 0.1$, corresponds to the viscoelastic deformation of the material. Since the testing temperature, T = 160 °C, is much higher than the glass transition temperature, $T_g = -75$ °C, deformation is mostly supported by shear and extensional processes in the rubberlike amorphous domains between the crystalline lamellae. At the yield point, for a stress $\sigma_y \approx 10$ MPa, the onset of plastic flow is observed by a knee in the stress-strain curve. No strain softening is observed, unlike in nominal load/extension curves at constant crosshead speed. This phenomenon, discussed in detail elsewhere (G'Sell *et al.*, 1992), is due to the fact that the true stress definition adopted here takes into account the current decrease in cross-sectional area while nominal definition of stress, $\sigma_N = F / S_o$, refers to the original geometry of the sample. It is thus evident that the stress drop observed in conventional test corresponds merely to an artefact due to sudden necking of the specimen. In this work, strain interpolation in the RVE gives access to the true stress/strain response which exhibits no softening at yield. One can note that the yield process in POM at 160 °C involves some viscous processes, since σ_y depends on strain-rate. According to standard procedure, the strain-rate sensitivity coefficient, $m = [\partial \ln \sigma_y / \partial \ln \dot\varepsilon]_\varepsilon$, is equal to about 0,09, a usual value for most crystalline polymers between T_g and T_m.

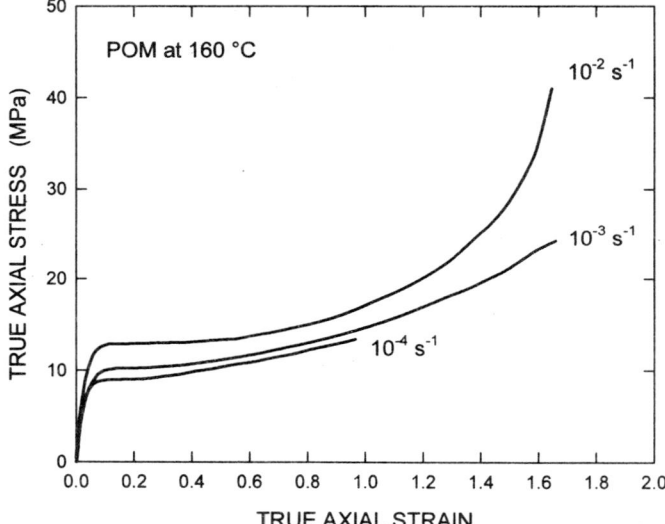

Figure 4. *Stress/strain curves of the POM at constant strain rate*

The interpretation of yield processes has been the object of active scientific debates in the last decade. Many authors, after the pioneering works of Schultz (Schultz, 1974) and Haudin (Haudin, 1982), agree on a plasticity scheme controlled by the activation of glide in the crystalline lamellae favourably oriented in terms of the resolved shear stress on the easy glide planes of the hexagonal structure (prismatic systems { $\bar{1}100$ } < 0001 >). In a microstructure with crystallites initially oriented in any direction, the critical resolved shear stress, τ_c, would be equal to half the macroscopic yield stress, that is $\tau_c \approx 5$ MPa. This model is supported by the evolution of the flow stress at higher deformation. It is seen in Figure 4 that an auto-accelerating strain-hardening occurs while stretching up to large deformation. As demonstrated elsewhere (Dahoun *et al.*, 1994), this consolidation contrasts with the nearly constant flow stress when the same material is deformed under simple shear.

A detailed discussion of this influence of strain path has shown (Dahoun, 1992) that hardening is intimately connected with the development of a crystal orientation texture in the microstructure. In the case of uniaxial tension, plastic tilt of the shearing lamellae brings the chain axis <0001> closer to the tensile axis as strain increases. Consequently, the Schmid factor for the easy glide systems decreases progressively and the continuation of plastic glide requires an increase of the applied stress. More difficult transverse slip systems ($\{\bar{1}100\}$ <11$\bar{2}$0>) may relay the prismatic systems but, nevertheless, true stress continues increasing until rupture occurs. Conversely, for simple shear testing, plastic tilt turns the chain axis towards the shear direction and tends to level-off the stress evolution. Quantitative prediction of these phenomena by means of a self-consistent simulation including the amorphous chain contribution (G'Sell et al., 1999) supports this model. Here in POM, it is seen that strain can be increase in tension up to $\varepsilon \approx 1.7$, that is for an extension ratio of about $\lambda \approx 5.5$ before rupture. It seems that, for such a large elongation a certain saturation of the strain-hardening is observed at smaller strain-rates. This is presumably due to some slippage of the "tie molecules" originally joining the crystallites together and consequently to the permanent activation of creep at long times. This loss of cohesiveness of the polymeric "network" is certainly influenced by the high testing temperature, 160 °C, at which it is shown by DSC that a given fraction of the smallest crystallites are already melted (Dahoun, 1992).

The evolution of volume strain during the stretching of POM is displayed in Figure 5. It is remarkable that damage processes are very active, since volume strain reaches values as high as $\varepsilon_v \approx 0.4$ while axial true strain attains $\varepsilon \approx 1.6$. Overall, that means that non-isochoric processes participate by 25% to the total deformation of the material. We are thus far from the classical approach which considers that, as in metals, plasticity of polymers is controlled by simple shear processes. Apart from the curve for $\dot{\varepsilon} = 10^{-2}$ s^{-1} which, for unidentified reasons, lags at zero volume change until it upturns after yield point, the behaviour in the elastic range corresponds to immediate dilatation with $D = d\varepsilon_v / d\varepsilon \approx 0.25$. As pointed out before, Poisson's ratio is then $\nu = (1-D)/2 \approx 0.37$, in good agreement with reported values for polymers. Dilatation in the elastic stage is generally not interpreted in terms of damage, but rather from the preferential deformation of cohesive bonds between molecular groups along the tensile axis. This effect is not specific to polymers; it is even larger in metals for which values of $\nu \approx 0.3$ are usually observed. Here, if we note that stress goes on increasing after yield, the contribution of this elastic dilatation, ε_v^{el}, becomes larger at higher strains. It can be estimated by the relation $\varepsilon_v^{el} = (1-2\nu)(\sigma/E)$, where $E \approx 340$ MPa, as assessed from the initial slope of the $\sigma(\varepsilon)$ curves in Figure 4. From the above relationship, we can estimate the contribution of elastic dilatation to the total volume strain. For the reference case at $\dot{\varepsilon} = 10^{-3}$ s^{-1}, it increases up to $\varepsilon_v^{el} = 0.019$ when $\varepsilon_v = 0.385$ at the ultimate point ($\varepsilon = 1.65$). Consequently, the elastic dilatation can be regarded as a secondary phenomenon with respect to the major contribution of damage. Unexpectedly, in the major deformation domain, nearly no influence of the strain-rate in noted on the value of volume strain. Only, in the experiment at $\dot{\varepsilon} = 10^{-2}$ s^{-1}, a particular phenomenon is displayed at strains above 1.4, where it is noted that the volume strain decreases, while true stress increases steeply. This effect can be ascribed to an enhanced orientation process under this fast strain rate. Instead of flowing apart, the amorphous POM molecules are supposed to uncoil and align themselves parallel to the tensile axis. Since oriented chains occupy less space than disordered coils, it is not unlikely that volume strain declines while oriented covalent bonds increase strain-hardening.

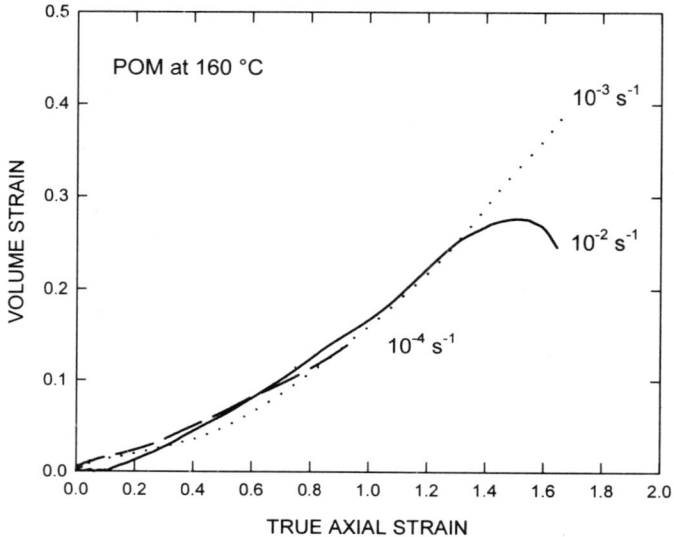

Figure 5. *Assessment of volume strain evolution during the same tests as in Figure 4*

Evidence of density decrease in POM subjected to large deformation has been published several times previously (Gezovitch and Geil, 1971; Choy et al., 1983; Hope et al., 1981; Komatsu, 1991; Dahoun, 1992). The influence of this effect on the management of the industrial die-drawing process is severalfold. Firstly, the modelling of solid-state flow in the die, which is essential for parameter adjustment, should be revised thoroughly. In their detailed mechanical model, Motashar et al. (Motashar et al., 1993) have taken into account many intrinsic properties of the material (yield stress, strain hardening, strain-rate sensitivity, pressure dependence, etc.) and of the polymer-die interaction (geometrical constraint, friction coefficient). However, isochoric deformation was kept as a valid approximation. The present results show that density variations should be introduced in such a model. However, more research effort should be devoted to fully understand the volume strain effect in a 3-dimensional frame. This is because axial true strain is certainly not the only sensible parameter. G'Sell et al. (G'Sell et al., 1997) have compared the respective evolution of POM microstructure after tension and shear deformation paths up to equivalent strains. They showed that while the density of POM decreased by about 30% from its initial value after being deformed in tension for a true strain $\varepsilon = 1.8$ at 100 °C, the corresponding variation after shear is only 2% under shear for the same equivalent strain-rate. Since the die-stretching model of Motashar et al. (Motashar et al., 1993) showed that transverse deformation is negative due to convergent stream in the conical die, the 3-D plastic response of the polymer is somehow intermediate between uniaxial tension behaviour and simple shear flow, so that density variations are probably smaller in the industrial process than in the experiments reported here. Whatever the importance of damage phenomena, they deserve to be characterised furthermore. This is the reason why the microscopic observation of damage processes during stretch forming of POM will be the next step of this cooperation work. Also, it will be interesting to study the influence of deformation damage on the final properties of the stretched monofilaments. In his systematic study of POM stretching at 100 °C, Dahoun (Dahoun, 1992) showed that the viscoelastic compliance of the polymer (still measured at 100 °C) was increased dramatically with respect to the undeformed material. Although die-stretched products are normally devoted to room-temperature applications (so that the viscoelastic contribution is certainly less critical), this negative influence of deformation damage on the final properties should be carefully evaluated, and introduced in the optimisation problematic of the process development.

4. Conclusions

The mechanical behaviour of poly(oxymethylene) under uniaxial stretching at 160°C was characterised by a novel video-controlled tensile testing technique capable of recording true stress, true strain and volume strain simultaneously. A major point in this technique is that plastic behaviour is assessed locally within a small representative volume element in the centre of the neck. Consequently, stress/strain curves and volume strain variation were determined at three different strain-rates without perturbation due to the development of necking. The main result of these experiments is that non-isochoric processes play an important role in the total deformation of the material, since volume strain contributes 25% to the total strain undergone by the polymer upon stretching. This result has several consequences: i) it indicates unambiguously that slip processes in crystalline lamellae are not the only mechanisms involved in the plastic deformation of POM, ii) it invites us to identify more specifically the microstructural mechanisms responsible for the dilatation of the material al large strains and, iii) it proves the necessity to introduce a detailed 3-D formulation of damage kinetics into the mechanical modelling.

5. References

Coates P.D., Ward I.M., "The plastic deformation behaviour of linear polyethylene and poly(oxymethylene)", *J. Mater. Sci.*, Vol. 13, 1978, pp. 1957–1970.

Coates P.D., Ward I.M., "Drawing of polymers through a conical die", *Polymer*, Vol. 20, 1979, pp. 1553–1560.

Coates P.D., Ward I.M., "Die drawing: solid phase drawing of polymers through a converging die", *Polym. Eng. Sci.*, Vol. 21, 1981, pp. 612–618.

Choy C.L., Leung W.P., Huang C.W., "Elastic moduli of highly oriented poly(oxymethylene)", *Polym. Eng. Sci.*, Vol. 16, 1983, pp. 910–922.

Dahoun A., Déformation plastique et textures cristallines induites dans les polymères semi-cristallins, *Doctorat de l'INPL*, Nancy, France, 10th December 1992.

Dahoun A., G'Sell C., Canova G.R., Philippe M.J., "Development of orientation and damage during the plastic deformation of semi-crystalline Poly(oxymethylene)", *Textures of Materials – Materials Science Forum*, Vol.157–162, Part 1, 1994, pp. 645–652.

Gibson A.G., Ward I.M., "High stiffness polymers by die drawing", *Polym. Eng. Sci.*, Vol. 20, 1980, pp. 1229–1235.

G'Sell C., Jonas J.J., "Determination of the plastic behaviour of solid polymers at constant true strain rate", *J. Mater. Sci.*, Vol. 14, 1979, pp. 583–591.

G'Sell C., Hiver J.M., Dahoun A., Souahi A., "Video-controlled tensile testing of polymers and metals beyond the necking point", *J. Mat. Sci.*, Vol. 27, 1992, pp. 5031–5039.

G'Sell C., Dahoun A., Favier V., Hiver J.M., Philippe M.J., Canova G.R., "Microstructure transformation and plastic behaviour of isotactic polypropylene under large plastic deformation", *Polym. Eng. Sci.*, Vol. 37, 1997, pp. 1702–1711.

G'Sell, C., Hiver, J.M., "Dispositif de caractérisation optique du comportement mécanique local d'une structure pouvant présenter des déformations finies homogènes", French Patent pending n° 01054210000, INPI, Paris, France, 23rd April 2001.

G'Sell C, Dahoun A., Royer F.X., Philippe M.J., "Influence of the amorphous matrix on the plastic hardening at large strain of semi-crystalline polymers", *Modell. Simul. in Mater. Sci. and Eng.*, Vol.7, 1999, pp. 817–828.

G'Sell C., Hiver J.M., Dahoun A., "Experimental characterization of deformation damage in solid polymers under tension and its interrelation with necking", *Int. J. Sol. Struct.*, 2002, in press.

Gezovitch D.M., Geil, P.H., "Deformation of poly(oxymethylene) by rolling", *J. Mater. Sci.*, Vol. 6, 1971, pp. 509–530.

Haudin J.M., "Plastic deformation of semi-crystalline polymers", *Plastic deformation of amorphous and semi-crystalline materials*, ed. by Escaig, B. and G'Sell, C, Les Editions de Physique, Les Ulis, Paris, France, 1982.

Hope P.S., Richardson A., Ward, I.M., "Manufacture of ultrahigh-modulus poly(oxymethylene) by die-drawing", *J. Appl. Polym. Sci.*, Vol. 26, 1981, pp. 2879–2896.

Komatsu T., Enoki S., Aoshima A., "The effect of pressure on drawing of poly(oxymethylene): drawn fiber properties and structure", *Polymer*, Vol. 32, 1991, pp. 1988–1993.

Kukureka S.N., Craggs G., Ward I.M., "Analysis and modelling of the die drawing of polymers", *J. Mater Sci*, Vol. 27, 1992, pp. 3379–3388.

Motashar F.A., Unwin A.P., Craggs G., Ward I.M., "Analytical and experimental study of the die-drawing of circular rods through conical dies", *Polym. Eng. Sci.*, Vol. 33, 1993, pp. 1288–1298.

Richardson A., Parsons B., Ward I.M., "Production and properties of high stiffness polymer rod, sheet and thick monofilament oriented by large-scale die drawing", *Plast. Rubber Process. Appl.*, Vol. 6, 1986, pp. 347–361.

Schultz, J, *Polymer materials science*, Prentice Hall, Englewood Cliffs, N.J., 1974.

Tarayia A.K., Nugent M., Sweeney J., Coates P.D., Ward I.M., "Development of continuous die drawing production process for engineered polymer cores for wire ropes", *Plast. Rubbers, Compos.*, Vol. 29, 2000, pp. 46–50.

Wunderlich B., *Macromolecular physics*, Vol. 1, Academic Press, New York, NY, USA, 1973, p.118, pp. 388 & 316.

Chapter 17

Mathematical Modelling of Dynamic Processes of Irreversible Deforming, Micro- and Macro-fracture of Solids and Structures

Alexey B. Kiselev and Alexander A. Luk'yanov
Mechanics and Mathematics Faculty, Moscow M.V. Lomonosov State University, Moscow, Russia

1. Models of mechanics of continual fracture

It is common practice to recognize the following three basic types of dynamic fracture: viscous, brittle, and with formation of zones of adiabatic shear.

Viscous fracture (observed in metals like aluminum and copper as well as in solid fuels and explosive) is characterized by formation and evolution of near-spherical pores during a process of plastic deformation.

It is typical of the brittle fracture of a body that arbitrary oriented coin-shaped microcracks (capable to grow during deformation) are formed in a great number of steels.

If the rate of deformation is high, then process of plastic flow is adiabatic. In a number of cases, the liberated heat is concentrated in thin domains whose thickness ranges up to several tens of microns. These domains are located along the surfaces of maximal tangential stresses; this leads to a considerable increase of characteristics of the plastic flow along these surfaces. In particular, such fractures with formation of zones of deformation shear are observed in still cylinders loaded by an explosion and in cases of punching the barriers by percussion mechanisms with flat front acts (forcing a "plus" out of a barrier).

Since a great number of the above-mentioned are formed in the process of dynamical deformation, it is difficult to consider each of such microfracture individually. In this connection, in recent years some approaches have been developed, whereby certain internal variables characterizing the evolution of microdamages are introduced into determining equations.

This line of investigation is called *mechanics of continual (or scattered) fracture.*

The development of mechanics of continual fracture was originated in the papers of (Kachanov 1958) and (Rabotnov 1959), dealing with theory of creeping of materials, where one scalar damage parameter was introduced. A short time later, tensor measures of damage were proposed (Il'ynshin 1967). Attempts to introduce these tensors are being undertaken now (Astaf'ev *et al.*, 2001).

The introduction of damage parameters into the system of internal variables and the usage of thermodynamic principles of continuum mechanics make possible the construction of thermodynamically correct coupled models of damageable solids (Coleman *et. al.*, 1967; Kiselev *et al.*, 1990, 1992a, 1994, 1998; Kiselev 1997, 1998).

In particular, the thermoelastovicsoplastic model with two damage parameters was constructed, permitting description of microfracture in domains of intensive tension and formation of zones of adiabatic shear (Kiselev 1997, 1998). Since we describe viscous fracture by the formation of spherical micropores and the fractures with formation of shear zones, we choose with respect to the damage parameters the following two invariants of the tensor ω_{ij}: the scalars $\omega = \omega_{kk}/3$ (volume damage) and $\alpha = \sqrt{\omega'_{ij}\omega'_{ij}}$ (intensity of the damage tensor deviator $\omega'_{ij} = \omega_{ij} - \omega\delta_{ij}$). We shall assume that in domains of intensive tension the parameter ω describes the accumulation of micropore type damage (which disappear under compression) and the parameter α describes shear fracture. As in classical theory (Kachanov 1958; Rabotnov 1959), we shall interpret the parameter ω as a relative decrease of the effective load-bearing elementary area due to formation of micropores inside the specimen. The parameter ω may be considered as a volume content of micropores in the material. In the damage-free material we have $\omega = \alpha = 0$; if damage is accumulated, then ω and α increase in such a manner that they remain less than 1.

The system of constitutive equations for a model of damageable thermoelastoviscoplastic medium is as follows:

$$\varepsilon_{kk} = \frac{\sigma}{K} + \alpha_v(T-T_0) + \Lambda\int_0^\omega \frac{\partial\varphi}{\partial\sigma}d\omega, \quad e^e_{ij} = \frac{S_{ij}}{2\mu} + A\int_0^\alpha \frac{\partial\varphi}{\partial\sigma_{ij}}, \qquad [1]$$

$$\varepsilon^p_{ij} = \frac{S_{ij}}{2\eta}\frac{S_u - \sqrt{\frac{2}{3}}Y}{S_u}H(S_u - \sqrt{\frac{2}{3}}Y),$$

$$\dot\omega = \varphi(\omega,\sigma) = B\left(\frac{\sigma}{1-\omega} - \sigma_*\right)H\left(\frac{\sigma}{1-\omega} - \sigma_*\right) +$$

$$+ \omega\frac{\sigma - \sigma^+}{4\eta_0}H(\sigma - \sigma^+) + \omega\frac{\sigma - \sigma^-}{4\eta_0}H(\sigma^- - \sigma),$$

$$\sigma^+ = -\frac{2}{3}Y_0\ln\omega, \quad \sigma^- = -\sigma^+,$$

$$\dot\alpha = \psi(\omega,\alpha,S_u) = C\left(\frac{S_u}{(1-\omega)(1-\alpha)} - S_u^*\right)H\left(\frac{S_u}{(1-\omega)(1-\alpha)} - S_u^*\right),$$

$$\eta = \eta_0(1-\omega)(1-\alpha), \quad Y = Y_0(1-\omega)(1-\alpha),$$

$$\rho c_\sigma \dot T + \alpha_v \dot\sigma T = S_{ij}\dot\varepsilon^p_{ij} + \Lambda\dot\omega^2 + A\dot\alpha^2 - \text{div }\vec q,$$

$$\vec q = -\kappa\,\text{grad}\,T, \quad S_u = \sqrt{S_{ij}S_{ij}}.$$

Here $\sigma_{ij}, \varepsilon^e_{ij}, \varepsilon^p_{ij}$ are the components of the stress tensor, elastic and nonelastic (viscoplastic) deformation tensors, respectively ($\varepsilon_{ij} = \varepsilon^e_{ij} + \varepsilon^p_{ij}$; $\varepsilon^p_{kk} = 0$); T is the absolute temperature; $\vec{q}$ – is a heat flux; ρ is density; A, D, C, Λ, σ_*, S_u^* – constants of materials, connected with damage parameters ω and α; K_0, μ_0, η_0, Y_0 – volume module, shear module, dynamic viscosity and static yield of elasticity for an undamaged material; c_σ is the heat conductivity at constant stress; α_v is the coefficient of cubic expansion; κ is the coefficient of heat conduction; $H(x)$ is Heaviside function; the dot over symbols indicates the material derivative with respect to tine.

The kinetic equation for the volume damage ω consists of three terms. The first has the form of the Tuler – Bucher equation and describes the stage of formation and initial growth of the volume damage ω. Then, as ω is accumulated, the second term describing the viscous growth in domains of tension of the material comes into play (Kiselev et al., 1992b). The third term describes the viscoplastic flow in pores when the material is compressed. Note that the equation for ω is taken without the dynamical problem on a single spherical pore of inner radius a and other radius b in a viscoplastic incompressible material.

We assume that the yield limit Y and shear modulus μ depend on temperature, pressure, density, accumulated plastic strain, as in the model of Steiberg-Guinan (Wilkins, 1984):

$$Y_0 = Y_{00}(1 + \beta \varepsilon_u^p)(1 - b\sigma(\frac{p_0}{\rho})^{1/3} - h(T - T_0)), \qquad [2]$$

$$Y_{00}(1 + \beta \varepsilon_u^p) \leq Y_{max}, \quad Y_{00} = 0 \text{ for } T > T_m,$$

$$T_m = T_{m0}(\frac{p_0}{\rho})^{2/3} \exp(2\gamma_0(1 - \frac{p_0}{\rho})),$$

$$\mu_0 = \mu_{00}(1 - b\sigma(\frac{p_0}{\rho})^{1/3} - h(T - T_0)),$$

where $\varepsilon_u^p = \sqrt{\frac{2}{3}\varepsilon_{ij}^p \varepsilon_{ij}^p}$ is intensity of tensor of plastic deformation; T_m is melting temperature of material; Y_{00}, μ_{00}, T_{m0}, β, h, b, γ_0 are materials constants. It is accepted that

$$\sigma_* = \sigma_*^0 \frac{Y_0}{Y_{00}}, \quad \eta_0 = \eta_{00}\frac{\mu_0}{\mu_{00}}, \quad S_u^* = S_{u0}^* \frac{Y_0}{Y_{00}} \qquad [3]$$

In formulas [2]–[3] we denote by two zeros the parameters of the undeformed material: $\rho = \rho_0$, $\sigma = 0$, $T = T_0$.

This model develops the model for elastoviscoplastic medium, Socolovsky–Malvern type, and takes into account the formation and accumulation of damage in intensive tensor domains, their disappearance under compression as well as the heating effects and the accumulation of damage in intensive tension domains, their disappearance under compression as well as the heating effects and the accumulation of damage under shear. Mechanical, structural, and heat processing are mutually dependent.

The evolution of intensive plastic flow and accumulation of microstructural damages may be considered as a process of prefracture of the material. The entropy criterion of limiting specific dissipation (Kiselev et al., 1990):

$$D = \int_0^{t_*} \frac{1}{\rho}(d_M + d_F + d_T)dt = D_*, \qquad [4]$$

$$d_M = S_{ij}\dot{\varepsilon}_{ij}^p, \quad d_F = \Lambda\dot{\omega}^2 + A\dot{\alpha}^2, \quad d_T = \kappa\frac{(grad\, T)^2}{T},$$

is proposed as the criterion of the beginning of macrofracture (i.e., the beginning of formation of cracks (new free surfaces) in the material). Here t_* is the time of the beginning of fracture; D_* is a constant of the material (the limiting specific dissipation); d_M, d_F and d_T are mechanical dissipation, dissipation of continuum fracture and thermal dissipation.

If zones of large stress appear in the body (as, for example, in the problem of plane collision of plates with spallation fracture (Kiselev et al., 1990), then the major contribution to dissipation [4] is made by d_M and by term $\Lambda\dot{\omega}^2$ from d_F. As for the developed shearing plastic flow with formation of zones of adiabatic shear, the major contribution to dissipation D is made by d_M, d_T and by the term $A\dot{\alpha}^2$ from d_F (as in the problem of forcing a "plug" out of a barrier by a percussion mechanism with a flat front cut) (Fomin et al., 1999)

When criterion [4] is fulfilled at some point of the material, a microcrack should be formed there, *i.e.*, a new free surface that will spread over the body. Thus, the problem of calculation of the further deformation becomes self-dependant in the framework of computational mechanics of deformable solids.

2. The methods for determination of material constants of damageable media

Models for damageable media contain some "unstandart" constants, connected with damage parameters and subjected to determination. As such, the models with single damage parameter (Kiselev et al., 1990, 1992a) contain three such constants. In addition, these models contain a fourth unknown constant – a constant of limiting specific dissipation. For determination of these constants we use a method based on comparison the results of physical and numerical experiments of the problem of flat collision of two plates with spallation destruction in a plate-target (Kiselev et al., 1990, 1992a). Note that experiments with spallation destruction are today the most informative and detailed for constructing dynamic constitutive equations for materials under high parameters (Kanel et al., 1996). For a model of porous medium (Kiselev et al., 1992a) intended to describe behaviour of solid fuels and explosives, we used the problem of compression of a spherical microscopical pore filled with gas medium (Kiselev et al., 1992b). The model with two damage parameters (see [1], [4]) contained seven such constants: A, B, C, Λ, σ_*, S_u^*, D_*. For determining these constants under quasidynamical deformation the modelling the method, based on numerical and physical modelling of processes of quasidynamical twisting and tension of thin-walled tubular samples with destruction and with the following mathematical data handing, was proposed (Kiselev et al., 1998). However, we do not know the published results of experiments, which make possible to determine constants for real materials.

For dynamical loading of materials with destruction there are many experiments involving flat collision of the plates with spallation destruction (Kanel et al., 1996), which make possible to determine the constants of damageable media [1], [4].

Therefore we consider the problem of flat collision of two thin plates. Since the thickness of the plates is small compared with the size and the characteristic time of the process is the time of several runs of elastic wave across the whole thickness of the plate target, the problem may be solved using the one-dimensional mathematical formulation (a uniaxial deformed state) and the adiabatic approximation ($div\,\vec{q} = 0$).

In this case, the equations of mass, momentum and internal energy are written in a Cartesian coordinate system $Oxyz$ (the x- axis is perpendicular to the plate surface) as follows:

$$\frac{\dot{\rho}}{\rho} = -\dot{\varepsilon}, \quad \dot{v} = \frac{1}{\rho}\frac{\partial(\sigma+S)}{\partial x}, \quad \rho c_\sigma \dot{T} + \alpha_v \dot{\sigma} T = \frac{3}{2}S\dot{\varepsilon}^p + \Lambda\dot{\omega}^2 + A\dot{\alpha}^2,$$

Here $v = v_x$ is the velocity, $\dot{\varepsilon} = \varepsilon_{xx} = \frac{\partial v}{\partial x}$, $\dot{\varepsilon}^p = \dot{\varepsilon}^p{}_{xx}$, $S = S_{xx}$. The rest of designations are indentical to the ones introduced above. Besides, we are take into account the following $S_{yy} = S_{zz} = -S_{xx}/2$, $\dot{\varepsilon}^p_{yy} = \dot{\varepsilon}^p_{zz} = -\dot{\varepsilon}^p_{xx}/2$, since $S_{kk} = 0$ and $\dot{\varepsilon}^p_{kk} = 0$. Constitutive equations of model [1] are given by

$$\sigma = K(\varepsilon - \alpha_v(T-T_0) + B\Lambda \ln(1-\omega) - \frac{\omega^2}{4\eta_0}), \qquad [5]$$

$$\dot{\varepsilon}^e = \frac{1}{3}\dot{\varepsilon} + \frac{\dot{S}}{2\mu} + \sqrt{\frac{3}{2}}\frac{AC}{(1-\omega)(1-\alpha)}\dot{\alpha}\,sign\,S,$$

$$\dot{\varepsilon}^P = \frac{S}{2\eta}(1 - \frac{2}{3}\frac{Y}{|S|})H(|S| - \frac{2}{3}Y), \quad \dot{\varepsilon} = \dot{\varepsilon}^e + \dot{\varepsilon}^P,$$

$$\dot{\omega} = \varphi(\omega, \sigma) = B\left(\frac{\sigma}{1-\omega} - \sigma_*\right)I\cdot\left(\frac{\sigma}{1-\omega} - \sigma_*\right) +$$

$$+ \omega\frac{\sigma - \sigma^+}{4\eta_0}H(\sigma - \sigma^+) + \omega\frac{\sigma - \sigma^-}{4\eta_0}H(\sigma^- - \sigma),$$

$$\sigma^+ = -\frac{2}{3}Y_0 \ln \omega, \quad \sigma^- = -\sigma^+,$$

$$\dot{\alpha} = \psi(\omega, \alpha, S_u) = C\left(\frac{S_u}{(1-\omega)(1-\alpha)} - S_u^*\right)H\left(\frac{S_u}{(1-\omega)(1-\alpha)} - S_u^*\right),$$

Initial conditions: for the striker

$$v = V_0, \rho = \rho_{01}, \sigma = S = 0, T = T_0 \quad (-h_1 \leq \overset{o}{x} \leq 0)$$

and for the target:

$$v = 0, \rho = \rho_{02}, \sigma = S = 0, T = T_0 \quad (0 \leq \overset{o}{x} \leq h_2)$$

Here $\overset{o}{x} = x|_{t=0}$ is the initial Lagrangian coordinate; h_1 and h_2 are the striker and target thicknesses, respectively. Boundary conditions at the contact surface $\overset{o}{x} = 0$: $v^+ = v^-$, $(\sigma + S)^+ = (\sigma + S)^-$ for compressive forces $(\sigma + S)^+ = (\sigma + S)^- < 0$; and condition of the free surface: $(\sigma + S) = 0$ otherwise.

In the same way as for one of the contact surfaces, a boudary condition is set in the cross-section $x = x^*$ where the criterion of destruction [4] is realized.

The problem is solved on the Lagrangian computational grid using the explicit finite-difference scheme. The algorithm to construct the surface of spallation destruction in the plate based on the procedure of transforming the Lagrangian computational grid to the destruction surface and the procedure of converting the state parameters into a new grid are described in (Fomin et al., 1999).

The striker is made of aluminum and the target is made of titanium. The thickness is $h_1 = 2$ mm for the striker, and $h_2 = 10$ mm for the target as in experiments (Kanel et al., 1996). The constants of materials is taken from (Wilkins 1984).

206 Prediction of Defects in Material Processing

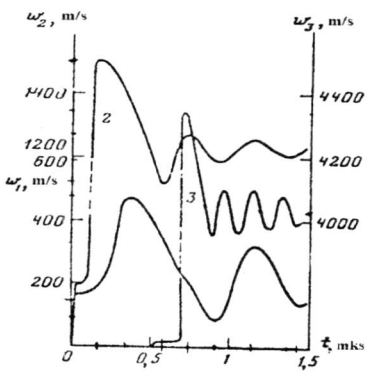 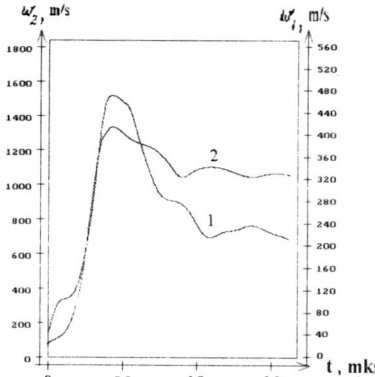

Figure 1. *Experimental velocity of the target rear surface* ω

Figure 2. *Dependence of velocity of the target rear surface* ω

Figure 1 shows the experimental velocity of the target rear surface $w = v|_{x=h_2}$ as a function of the time for impact velocities $V_0 = 660 \, m/s$ (curve 1), $V_0 = 1900 \, m/s$ (curve 2) and $V_0 = 5300 \, m/s$ (curve 3) (Kanel et. al., 1996). In addition, these experiments are given information that residual temperature in plate–target is $\approx 1100 \, K$.

The model parameters A, B, C, Λ, σ_*, S_u^*, D_* are chosen from the condition of the best agreement between calculation results and experimental data (Kanel et al., 1996): for aluminum $A = 100 \, Pa \cdot s$, $\eta_{00} = 800 \, Pa \cdot s$, $B = 1.034 \cdot 10^{-3} \, (Pa \cdot s)^{-1}$, $C = 4.225 \cdot 10^{-4} \, (Pa \cdot s)^{-1}$, $\Lambda = 64.43 \, Pa \cdot s$, $\sigma_* = 0.145 \, GPa$, $S_u^* = 0.746 \, GPa$; for titanium $A = 1800 \, Pa \cdot s$, $\eta_{00} = 43 \, Pa \cdot s$, $B = 4.225 \cdot 10^{-4} \, (Pa \cdot s)^{-1}$, $C = 4.225 \cdot 10^{-6} \, (Pa \cdot s)^{-1}$, $\Lambda = 197.23 \, Pa \cdot s$, $\sigma_* = 1.065 \, GPa$, $D_* = 50 \, kJ/kg$, $S_u^* = 0.746 \, GPa$.

Figures 2–8 represent the results of calculations. Figure 2 shows the dependence of velocity of the target rear surface w as a function of the time for impact velocities $V_0 = 660 \, m/s$ (curve 1) and $V_0 = 1900 \, m/s$ (curve 2). As we can see, results of experiments (Figure 1) and results of calculations (Figure 2) agree rather well.

Figures 3–6 represent the distribution of values of deformation ε, damage parameter ω, temperature T and dissipation D in the moments of realization

Mathematical Modelling of Dynamic Processes 207

destruction criterion [4] for these two impact velocities ($t_1^* = 2.24\ mks$, $t_2^* = 2.54\ mks$). Figures 7, 8 represents the distribution of values of mechanical dissipation D_M and dissipation of continuum fracture D_F in the moments of realization destruction criterion [4]. The contribution of termal dissipation D_T to full dissipation D is small in comparison with other two dissipations in the collision of two plates problem. Also, the influence of the second damage parameter α is small.

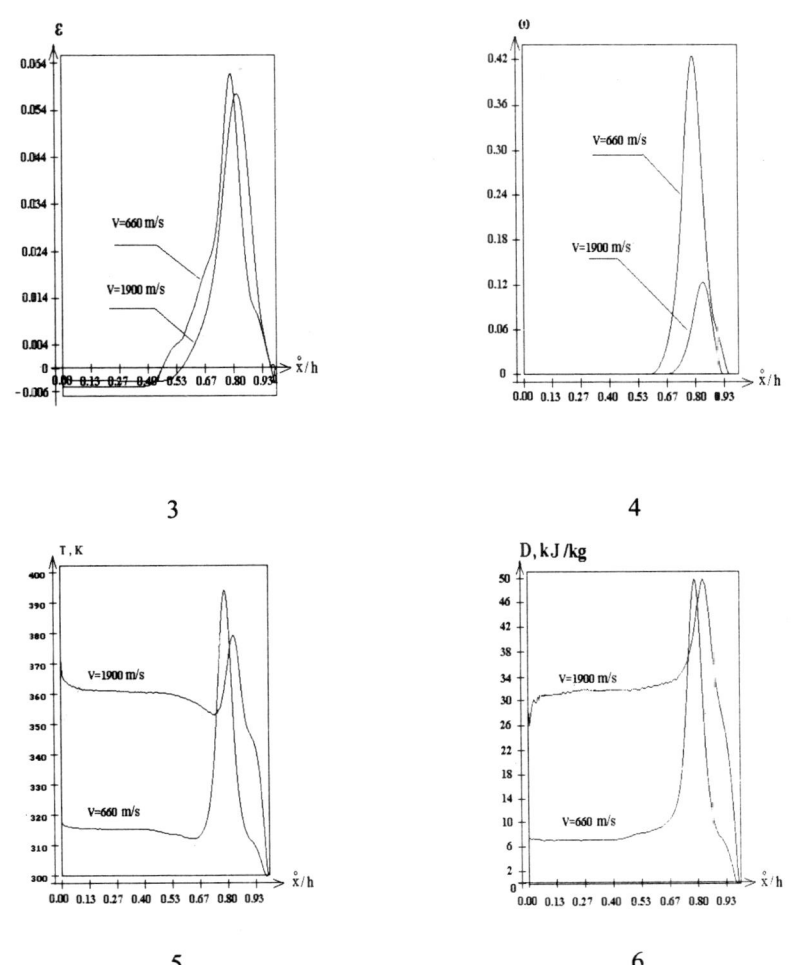

Figures 3–6. *Distribution of values of deformation ε*

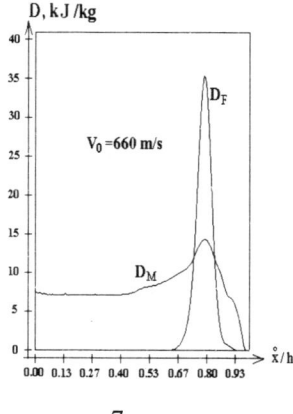 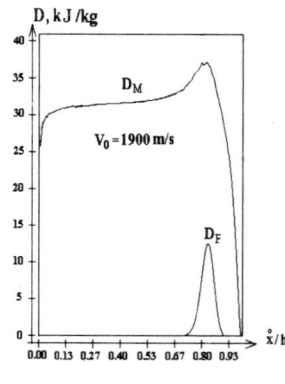

7   8

Figures 7–8. *Distribution of values of mechanical dissipation D_M and dissipation of continuum fracture D_F*

The results represented of numerical investigations lead to the conclusions:

– The maxima of deformation ε, damage parameter ω, temperature T and dissipation D (as its components D_M and D_F) are reached in the same cross-section as where the destruction criterion [4] is realized.

– As V_0 increases, the limiting value of damage parameter ω decreases appreciably, and therefore the limiting value of ω cannot be a destruction criterion.

– As V_0 increases, the maximum tensile deformation ε in the destruction cross-section practically coincides. Therefore, the second classical strength theory (the maximum deformation theory) may be used as a destruction criterion in the problem of spallation destruction.

– As V_0 increases the contribution of continuum of fracture D_F to full dissipation D is decreased near the zone of spallation destruction.

– The temperature near the spallation destruction zones decreases, however, and naturally, the average temperature in plate–target increases. This effect is connected with real decreases of D_F's contribution to full dissipation D and then V_0 is increased.

This mathematical model of damageable thermoelastoviscoplastic medium to make possible a correct description of real peculiarities of dynamical deforming and microfracture of solids and destruction criterion of limit specific dissipation is applicable to the description of spallation destructions.

Acknowledgements

The authors would like to thank Russian Foundation of Basic Research (grants No. 00-01-00245 and No. 00-15-99060) and INTAS (grant No. 00-0706) for financial support.

3. References

Astaf'ev V.I., Radaev Yu.I., Stepanova L.V., *Nonlinear fracture mechanics*, Samara: Samara Univ. Publ., 2001 (in Russian).

Coleman B.D., Gurtin H.E., "Thermodynamics with internal state variables", *J. Chem. Phys.*, Vol. 47, No. 2, 1967, pp. 597–613.

Il'yushin A.A., "On some theory of long strength", *Mechanics of Solids*, Vol. 2, No. 3, 967, p. 21–35.

Kachanov L.M., "On destruction in creeping of materials", *Izv. Acad. Nauk USSR, Section of Eng. Sci.*, Vol. 22, No. 8, 1958, pp. 26–31.

Kanel G.I., Razorenov S.V., Utkin A.V., Fortov V.E., *Shock-wave phenomena in condensed media*, Moscow: Janus-K, 1996 (in Russian).

Kiselev A.B., "Mathematical modelling of dynamical deforming and microfracture of damageable thermoelastoviscoplastic medium", in M. Predeleanu, P. Gilormini, eds., *Studies in Applied Mechanics 45: Advanced Methods in Materials Processing Defects*, Amsterdam, Elsevier, 1997, pp. 43–50.

Kiselev A.B., "Mathematical modelling of dynamical deformation and combined microfracture of a thermoelastoviscoplastic medium", *Moscow Univ. Mech. Bull.*, Vol. 53 No. 6, 1998, pp. 32–40.

Kiselev A.B., Yumashev M.V., "Deforming and fracture under impact loading. The model of thermoelastoplastic medium", *J. Appl. Mech. Tech. Phys.*, Vol. 31, No. 5, 1990, pp. 116–123.

Kiselev A.B., Yumashev M.V., "A mathematical model of deformation and fracture of a solid fuel under impact loading", *J. Appl. Mech. Tech. Phys.*, Vol. 33, No. 6, 1992a, pp. 126–134.

Kiselev A.B., Yumashev M.V., "Numerical investigation of micropore shock compression in a thermoelasto-viscoplastic material", *Moscow Univ. Mech. Bull.*, Vol. 47, No. 1, 1992b, pp. 31–36.

Kiselev A.B., Yumashev M.V., "Numerical investigation of dynamic processes of deforming and microfracture of damageable thermoelastoplastic medium", *Moscow Univ. Mech. Bull.*, Vol. 49, No. 1, 1994, pp. 14–24.

Kiselev A.B., Yumashev M.V., Volod'ko O.V., "Deforming and fracture of metals. The model of damageable thermoelastoviscoplastic medium", *J. of Materials Processing Technology*, Vol. 80–81, 1998, pp. 585–590.

Rabotnov Yu.N., "Mechanism of long fracture", *Problems of strength of materials and constructions,* Moscow, USSR Academy of Sci. Publ., 1959, pp. 5–7.

Fomin V.M., Gulidov A.I., Kiselev A.B. *et al., High-speed interaction of bodies,* Novosibirsk, Sibirian Branch of RAS, Russia, 1999 (in Russian).

Wilkins M.L., "Modelling the behavior of materials", *Structural impact and crushworthiness: Proc. Int. Conf.,* London, 16–20 July 1984, London, New York, 1984, pp. 243–287.

Chapter 18

Numerical Simulation of Ductile Damage in Metal Forming Processes: A Simple Predictive Model

Part 1: Theoretical and Numerical Aspects

Khémaïs Saanouni, Philippe Lestriez and Abdelhakim Cherouat
Université de Technologie de Troyes, France

Jean-François Mariage
Université de Technologie de Troyes, France, and CETIM St Etienne, France

1. Introduction

When materials are formed by large plastic deformation processes, they experience large plastic deformations leading to the onset of internal or surface micro-defects as voids and micro cracks. When these micro-defects initiate and grow inside the plastically deformed metal, the thermomechanical fields are deeply modified, leading to significant modifications in the deformation process. On the other hand, the coalescence of these defects (micro-voids) during the deformation can lead to the initiation of macro cracks or damaged zones, inducing irreversible damage inside the part formed and consequently its loss. Accordingly, it is very important for engineers that the virtual metal forming tools allow the possibility to predict damage occurrence during the process simulation. This provides a helpful way to improve and optimise, in situ, the process plan in order to avoid this damage occurrence. In fact, the unavailability of the damage prediction possibility provokes a certain amount of scepticism with the users of these codes. Fortunately, an increasing number of works as well as special sessions of scientific conferences are devoted to the numerical modelling of damage occurrence during metal forming processes. Taking into account that ductile damage in metal forming necessitates not only the development of a continuum damage theory, but also its coupling with the other thermomechanical field under concern. Whichever the damage theory used, two principal methods of damage calculations arise. The first one, called the uncoupled approach, aims to calculate the damage distribution without taking into account its effect on the other thermomechanical fields. Generally, this is achieved by post-processing the finite element analysis for a given time step to calculate the damage distribution using the stress and strain fields. This approach has been used in many situations by many authors to predict zones where local failure has taken place inside the deformed work piece (Cordebois, 1982; Gelin *et al.*, 1985; Hartley *et al.*, 1989; Liu and Chung, 1990; Zhu and Cescotto, 1991; Clift 1992). In much other work the damage is used to find the forming limit strain and fraction limit in metal forming (Cordebois, 1982; Cordebois and Ladeveze, 1985; Gelin *et al.*, 1985 among many others). It is worth noting that in these uncoupled formulations, the effects of damage on the elastic, thermal, plastic and hardening behaviours are neglected. Consequently, the redistribution of these thermomechanical fields generated by the damage initiation and growth is not taken into account.

The second method, is called the fully coupled approach, because the damage effect is introduced directly into the overall constitutive equations and affects all the thermomechanical fields according to the coupling theory. Different kinds of this coupled approach have been employed by many authors to predict the damage occurrence in metal forming. The widely used models are based on Gurson's damage theory based of the void volume fraction evolution and its effect on the plastic yielding. Hence, only the effect of damage on the plastic behaviour is taken into account, leaving the elastic behaviour completely insensitive to the damage occurrence (Aravas, 1986; Onate and Kleiber, 1988; Gelin, 1990; Bontcheva and Iankov, 1991; Brunet *et al.*, 1996; Picart *et al.*, 1998). Other works are based on the continuum damage mechanics theory (CDM) (Lee *et al.*, 1985; Mathur and Dawson, 1987; Zhu *et al.*, 1992; Zhu et Cescotto, 1995; Bezzina and Saanouni, 1996; Saanouni and Franqueville, 1999; Hammi, 2000; Saanouni and Hammi, 2000, Saanouni *et al.*, 2000; Saanouni *et al.*, 2001, Borouchaki *et al.*, 2002, Cherouat and Saanouni, 2002). These fully coupled approaches have shown their ability to "optimise" the process plane, not only to avoid the damage occurrence, but also to enhance the damage in order to simulate any metal cutting processes (Abdali *et al.*, 1995; Bezzina and Saanouni, 1996; Homsi *et al.*, 1996; Saanouni and Franqueville, 1999; Brokken, 1998; Saanouni and Hammi, 2000; Saanouni *et al.*, 2000; Hambli, 2001; Cherouat and Saanouni, 2002).

The present work is devoted to the "simplified" modelling compared to the "advanced" modelling developed in our laboratory during some six years. In Part 1 of the work, the constitutive equations used in Forge2® and Forge3® codes are modified to include the ductile damage effect. This latter is modelled according to the Lemaitre damage model (Lemaitre and Chaboche, 1990; Lemaitre, 1992). The associated numerical aspects are then discussed in the light of the algorithms used in these codes. Applications will be presented in the second part of this work.

2. Simple fully coupled constitutive equations

The fully coupled approach, where the damage process is properly incorporated into the constitutive relations is expected to achieve better localisation on the onset of necking and failure (Saanouni *et al.*, 1994, Hammi, 2000). If the influence of damage is added to standard linear elasticity, the classical stress-strain relation of elasticity-based damage mechanics is obtained:

$$\underline{\sigma} = (1-D)\left[\lambda tr\underline{\varepsilon}^e \underline{1} + 2\mu\underline{\varepsilon}^e - 3K\alpha(T-T_0)\underline{1}\right] \quad \text{with} \quad K = \frac{3\lambda + 2\mu}{3} \quad [1]$$

where $\underline{\sigma}$, $\underline{\varepsilon}^e$ and T denote respectively the Cauchy stress tensor, the linear elastic strain tensor and the absolute temperature. K is the bulk modulus while λ and μ are the classical Lame constants, α is the coefficient of thermal expansion and T_0 is the reference temperature. The fully isotropic rate formulation used in Forge2® and Forge3® codes assumes the small strain hypothesis for the constitutive equations, justified by the fact that the applied load increments are still very small. Accordingly, the total strain rate tensor $\underline{\dot{\varepsilon}}$ is additively partitioned:

$$\underline{\dot{\varepsilon}} = \underline{\dot{\varepsilon}}^e + \underline{\dot{\varepsilon}}^p \quad [2]$$

with $\underline{\dot{\varepsilon}}^e$ and $\underline{\dot{\varepsilon}}^{pl}$ are respectively the elastic and plastic strain rate components. The stress and total strain tensors can be decomposed into deviatoric and spherical parts:

$$\underline{S} = \underline{\sigma} - P_H \underline{1}, \quad P_H = \frac{1}{3} \text{tr}\underline{\sigma} \quad \text{and} \quad \underline{e} = \underline{\varepsilon} - \varepsilon_H \underline{1}, \quad \varepsilon_H = \frac{1}{3} \text{tr}\underline{\varepsilon} \qquad [3]$$

Applied to equation [1] one gets:

$$\underline{S} = 2\mu(1-D)\underline{e}^e \quad \text{and} \quad P_H = -3K(1-D)[\varepsilon_H - \alpha(T-T_0)] \qquad [4]$$

Assuming a fully isotropic material behaviour; the isotropic hardening is characterised by the accumulated plastic strain p solution of:

$$\dot{p} = \sqrt{\frac{2}{3} \underline{\dot{\varepsilon}}^p : \underline{\dot{\varepsilon}}^p} \qquad [5]$$

The damage effect leads to a continuous reduction of the yield surface size in the softening stage. Hence, the classical von Mises yield criterion $f(\underline{S}, p, D, T)$ is used for damage effect:

$$f = \sigma_{eq} - (1-D)\sigma_y(p,T) \leq 0 \qquad [6]$$

where $\sigma_y(p,T)$ represents the size of the yield surface and the invariant σ_{eq} is the equivalent stress defined by:

$$\sigma_{eq} = \sqrt{\frac{3}{2} \underline{S} : \underline{S}} \qquad [7]$$

In this simple formulation, the yield surface radius $\sigma_y(p,T)$ is only a function of the cumulated plastic strain p in order to recover the classical time independent plasticity theory. The extension of the present model to the time dependent plasticity (or viscoplasticity) is simply obtained by considering that the yield surface size is sensitive to the accumulated plastic strain rate $\dot{p}$ (equation [5]), i.e. $\sigma_y(p,\dot{p},T)$ (see Chenot et al., 1998; Gay, 1995). By using equation [7], the yield function (equation [6]) transforms to:

$$\bar{f} = \underline{S} : \underline{S} - \frac{2}{3} \sigma_y^2(p)(1-D)^2 = 0 \qquad [8]$$

For the case of time independent plasticity, the deviatoric plastic strain rate tensor is obtained from $\bar{f}$ by using the normality rule:

$$\underline{\dot{\varepsilon}}^p = \underline{\dot{e}}^p = \dot{\delta}\frac{\partial \bar{f}}{\partial \underline{S}} = 2\dot{\delta}\underline{S} \qquad [9]$$

in which the plastic multiplier $\dot{\delta}$ is deduced from the consistency condition:

$$\dot{\bar{f}} = \frac{\partial \bar{f}}{\partial \underline{S}}:\dot{\underline{S}} + \frac{\partial \bar{f}}{\partial \sigma_y}\frac{\partial \sigma_y}{\partial p}\dot{p} + \frac{\partial \bar{f}}{\partial D}\dot{D} + \frac{\partial \bar{f}}{\partial \sigma_y}\frac{\partial \sigma_y}{\partial T}\dot{T} = 0 \qquad [10]$$

The ductile damage evolution is described by Lemaitre's model (Lemaitre, 1992):

$$\dot{D} = \left(\frac{Y}{S}\right)^s \frac{\dot{p}}{(1-D)^\beta} = \hat{Y}\dot{p} \quad \text{with} \quad \hat{Y} = \left(\frac{Y}{S}\right)^s \frac{1}{(1-D)^\beta} \qquad [11]$$

in which Y is the thermodynamic force associated with the damage variable D and given by:

$$Y = \frac{\sigma_{eq}^2}{2E(1-D)^2}\left[\frac{2}{3}(1+\nu) + 3(1-2\nu)\left(\frac{P_H}{\sigma_{eq}}\right)^2\right] - 3K\alpha(T-T_0)\varepsilon_H^e \qquad [12]$$

The consistency condition equation [10] is written:

$$\underline{S}:\underline{\dot{e}} - \frac{4}{3}\sigma_y^2(1-D)^2\left[1 + \frac{1}{3\mu}\frac{\partial \sigma_y}{\partial p}\right]\dot{\delta} - \frac{(1-D)\sigma_y}{3\mu}\left[\frac{\partial \sigma_y}{\partial T} - \frac{1}{\mu}\frac{\partial \mu}{\partial T}\sigma_y\right]\dot{T} \qquad [13]$$

from which the plastic multiplier can be expressed in term of the deviatoric strain rate tensor as follows:

$$\dot{\delta} = \frac{1}{H_{pd}}\langle \underline{S}:\underline{\dot{e}} - H_{pT}\dot{T}\rangle \qquad [14]$$

where the tangent plastic modulus H_{pd} and the thermal tangent "modulus" H_{pT} are defined by:

$$H_{pd} = \frac{4}{3}\sigma_y^2(1-D)^2\left[1 + \frac{1}{3\mu}\frac{\partial \sigma_y}{\partial p}\right] \text{ and } H_{pT} = \frac{(1-D)\sigma_y}{3\mu}\left[\sigma_y' - \frac{1}{\mu}\frac{\partial \mu}{\partial T}\sigma_y\right] \qquad [15]$$

On the other hand the time derivatives of the deviatoric stress tensor and the hydrostatic pressure (equation [4]) give:

$$\underline{\dot{S}} = \underline{\underline{L}}_s:\underline{\dot{e}} + \underline{L}_T\dot{T} \qquad \dot{P}_H = \underline{L}_p:\underline{\dot{\varepsilon}} + L_T\dot{T} \qquad [16]$$

in which the fourth order symmetric "deviatoric" elastoplastic tangent operator is:

$$\underline{\underline{L}}_s = 2\mu(1-D)\underline{\underline{1}} - \frac{4}{H_{pd}}\left(\mu(1-D) + \frac{\sigma_y}{3}\hat{Y}\right)\underline{S} \otimes \underline{S} \qquad [17]$$

The second order symmetric "thermal" tangent operator given by:

$$\underline{L}_T = \frac{4H_{pT}}{H_{pd}}\left(\mu(1-D) + \frac{\sigma_y}{3}\hat{Y} + \frac{1}{\mu}\frac{\partial \mu}{\partial T}\right)\underline{S} \qquad [18]$$

The symmetric second order tangent operator $\underline{L}_P$ as well as the scalar $\underline{L}_T$ defining the hydrostatic stress rate are:

$$\underline{L}_P = -\frac{1}{3}\left[K(1-D)\underline{1} + \frac{4}{3}\frac{P_H \sigma_0 \hat{Y}}{H_{pd}}\underline{S}\right] \qquad [19]$$

$$L_T = \left[P_H \frac{\frac{\partial K}{\partial T}}{K} + K(1-D)\left\{(T-T_0)\frac{\partial \alpha}{\partial T} + \alpha\right\} - \frac{4}{3}P_H \sigma_y \hat{Y}\frac{H_{pT}}{H_{pd}}\right] \qquad [20]$$

From these equations the accumulated plastic strain rate equation [5] can be deduced:

$$\dot{p} = \frac{4}{3}\dot{\delta}(1-D)\sigma_y = \frac{\underline{S}:\dot{\underline{e}}}{(1-D)\sigma_y\left(1 + \frac{1}{3\mu}\frac{\partial \sigma_y}{\partial p}\right)} \qquad [21]$$

Finally, let us note that when the hypothesis of dissipation uncoupling is assumed, the volumetric thermomechanical dissipation can be decomposed and written, in this simple case, in the following form:

$$\wp^v = \wp_m^v + \wp_{th}^v \geq 0 \Leftrightarrow \wp_m^v = \underline{\sigma}:\dot{\underline{\varepsilon}}^p - Y\dot{D} \geq 0 \text{ and } \wp_{th}^v = -\frac{\mathbf{q}}{T}\cdot\mathbf{grad}(T) \qquad [22]$$

where q is the heat flux vector. The temperature distribution inside the deformed part an be deduced from the energy balance (first law of thermodynamics) together with ıe Fourier law of heat flow including the plastic and damage dissipation (the body eat source and the thermoelastic coupling being neglected):

Numerical Simulation of Ductile Damage in Metal Forming   217

$$\rho C_v (1-D)\dot{T} = \text{div}(k\text{grad}(T)) + \underline{\sigma}:\underline{\dot{\varepsilon}}^p - Y\dot{D} = 0 \qquad [23]$$

where ρ is the material density, k is the heat conduction factor and C_V is the specific heat, both associated with the current deformed configuration. Appropriate mixed initial and limit thermal conditions should be added to the heat equation (equation [23]) in order to specify the heat exchange nature between the part and the tools.

Equations [1] to [23] together with the classical equilibrium equations define the initial and boundary value problem (IBVP) governing the material motion and the temperature field in the part during its forming.

3. Numerical aspects

The coupled non linear IBVP defined above should be solved numerically using the classical time and space discretization using both the finite difference (for time) and the finite element (for space) methods. Practically, the thermal problem (defined by the heat flow equation) is solved using the mechanical fields at the beginning of the time step to obtain the temperature field. This latter is injected in the mechanical problem, which is then solved in order to calculate the mechanical fields at the end of the time step.

3.1. *Finite element formulation*

The problem can be represented by the deformation of a given 2D or 3D domain (Ω) between one fixed tool defined by the boundary Γ_0 and another tool defined by Γ_V moving with the velocity field V_{tool}, both assumed to be rigid bodies. The material (Ω) is assumed to be homogeneous, isotropic and its behaviour modelled by equations [1] to [23] accounting for the coupled thermo-elasto-(visco)plastic and ductile damage phenomena.

The contact boundaries Γ_0 and Γ_V between the tools and the workpiece are governed by the contact conditions. On these boundaries, the non penetration condition is given by:

$$(v_i - v_i^{tool}).n_i = v_i^s.n_i = 0 \qquad [24]$$

The friction behaviour is described by Coulomb friction law which is a non linear relation between the shear stress τ_i and the relative tangential velocity difference v_i^s at the tool part interfaces Γ_0 and Γ_V:

$$\tau_i = -\xi_f \left\| v_i^s \right\|^{q-1} v_i^s \qquad [25]$$

where ξ_f and q are temperature dependent friction coefficients.

218   Prediction of Defects in Material Processing

The weak variational forms associated with both the equilibrium equations and the heat equation can be written under the following three integral functionals (the superscript (*) indicates the kinematically admissible fields):

$$\begin{cases} \int_\Omega \underline{S}:\underline{\dot{\varepsilon}}^* dV - \int_\Omega P_H \text{tr}\underline{\dot{\varepsilon}}^* + \int_\Omega \rho f_i v_i^* dV - \int_{\partial\Omega} F_i v_i^* dS - \int_{\partial\Omega} \tau_i v_i^* dS = 0 \\ \int_\Omega P_H^* (P_H + K\text{tr}\underline{\dot{\varepsilon}})dV = 0 \\ \int_\Omega T^* \rho C_v \dot{T} dV - \int_\Omega T^* \wp_m^v dV - \int_{\partial\Omega} T^* q_f dS + \int_\Omega \nabla T^* (k\nabla T)dV + \int_{Sq} T^* \overline{q} dS = 0 \end{cases} \quad [26]$$

where $\overline{q} = q_c + q_r + q_s$ on s_q with $q_c = h(T - T_e)$, $q_r = \zeta\varphi(T^4 - T_e^4)$, h is the convective thermal exchange coefficient, Te the external temperature, ζ is the emission factor, φ the Stephan constant and q_S defines the imposed heat flux.

Using the classical displacement (velocity) based F.E. method, with isoparametric elements, the above system equation [26] can be written with respect to the reference element Ω_r under the following discrete integrals:

$$\begin{cases} R_V(V) = \int_{\Omega_r}\{S\}[B]J_v dV_0 + \int_{\Omega_r}[B^T][I][\overline{N}]\{P_H\}J_v dV_0 + \int_{\Omega_r}\rho\{f\}[N]^T J_v dV_0 \\ \qquad - \int_{\partial_c\Omega_r}\{\xi_f\|v^s\|^{q-1} v^s\}[N]^T J_s dS - \int_{\partial_c\Omega_r}[N]^T \{T\}J_s dS = 0 \\ R_P(P) = \int_{\Omega_0}[\overline{N}]^T [I]^T [B]\{V\}J_v dV_0 + \frac{1}{\Delta t K}\int_{\Omega_0}[\overline{N}]^T [\overline{N}]\{\dot{P}\}J_v dV_0 = 0 \qquad [27] \\ R_T(T) = \int_{\Omega_0}\rho C_v \left[\overline{\overline{N}}\right]^T \left[\overline{\overline{N}}\right]\{\dot{T}\}J_v dV_0 - \int_{\Omega_0}\{\wp_m^v\}\left[\overline{\overline{N}}\right]^T J_v dV_0 + \int_{S_{q0}^-}\{\overline{q}\}\left[\overline{\overline{N}}\right]^T J_s dS_0 \\ \qquad - \int_{d\Omega_0}\{q_f\}\left[\overline{\overline{N}}\right] J_s dS_0 + \int_{\Omega_0}[B]^T k\{T\}[B]J_v dV_0 = 0 \end{cases}$$

in which the following linear interpolation functions have been used:

$$\dot{U}_j = \sum_k N_k \dot{U}_j^k, \quad P_{Hj} = \sum_k \overline{N}_k P_{Hj}^k, \quad \dot{T}_j = \sum_k \overline{\overline{N}}_k \dot{T}_j^k \quad [28]$$

where [B] is the geometric or strain displacement matrix in the current configuration, [N], $[\overline{N}]$ and $\left[\overline{\overline{N}}\right]$ are respectively the matrices of the nodal interpolation functions for displacement, pressure and temperature. Numerical experience shows that it will be enough to take the same linear interpolation for the three fields, so that

$[N] = [\overline{N}] = [\overline{\overline{N}}]$. The scalars J_v and J_s are respectively the determinant of the volume and boundary Jacobian matrices.

The algebraic system defined in equation [27] is highly non-linear with respect to the displacement field since the stress and the friction force are displacement dependent. It can be classically linearized using the implicit iterative Newton-Raphson scheme for a typical time (or load) increment $[t_n, t_{n+1}]$ with $t_{n+1} = t_n + \Delta t$. This allows one to compute the velocity, the pressure and the temperature increments after the convergency is reached at a given iteration. To do this, one needs to compute the tangent stiffness matrix by taking the derivatives of each term in the integrals of equation [27] with respect to the problem unknowns, namely: displacement, temperature and pressure increments. In obtaining these derivatives, the terms including the derivatives of J_v and J_s and the contact forces are omitted for the sake of simplicity. However, the term containing $\partial S/\partial \varepsilon$ is called the incremental tangent elastoplastic matrix and should be consistent with the time discretization scheme used for the deviatoric stress calculations. This will be briefly outlined in the next sections, limited to the fully isothermal case (thermal effect and heat equations will be dropped).

3.2. Deviatoric stress computation

For the integration of the fully coupled damage elasto-viscoplastic constitutive equations, the so-called backward Euler scheme is used since it contains the property of absolute stability and the possibility of appending further equations to the existing system of non linear equations. Due to the strong non-linearity of the constitutive equations an accurate numerical tool has to be associated with the previous model. Let I be the time interval [0,T], with the partition $I = \bigcup_{n=1}^{N}[t_n, t_{n+1}]$. Given the mechanical state $q_n = (\underline{s}^n, P_H^n, p^n, D^n)$ at time t_n and the prescribed strain increment $\Delta\underline{\varepsilon}$ (such as $\underline{\varepsilon}^{n+1} = \underline{\varepsilon}^n + \Delta\underline{\varepsilon}$), the integration problem amounts to calculating the state $q_{n+1} = (\underline{s}^{n+1}, P_H^{n+1}, p^n, D^n)$. Two steps have to be considered:

1/ Elastic prediction: First, the increment is assumed elastic, i.e., $\Delta p = \Delta D = 0$. One checks whether the mechanical state $q_{n+1} = (\underline{S}^{trial}, P_H^{trial}, p^n, D^n)$ fulfils the condition $f(\underline{S}^{trial}, P^n, D^n) < 0$. If yes, the elastic prediction coincides with the solution of the problem for this load increment and we have:

$$\underline{S}^{n+1} = \underline{S}^n + 2(1-D^n)\mu\Delta\underline{e} \quad , \quad p^{n+1} = p^n \text{ and } D^{n+1} = D^n \tag{29}$$

Otherwise (*i.e.* if $f(\underline{S}^{trial}, p^n, D^n) \geq 0$) the mechanical state has to be corrected in order to determine the actual state q_{n+1},

2/ Plastic-damage corrector: This step consists of a plastic corrector phase, where the trial deviatoric stress and the corresponding accumulated plastic strain and damage variables are "returned back" to the yield surface in order to have $f(\underline{S}^{trial}, p^n, D^n) = 0$. This can be done by solving, iteratively (Newton-Raphson method) the coupled constitutive equations discussed in section 2. However, it will be very helpful to rearrange the differential equations of the problem in order to reduce the size of the algebraic system to be solved. Therefore, expressing the deviatoric stress tensor under the incremental form gives:

$$\underline{S}^{n+1} = \underline{S}^n + \Delta\underline{S} = \frac{\Delta\underline{e} + \dfrac{\underline{S}^n}{2\tilde{\mu}}}{\dfrac{1}{2\tilde{\mu}} + \dfrac{\Delta D}{2\tilde{\mu}(1-D)} + 2\Delta\delta} \qquad [30]$$

in which $\tilde{\mu} = (1-D)\mu$. By combining the equations [29] and [30] and saving the damage evolution equation one may obtain the following system of two scalar equations f_1 and f_2 with only two unknowns Δp and ΔD:

$$\begin{cases} f_1 = (1-D+\Delta D)\sigma_y(p^n + \Delta p) + 3\tilde{\mu}\Delta p \\ \qquad - \sqrt{6\tilde{\mu}^2 \Delta\underline{e} : \Delta\underline{e} + 6\tilde{\mu}\underline{S}^n : \Delta\underline{e} + \dfrac{3}{2}\underline{S}^n : \underline{S}^n} = 0 \\ f_2 = \Delta D - \Delta p \dfrac{1}{(1-D)^\beta}\left(\dfrac{Y}{S}\right)^s = 0 \end{cases} \qquad [31]$$

This simple system is iteratively solved thanks to the Newton-Raphson scheme to determine Δp and ΔD at the time t_{n+1}. The knowledge of Δp and ΔD enables the updating of the hardening p and damage D variables as well as the deviatoric stress tensor by equation [30] at the end of the time step.

For the purpose of solution, let us write the system equation [31] formally as:

$$f_i(x_j) = 0 \qquad [32]$$

where $i = 1,2$ (number of unknown variables) and x_j ($j = 1,2$) is the "vector" of unknown variables. If x_j^k is the solution at the iteration k, when we have:

$$x_j = x_j^k + \delta x_j^k \qquad [33]$$

where δx_j^k is the field variation for the iteration under concern. The linearization of equation [32] using the Newton procedure leads to:

$$f_i(x_j) = f_i(x_j^k + \delta x_j^k) = f_i(x_j^k) + \frac{\partial f_i(x_j^k)}{\partial x_j^k}\delta x_j^k = 0 \text{ or } P_{ij}\delta x_j^k = -f_i(x_j^k) \quad [34]$$

where the components P_{ij} of the Jacobian material 2x2 matrix are given by:

$$P_{11} = \frac{\partial f_1}{\partial \Delta p} = (1 - D + \Delta D)\frac{d\sigma_y}{dp} + 3\tilde{\mu} \quad [35]$$

$$P_{12} = \frac{\partial f_1}{\partial \Delta D} = -3\mu\Delta p + \frac{(6\mu^2(1-D)\Delta\underline{e}:\Delta\underline{e} + 3\mu\underline{S}^n:\Delta\underline{e})}{\sqrt{6\tilde{\mu}^2\Delta\underline{e}:\Delta\underline{e} + 6\tilde{\mu}\underline{S}^n:\Delta\underline{e} + \frac{3}{2}\underline{S}^n:\underline{S}^n}} \quad [36]$$

$$P_{21} = \frac{\partial f_2}{\partial \Delta p} = -\frac{1}{(1-D)^\beta}\left(\frac{Y}{S}\right)^s - \frac{s}{S}\Delta p\frac{1}{(1-D)^\beta}\left(\frac{Y}{S}\right)^{s-1}\frac{dY}{dp} \quad [37]$$

$$P_{22} = \frac{\partial f_2}{\partial \Delta D} = 1 - \beta\Delta p\frac{1}{(1-D)^{\beta+1}}\left(\frac{Y}{S}\right)^s - \frac{s}{S}\Delta p\frac{1}{(1-D)^\beta}\left(\frac{Y}{S}\right)^{s-1}\frac{dY}{dD} \quad [38]$$

3.3. Consistent stiffness matrix

As discussed above (see equation [27]) under isothermal conditions, three contributions are required in the calculation of the tangent stiffness matrix: the stress contribution, the geometric contribution (via J_v and J_s) and the contact/friction contribution. As pioneered by Simo and Taylor (Simo and Taylor, 1985), the tangent stiffness matrix should be consistent with the time discretization scheme. In this work, we limit ourselves to the calculation of the deviatoric stress contribution. By the derivation of the discretized deviatoric stress tensor $\underline{S}$ given by equation [30] with respect to the total strain tensor, one gets:

$$\frac{\partial \underline{S}}{\partial \underline{\varepsilon}} = \frac{\partial \underline{S}}{\partial \underline{e}}\frac{d\underline{e}}{d\underline{\varepsilon}} + \frac{\partial \underline{S}}{\partial p}\frac{dp}{d\underline{\varepsilon}} + \frac{\partial \underline{S}}{\partial D}\frac{dD}{d\underline{\varepsilon}} \quad [39]$$

in which the terms $\frac{dp}{d\underline{\varepsilon}}$ and $\frac{dD}{d\underline{\varepsilon}}$ are obtained from the derivatives of the system given by equations [31]:

$$\begin{pmatrix} \dfrac{\partial f_1}{\partial p} & \dfrac{\partial f_1}{\partial D} \\ \dfrac{\partial f_2}{\partial p} & \dfrac{\partial f_2}{\partial D} \end{pmatrix} \begin{pmatrix} \dfrac{dp}{d\underline{\varepsilon}} \\ \dfrac{dD}{d\underline{\varepsilon}} \end{pmatrix} = \begin{pmatrix} -\dfrac{\partial f_1}{\partial \underline{e}} \dfrac{\partial \underline{e}}{\partial \underline{\varepsilon}} \\ -\dfrac{\partial f_2}{\partial \underline{e}} \dfrac{\partial \underline{e}}{\partial \underline{\varepsilon}} \end{pmatrix} \qquad [40]$$

This gives:

$$\frac{dP}{d\underline{\varepsilon}} = \frac{1}{\text{DESCRIM}} \left(-\frac{\partial f_1}{\partial \underline{e}} \frac{\partial f_2}{dD} + \frac{\partial f_2}{\partial \underline{e}} \frac{df_1}{dD} \right) \qquad [41]$$

$$\frac{dD}{d\underline{\varepsilon}} = \frac{1}{\text{DESCRIM}} \left(-\frac{\partial f_2}{\partial \underline{e}} \frac{\partial f_1}{dp} + \frac{\partial f_1}{\partial \underline{e}} \frac{df_2}{dp} \right) \qquad [42]$$

where

$$\text{DESCRIM} = \frac{\partial f_1}{\partial p} \frac{\partial f_2}{dD} - \frac{\partial f_1}{\partial D} \frac{df_2}{dp} \qquad [43]$$

4. Conclusion

In Part 1 of this work, a simple predictive ductile damage model has been coupled to a classical isotropic thermoelasto-(visco)plastic constitutive equation accounting for the isotropic non-linear hardening. Numerical implementation into industrial finite element codes (Forge2® and Forge3®) dedicated to the metal forming simulation has been made. This leads to a very practical numerical tool able to predict when and where damaged zones can take place inside the formed part. The model obtained will be applied to the simulation of the damage in some metal forming processes as presented in Part 2 of the paper.

Acknowledgements

The financial support of the FEDER (contract N° 99-2-50-059), the CRCA via the Pôle Mécanique et Matériaux Champardenais (PMMC) and the CIFRE N° 1.0.1591 is gratefully acknowledged.

5. References

Abdali A.K., Denkrid K. and Bussy P., Numerical Simulation of Sheet Cutting. in *Numiform'95*, Eds. Shen & Dawson, MAB, pp. 807–813, 1995.

Aravas, N., "The Analysis of Void Growth that Leads to Central Burst During Extrusion", *J. Mech. Phys. Solids*, 34:55–79, 1986.

Bezzina S. and Saanouni K., "Theorical and Numerical Modeling of sheet Metal Shear Cutting", *Mat-Tech '96*, Edt J. LU, ITT international, pp. 15–24, 1996.

Bontcheva N. and Iankov R., "Numerical Investigation of the Damage Process in Metal Forming", *Eng. Frac. Mech.*, 40:387–393, 1991.

Borouchaki H., Cherouat A., Saanouni K. and Laug P., «Remaillage en grandes deformation. Application à la mise en forme de structures 2D», *Revue européenne des éléments finis*, (to appear), 2002.

Brokken D., Brekemans W. and Baaijens F., "Numerical modelling of the metal blanking process", *J. Mat. P. Tech.*, 83, pp. 192–199, 1998.

Brunet M., Sabourin F. and Mguil-Touchal S., "The prediction of Necking and Failure in 3D Sheet Forming Analysis Using Damage Variable", *Journal de Physique III*, 6:473–482, 1996.

Chenot J.L., Fourment L., Coupez T., "Forge3® – A general tool for practical optimization of forging sequence of complex 3-D parts in industry", *ImechE*, C546/033, 1998.

Cherouat A., Saanouni K., "Numerical simulation of sheet metal blanking process using a coupled finite elastoplastic damage modelling", *Int. J. Forming Processes* (to appear), 2002.

Clift S.E., "Fracture in Forming Processes", *Numerical Modeling of Material Deformation Processes*, Eds. P. Hartley *et al.*, Springer, Berlin, pp. 406–418, 1992.

Cordebois J.P., Critères d'instabilité plastique et endommagement ductile en grandes déformations, Thèse d'Etat, Université Paris VI, 1983.

Cordebois J.P., Ladevèze P., "Necking criterion applied in sheet metal forming", *Plastic behavior of anisotropic solids*, Edt. Boehler J.P., Editions CNRS, 1985.

Gay C., Contribution à la simulation numérique 3-D du forgeage à froid, PhD thesis, ENSMP, Mars, 1995.

Gelin J.C., Oudin J. and Ravalard Y., "An Imposed Finite Element Method for the Analysis of Damage and Ductile Fracture in Cold Metal Forming Processes", *Annals of the CIRP*, 34(1):209–213, 1985.

Gelin J.C., "Finite Element Analysis of Ductile Fracture and defects Formations in Cold and Hot Forging", *Annals of the CIRP*, 39, pp. 215–218, 1990.

Hambli R., "Finite element model fracture prediction during sheet-metal blanking processes", *Eng. Fracture Mechanics*, 68, pp. 365–378, 2001.

Hammi Y., Simulation numérique de l'endommagement dans les procédés de mise en forme, Thèse de doctorat, Université de Technologie de Troyes, Avril 2000.

Hartley P., Clift S.E., Salimi J., Sturgess C.E.N and Pillinger I., "The Prediction of Ductile Fracture Initiation in Metalforming Using a Finite Element Method and Various Fracture Criteria", *Res. Mech.*, 28, pp. 269–293, 1989.

Homsi M., Morançay L. and Roelandt J.M., "Remeshing processes applied to the simulation of metal cutting", *Revue européenne des éléments finis*, 5, pp. 297–321, 1996.

Lee H., Peng K.E. and Wang J., "An anisotropic damage criterion for deformation instability and its application to forming limit analysis of metal plates", *Eng. Frac. Mech.*, 21(5), pp. 1031–1054, 1985.

Lemaitre J. and Chaboche J.L., *Mécanique des Milieux Solides*, Dunod, Paris, French edition 1985, Cambridge Univ. Press, English edition, 1990.

Lemaitre J., *A Course on Damage Mechanics*, Springer Verlag, 1992.

Liu T.S., and Chung N.L., "Extrusion Analysis and Workability Prediction Using Finite Element Method", *Computer and Structures*, 36, pp. 369–377, 1990.

Mathur K. and Dawson P., "Damage Evolution Modeling in Bulk Forming Processes", *Computational Methods for Predicting Material Processing Defects*, Edt, Predeleanu, Elsevier, 1987.

Onate E. and Kleiber M., "Plastic and Viscoplastic Flow of Void Containing Metal – Applications to Axisymmetric Sheet Forming Problem", *Int. J. Num. Meth. In Engng.* 25:237–251, 1988.

Picart P., Ghouati O. and Gelin J.C., "Optimization of Metal Forming Process Parameters with Damage Minimization", *J. Mat. Proc. Tech.*, 1998, 80–81:597–601.

Saanouni K., Forster C. and Ben Hatira F., "On the Anelastic Flow with Damage", *Int. J. Dam. Mech.*, 3:140–169, 1994.

Saanouni K. and Franqueville Y., "Numerical Prediction of Damage During Metal Forming Processes", *Numisheet 99*, Besançon, September, France, pp. 13–17, 1999.

Saanouni K., Nesnas K. and Hammi Y., "Damage modelling in metal forming processes", Int. *J. of Damage Mechanics*, Vol. 9, No. 3, pp. 196–240, July 2000.

Saanouni K., Hammi Y., "Numerical simulation of damage in metal forming processes", *Continuous Damage and Fracture*, Editor A. Benallal, Elsevier, ISBN. 2-84299-247-4, pp. 353–363, 2000.

Saanouni K., Cherouat A. and Hammi Y., "Numerical aspects of finite elastoplasticity with isotropic ductile damage for metal forming", *Revue européenne des éléments finis*, 2–3–4, pp. 327–351, 2001.

Simo J.C., Taylor R., "Consistent tangent operators for rate independent elastoplasticity", *Comput. Methods Appl. Mech. Eng.* 48, pp. 101–118, 1985.

Zhu Y.Y., Cescotto S., "The Finite Element Prediction of Ductile Fracture Initiation in Dynamic Metalforming Processes", *Journal de Physique III*, 1, pp. 751–757, 1991.

Zhu Y.Y., Cescotto S., Habraken A.M., "A Fully Coupled Elastoplastic Damage Modeling and Fracture Criteria in Metal forming Processes", *J. Met. Proc. Tech.*, 32: 197–204, 1992.

Zhu Y.Y. and Cescotto S., "A fully coupled elasto-visco-plastic damage theory for anisotropic materials», *Int. J. Molids and Structures*, 32 (11), pp. 1607–1641, 1995.

Chapter 19

Numerical Simulation of Ductile Damage in Metal Forming Processes: A Simple Predictive Model

Part 2: Some Applications

Khémaïs Saanouni, Philippe Lestriez and Abdelhakim Cherouat
Université de Technologie de Troyes, France

Jean-François Mariage
Université de Technologie de Troyes, France, and CETIM St Etienne, France

1. Introduction

Numerical simulation of metal forming processes (or virtual metal forming) has undergone an important development during the last decade. Its seems to be justified by a significant change in emphasis in the metal forming industry. In fact, the increasing cost of experimentation under along with the changing economic imperative of companies has focused development effort into process improvement and optimisation. This is facilitated by the increasingly affordable computing capability, the development of more and more efficient constitutive equations accounting, for the main (coupled) thermomechanical and metallurgical phenomena, maturing finite element methods (FEM) and their associated numerical and geometrical "tools" oriented to metal forming processes simulations. As indicated by the periodically organised conferences devoted to metal forming (ESAFORM, NUMIFORM, NUMISHEET...) and the growing number of special sessions dedicated to this field, virtual metal forming has revealed significant progress to simulate many processes accurately. Nowadays, many integrated virtual metal forming "tools" are available (Forge2®, Forge3®, Deform®, Marc®, Abaqus/Explicit, Pamstamp, etc.) to simulate accurately various processes by solving the main related nonlinearities, such as large elastoplastic deformation, thermal exchanges, hardening, evolving contact with friction, spring back, adaptive meshing in 2D and 3D situations. There is no doubt that the availability of such "virtual" tools allows the design of manufacturing processes to be integrated with the product design processes allowing rapid and cost effective production.

In our laboratory, for six years, extensive work has been developed in order to describe the ductile damage modelling in bulk and sheet metal forming. Two kinds of modelling are investigated; namely the "advanced" and the "simplified" approaches. Based on the thermodynamics of irreversible processes with state variables, the advanced approach aims to model the "multi-physics" coupling between main thermo-mechanical phenomena including the ductile damage in both isotropic and anisotropic cases (Saanouni *et al.*, 1994; Hammi, 2000; Saanouni and Hammi, 2000; Saanouni *et al.*, 2000; Saanouni *et al.*, 2001; Borouchaki *et al.*, 2002; Cherouat and Saanouni, 2002). These models have been implemented in Abaqus/standard and Abaqus/explicit using the available user subroutines (Umat, Vumat, Uele, Ufric). The simplified modelling uses very simple constitutive thermoelasto-(visco)plastic constitutive equations coupled to a simple isotropic damage model. These are implemented into the codes Forge2® and Forge3® devoted to the industrial use of virtual metal forming (Forge2®, Forge3® user's manual).

The present work is devoted to simplified modelling based on the formulation developed in Part 1 of this work. Some examples of metal forming processes are worked out in order to illustrate the ability of the proposed methodology to predict the damage effect. Comparisons between coupled and uncoupled calculations are sometimes proposed.

2. Implementation into Forge2 and Forge3

The constitutive equations discussed in section 2 of Part 1 and the corresponding numerical aspects (see section 3 of Part 1) have been implemented in the finite element codes Forge2® and Forge3®dedicated to the virtual metal forming simulation (see Forge2 and Forge3 Users Manuals 2001). Two implementation procedures related to the architecture of these codes have been investigated. The first is the one discussed above and called the "strong" coupled procedure only implemented into Forge2 code after the modification of many original subroutines in order to integrate the system defined by equation [31] of Part 1. The second solves only the equation f_1 of the system (equation [31] of Part 1) without damage effect in order to obtain Δp. After convergence Δp is used to calculate the damage increment ΔD without any iteration procedure applied to f_2 using available user's subroutine. Called "weak" coupling procedure, this latter has been implemented in both Forge2® and Forge3®codes. Note that comparison between the weak and the strong coupling procedures have been made using Forge2 and have shown that a small difference has been observed in the softening stage (Nesnas and Saanouni, 2000; Mariage and Saanouni, 2001). The implementation has been made in a manner that computations can be made with or without coupling. This is achieved according to the value of the flag Icoupl = 0 if no coupling (*i.e.* damage is calculated at each load increment but without any effect on the other thermomechanical fields) and 1 if there is coupling. This procedure is very helpful to show the superiority of the coupled computation compared to the uncoupled one against the prediction of the crack path.

For 2D problems Forge2® uses special three nodes triangular elements, while for 3D problems Forge3® uses special four nodes tetrahedral elements (see Chenot *et al*, 1998 for more details). A remeshing facility is provided by both Forge2® and Forge3® in order to improve the solution by adaptive meshing based on boundary curvature and internal mesh topology optimization (Coupez and Chenot, 1992).

When the damage variable reaches its critical value (D > 0.999) at a given gauss point giving a zero stress, this point is excluded from the integration procedure and supposed fully damaged with zero stiffness, zero stress and no more increase of strain and damage. When the overall gauss points inside a given element are fully damaged, the corresponding elementary stiffness matrix is zero. Hence, this element has no further contribution in the overall stiffness matrix and is extracted and killed from the structure and a new mesh of the overall structure is regenerated (Forge3® user manual).

3. Applications

To illustrate how the proposed coupled methodology can be used to predict the ductile damage occurrence during metal forming processes, some examples are worked out and discussed hereafter. All the presented examples are performed using

228   Prediction of Defects in Material Processing

both Forge2® and Forge3® codes with isothermal (constant temperature) time independent plasticity. The closed form of the isotropic hardening represented by the function σ_y as well as the overall material constants will be given for each material based on the available experimental results using uniaxial tension tests performed for materials under concern.

3.1. *Fracture of a round tensile bar*

The first example deals with the classical uniaxial tensile test of a 10 mm diameter and 30 mm long round specimen (see Figure 3). The material of the specimen is a commercial steel defined, at a constant temperature T = 800 °C, by the following constants:

– Isotropic elasticity: E = 70000 MPa, v = 0.3;

– Isotropic plasticity with isotropic hardening: $\sigma_y = \sqrt{3}K(p_0 + p)^n$ where n = 0.3 and K = 130 Mpa;

– Isotropic ductile damage: S = 0.21 MPa, s = 9.5 and β = 5 10^{-4}.

Two types of computations have been conducted using a 2D axisymmetric specimen and a 3D round tensile specimen (Figure 3). For each case the same controlled displacement loading path is applied with the constant velocity v = 1 s^{-1}. According to this very high displacement rate (similar to that used for the metal forming processes) the rate effect will be neglected by using a time independent plasticity model at constant temperature. Figure 1 shows the comparison of the experimental and the predicted engineering stress versus plastic strain for uncoupled and fully coupled cases both obtained in 2D and 3D specimens. One can observe the effectiveness of the chosen material constants to describe the stress-strain curve (in the coupled case) in comparison to the experimental one. The maximum stress reached before the softening induced by the damage effect take place is σ_{max}= 158.3 MPa for p = 37.6% (corresponding experimental values are σ_{max}= 161 MPa, p = 34,3%). The predicted plastic strain and the specimen fracture (material ductility) is found equal p_f = 50.3% (see Figures 1) and its corresponding experimental value is p_f= 60%.

Figure 2 shows the comparison between the coupled and the uncoupled calculations in terms of the force versus specimen elongation. Note the abrupt decrease of the force when the final fracture occurs after 5.3 mm of elongation.

Figure 3 shows the damage distribution inside the round bar for four different values of the imposed displacement u (mm). As can be verified on those images the fully damaged zones initiate at the specimen center and propagated towards the external surface in both 2D and 3D computations.

Numerical Simulation of Ductile Damage in Metal Forming   229

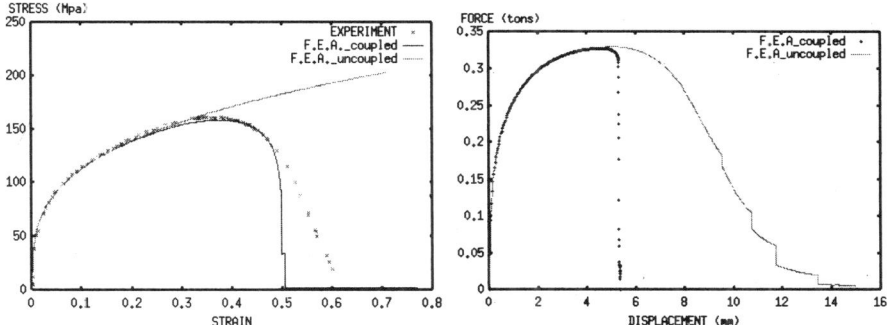

Figure 1. *Local stress-strain curves*   **Figure 2.** *Force-displacement curves*

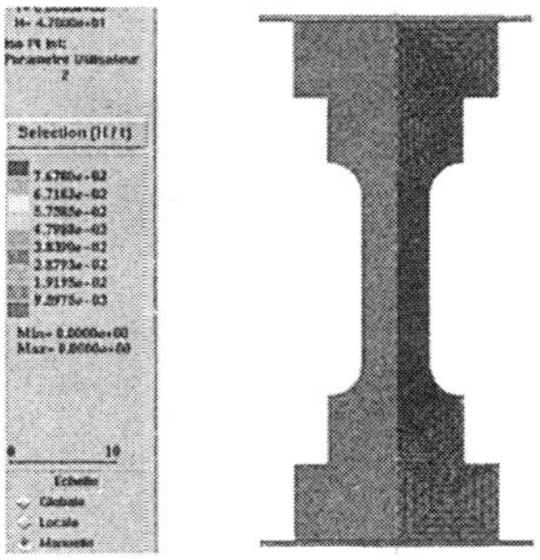

a1) u = 0.0 mm 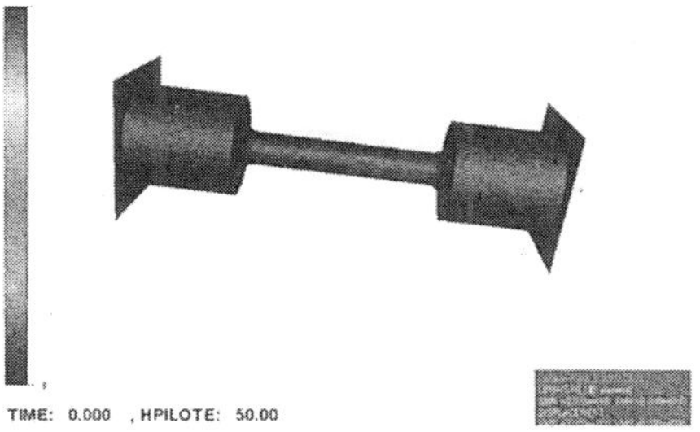   a2) u = 0.0 mm

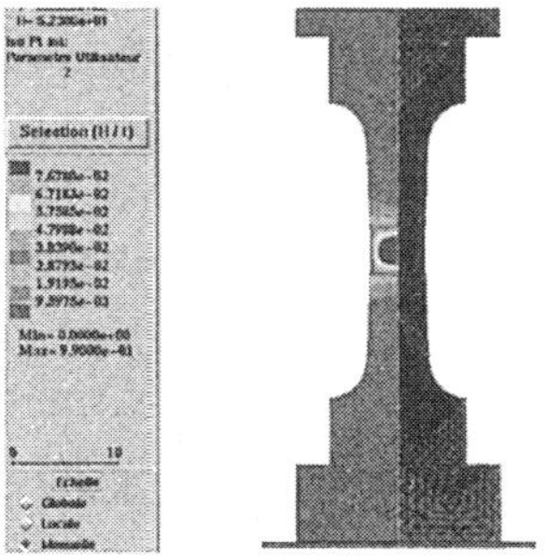

b1) u = 5.31 mm 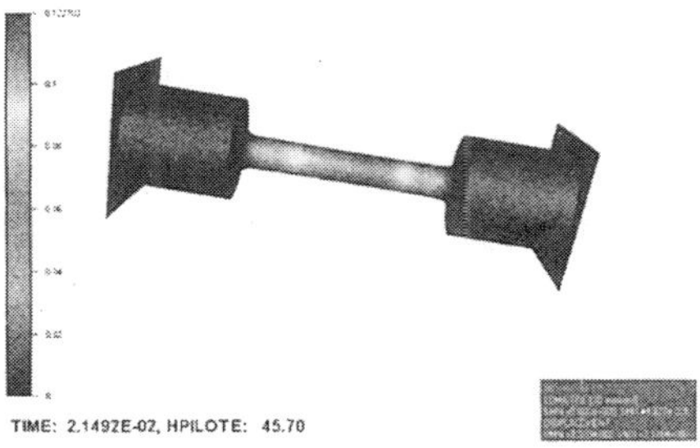   b2) u = 4.30 mm

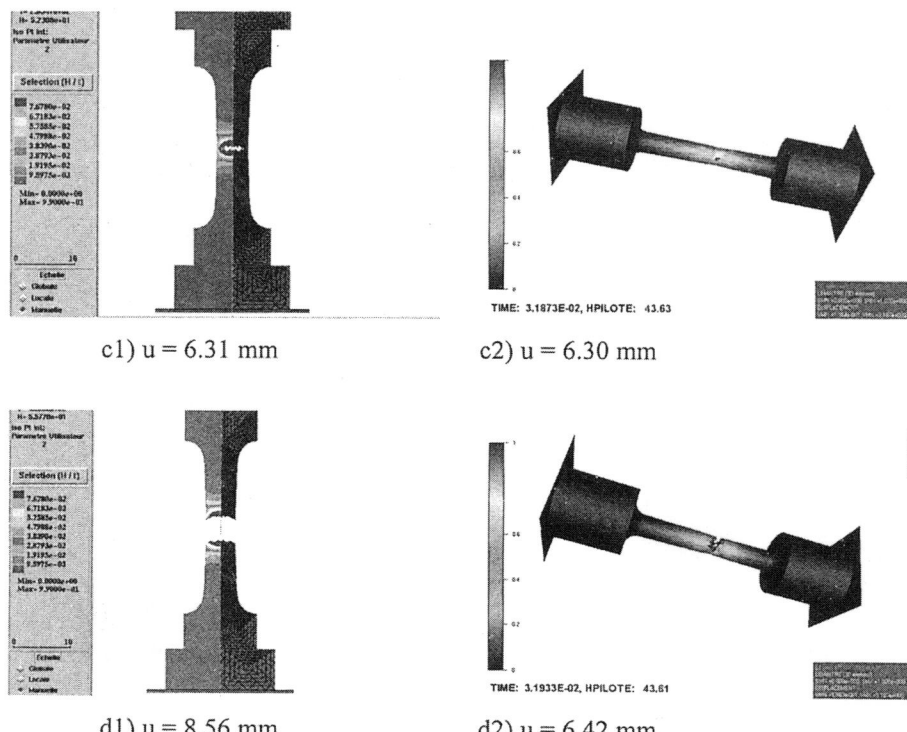

c1) u = 6.31 mm   c2) u = 6.30 mm

d1) u = 8.56 mm   d2) u = 6.42 mm

Figure 3. *Ductile damage distribution inside the round bar for different values of displacement*

3.2. *Tensile test of a plane strain plate*

In order to show the ability of the proposed methodology to predict the formation of shear bands, the classical uniaxial tensile test of a 10 mm large and 30 mm long strip is considered. The material is the same steel as before but taken at room temperature and defined by the following constants:

– Isotropic elasticity: E = 210000 MPa, ν = 0.3;

– Isotropic plasticity with isotropic hardening: $\sigma_y = \sqrt{3}K(1+ap^n)$ where n = 1.2 K = 231 Mpa;

– Isotropic ductile damage: S = 4 MPa, s = β = 1.

Only the 2D computation using Forge2 software is considered. Figure 4 shows the comparison of the predicted force-displacement curves for uncoupled and fully coupled cases obtained under a controlled displacement with the constant velocity v = 1 s^{-1}. From this figure, one can observe the strong effect of the softening induced by the damage occurrence giving a final fracture of the specimen around 3mm displacement, in the coupled case. In that case the maximum force is Fmax =8349 N reached for u = 1.12mm elongation, while for the uncoupled case we have Fmax = 8469 N and u = 1.42 mm. The local stress strain curve is similar to the one obtained for the round bar discussed above (see Figure 1).

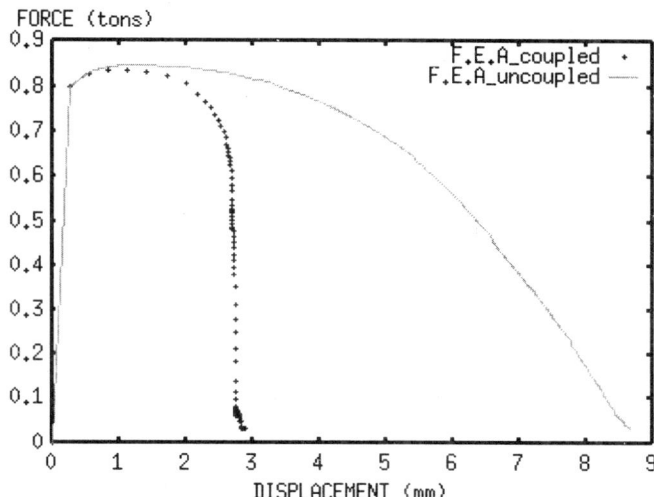

Figure 4. *Force-displacement curves for both coupled and uncoupled cases*

From Figure 5, where the damage maps at different displacement values are presented, one can note that the shear bands are realistically predicted. The voids (damage) initiate at the intersection of the two shear bands (Figure 5b) and propagate along only one shear band until final fracture of the specimen (Figures 5c and d). This damage localization inside a given path is controlled by the (infinitesimal) difference in the damage values which provokes the damage localization in points where the damage value is slightly higher then in their neighborhoods. Accordingly, the final macroscopic crack path is highly dependent in the time (Δt) and space (Δx) discretization steps. This very well known problem of mesh dependency due to the softening induced by the damage effect can be solved using an appropriate non-local or higher gradient formulation which is out of the scope of this paper.

232   Prediction of Defects in Material Processing

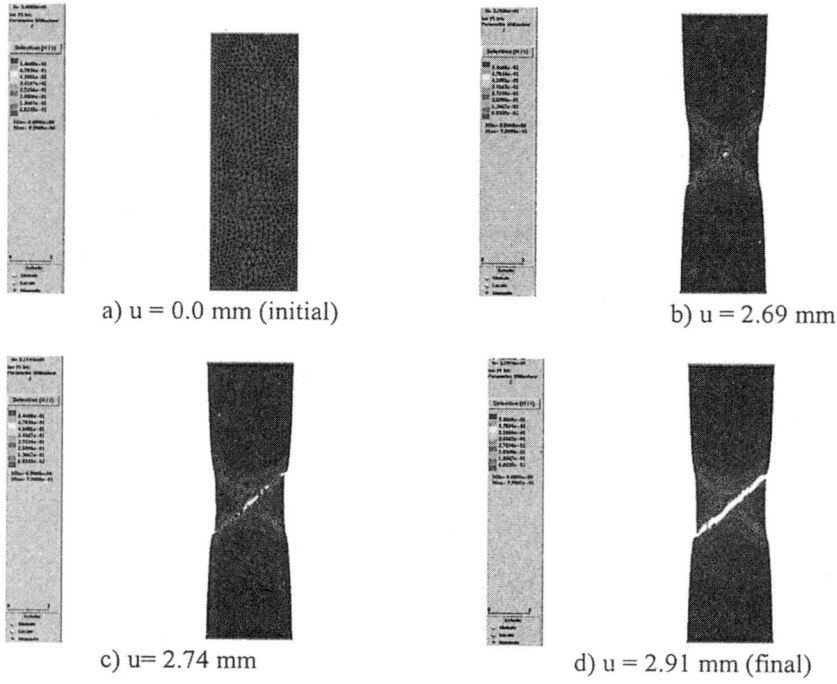

a) u = 0.0 mm (initial)

b) u = 2.69 mm

c) u = 2.74 mm

d) u = 2.91 mm (final)

Figure 5. *Ductile damage distribution in the specimen for different values of the controlled displacement*

3.3. Blanking of thin sheet

A 2D sheet metal blanking operation is presented in order to test the capability of the proposed model to predict the ductile damage occurrence during the cutting process. The initial sheet is circular of 10 mm diameter and 2.5 mm thickness. This sheet is clamped between the die and the blank-holder with a friction parameter (Coulomb friction) $\xi = 0.3$. The cylindrical punch has 4 mm diameter and moves vertically with 1 s^{-1} constant velocity. The friction between the punch and the sheet is taken to be similar to that between the sheet and the other tools *i.e.* $\xi = 0.3$. In order to study the effect of the tool wear, two different values of the punch angle radius and the die angle radius have been taken in order to model the "new tools" with 0.1 mm tool radii and the "worn tools" with 1 mm tool radii. The clearance (distance between the die and the punch) is 0.06 mm for the new tools and 0.13 mm for the worn tools. The material is the commercial steel used for the tensile test with the same constant values at constant room temperature (section 4.2). In absence of pertinent experimental data on this blanking process (*i.e.* force-stroke curves), we limit ourselves to the discussion of obtained numerical results. These are the predicted load-stroke curves and the initiation and the propagation of the macroscopic cracks for both new and worn tools.

Figure 6 shows four different plots of the damage distribution at four different punch displacement values for both new and worn tools. A clear difference between the new and worn tools appears concerning the location of the cutting initiation. With the new tools, two cracks initiate under the tools (punch and die) edges (Figure 6c1) and propagate towards each other until the final fracture of the sheet (Figure 6d1). The cutting scenario is quite different with worn tools: a single crack initiates at the sheet center (Figure 6c2) and propagates in the double directions of the die and punch edges until the final fracture occurs (Figure 6d2).

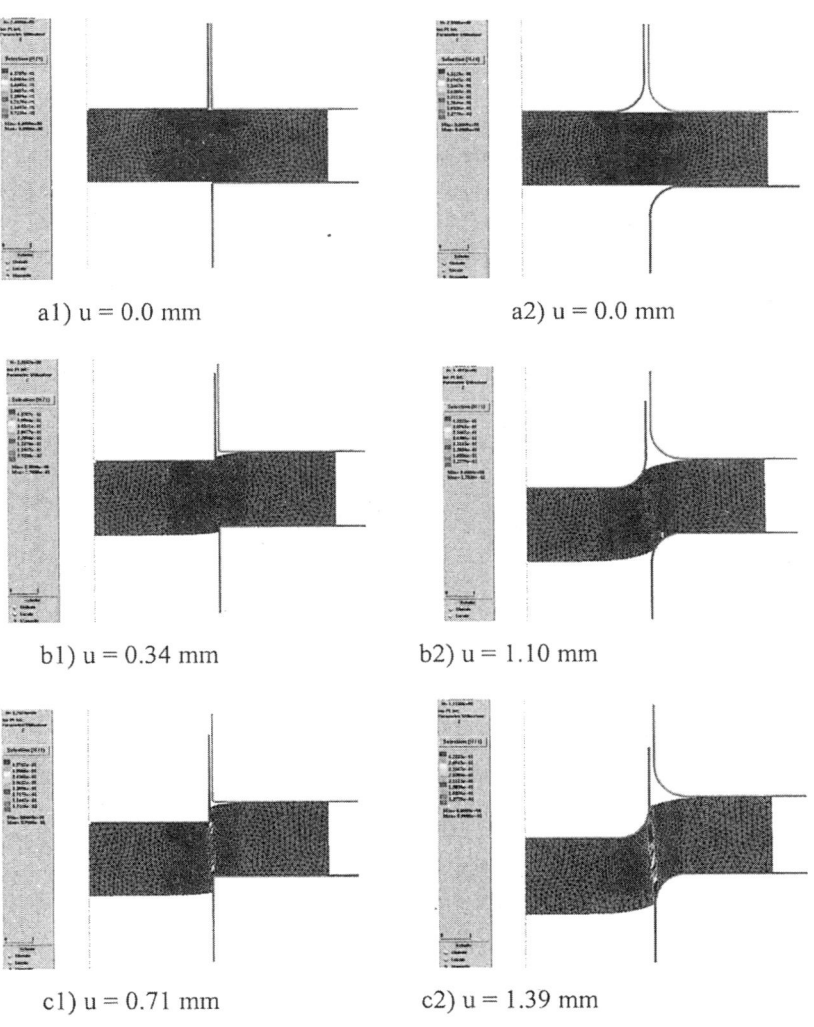

a1) u = 0.0 mm   a2) u = 0.0 mm

b1) u = 0.34 mm   b2) u = 1.10 mm

c1) u = 0.71 mm   c2) u = 1.39 mm

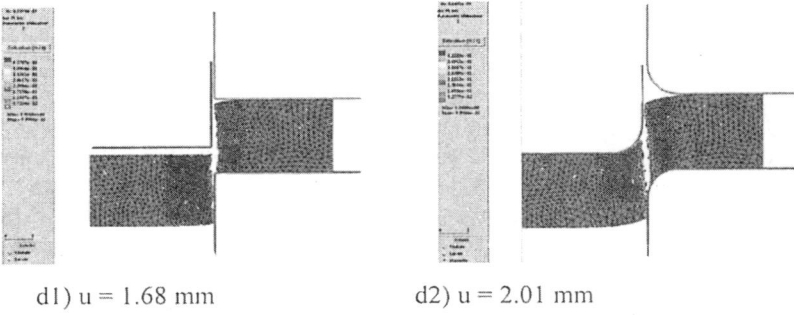

d1) u = 1.68 mm   d2) u = 2.01 mm

Figure 6. *Crack initiation and propagation with new and worn tools*

Figure 7 shows the blanking force versus punch displacement (or load-stroke) curves. It is very clear, as expected, that the cutting operation is easier with the new tools. In fact, with the new tools the maximum cutting force is 1111 N for u = 0.19 mm punch displacement while it was 1194 N for u = 1.04 mm punch displacement with worn tools. Also the cutting operation is terminated for 1.02 mm punch displacement for new tools, while it needs 1.5 mm with worn tools.

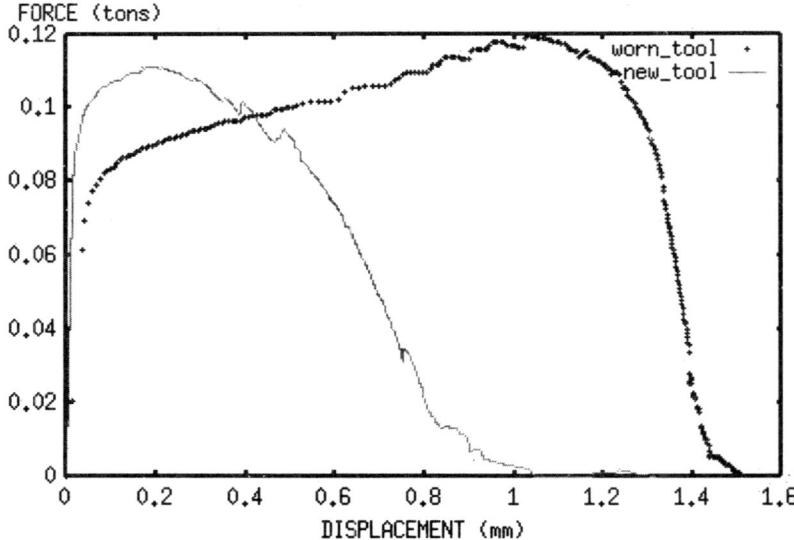

Figure 7. *Load-stroke curves for both new and worn tools*

3.4. Cutting of an axisymmetric wheel

This industrial example, experimentally investigated by CETIM, deals with a numerical simulation of drilling and deburring of a wheel obtained by hot forging and shown in Figure 8. According to the confidential character of this industrial problem no practical information can be given concerning the dimensions of the wheel, the kinematics of the tools, the material constants values and the experimental force-displacement curves. Hence, we limit ourselves to the presentation and discussion of the numerical results obtained when simulating the drilling and the deburring operations thanks to the coupled approach presented above.

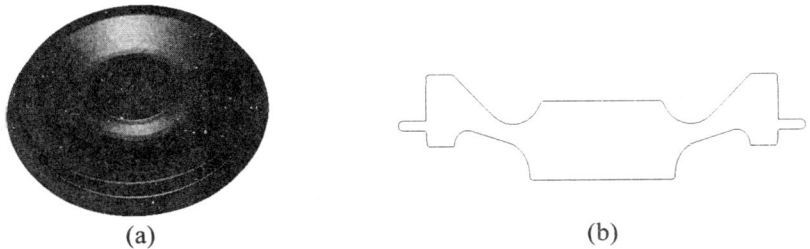

(a)   (b)

Figure 8. *Schematic representation of the wheel geometry a) 3D, b) 2D*

Figure 9 summarizes the simulation of the drilling (Figures 9a1, b1, c1, d1) and the deburring (Figures 9a2, b2, c2, d2) operations using the axisymmetric hypothesis (2D calculation). The drilling operation is terminated after 4.75 mm displacement of the punch (Figure 9d1), while the deburring operation needs 3.3 mm upper tool displacement (Figure 9d2). On the other hand, the force-displacement curves for both operations are shown in Figure 10. As expected, the drilling operation needs a maximum drilling force of 279 KN for u = 2.16 mm tool displacement; while the maximum deburring force is 128 KN for u = 0.83 mm tool displacement.

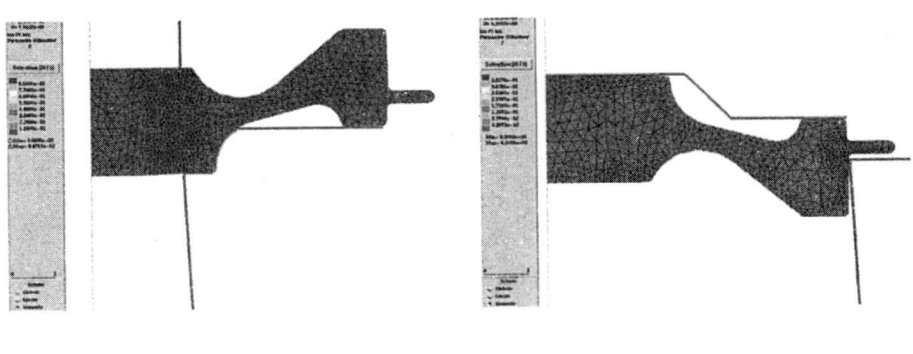

a1) u = 0.0 mm   a2) u = 0.0 mm

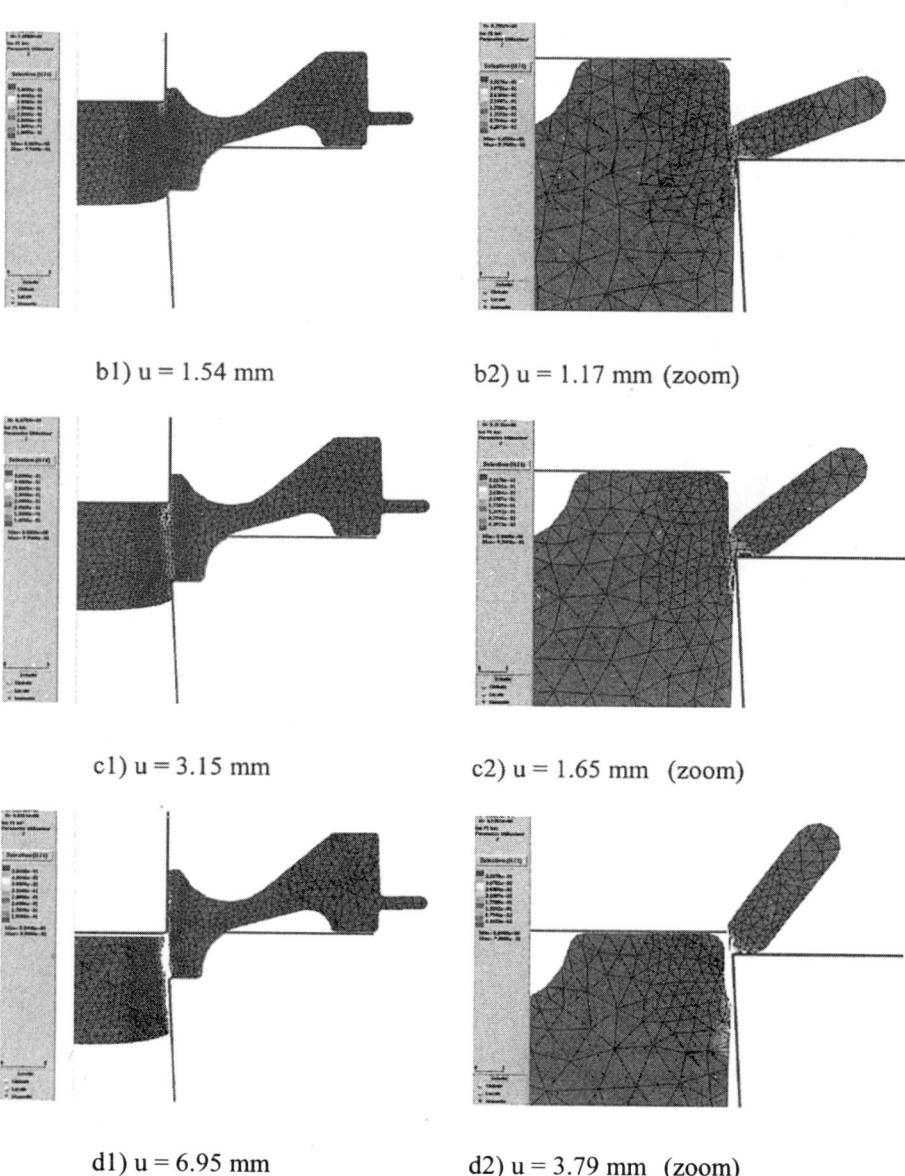

b1) u = 1.54 mm   b2) u = 1.17 mm (zoom)

c1) u = 3.15 mm   c2) u = 1.65 mm (zoom)

d1) u = 6.95 mm   d2) u = 3.79 mm (zoom)

Figure 9. *Damage distribution during the drilling (first column) and the deburring (second column) operations*

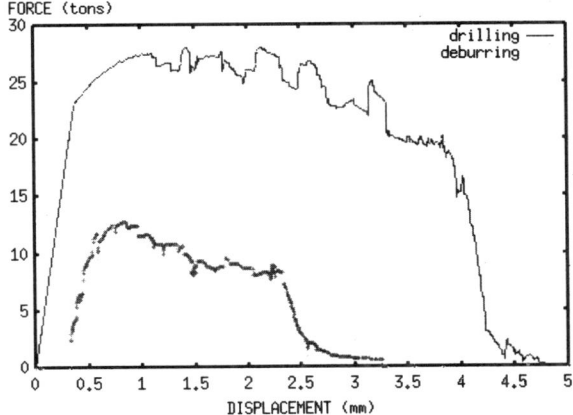

Figure 10. *Force-displacement curves for both drilling and deburring operations*

4. Conclusion

The implementation of the coupled elastoplastic/damage model for metal forming, developed in Part 1, into industrial finite element codes (Forge2® and Forge3®) dedicated to the metal forming simulation has been made. This provides a very helpful way to improve numerically any metal forming process in order to avoid or to enhance the ductile damage occurrence. Applications have been made to some cutting operations including the classical blanking of thin sheets as well as the drilling and deburring of a forged wheel. The results obtained attest the capability of the proposed approach to enhance any metal forming processes with respect to ductile damage occurrence.

Acknowledgements

The financial support of the FEDER (contract N° 99-2-50-059), the CRCA via the Pôle Mécanique et Matériaux Champardenais (PMMC) and the CIFRE N° 1.0.1591 is gratefully acknowledged.

5. References

Borouchaki H., Cherouat A., Saanouni K. and Laug P., "Remaillage en grandes deformation. Application à la mise en forme de structures 2D", *Revue euro. des éléments finis*, (to appear), 2002.

Chenot, J.L., Fourment L., Coupez T., "Forge3® – A general tool for practical optimization of forging sequence of complex 3-D parts in industry", *ImechE*, C546/033, 1998.

Cherouat A., Saanouni K., "Numerical simulation of sheet metal blanking process using a coupled finite elastoplastic damage modelling", *Int. J. Forming Processes* (to appear), 2002.

Coupez T. and Chenot J.L., "Large deformations and automatic remeshing", *Computational plasticity*», Pineridge Press, Swansea, Eds. D.R.J. Owen *et al.* pp. 1077–1088, 1992.

Hammi, Y., Simulation numérique de l'endommagement dans les procédés de mise en forme, Thèse de doctorat, Université de Technologie de Troyes, April 2000.

Mariage J.F. and Saanouni K., Simulation numerique de l'endommagement dans les procédés de Forgeage/Estampage, Rapport ANRT, UTT, 2001.

Nesnas K. and Saanouni K., Simulation numerique de l'endommagement dans les procédés de mise en forme avec Forge2, Rapport CETIM, UTT, 2000.

Saanouni, K., Forster C. and Ben Hatira F., « On the Anelastic Flow with Damage », *Int. J. Dam. Mech.*, 3:140–169, 1994.

Saanouni K., Nesnas K. and Hammi Y., "Damage modelling in metal forming processes", *Int. J. of Damage Mechanics*, Vol. 9, No. 3, pp. 196–240, July 2000.

Saanouni K., Cherouat A., Hammi Y., "Numerical aspects of finite elastoplasticity with isotropic ductile damage for metal forming", *Revue européenne des éléments finis*, 2–3–4, pp. 327–351, 2001.

Chapter 20

A Simplified Model of Residual Stresses Induced by Punching

Vincent Maurel, Florence Ossart and René Billardon
LMT-Cachan, ENS de Cachan, CNRS, Cachan, France

1. Introduction

Magnetic Fe-3%Si alloys are widely used in power engineering. These alloys are processed as 0.2 to 1.0 mm thick laminations, that are punched and stacked to build the magnetic circuit of most electrical machines. Punching the lamination induces plasticity localised in the vicinity of the cutting edges and long range residual stresses which alter the magnetic properties of the lamination through the local magneto-mechanical coupling existing in the material [BIL 99]. Hence, the accurate prediction of the magnetic behaviour of electrical machines requires both phenomena to be accounted for.

A first coupled model was developed, in which only the influence of plasticity on the magnetic behaviour of the material was considered [OSS 00]. A good agreement with experimental data was found in the case of a simple device with no residual stress (long rectangular stripe of constant width). But in other cases (for instance, the same rectangular stripe with circular holes), this model underestimates the influence of punching. Long range residual stresses of small magnitude may explain such discrepancies. The above-discussed application is an incentive-among others-to try and develop hereafter an analysis of the residual stresses induced by blanking.

2. Strain mechanisms activated during the blanking process

For most applications, the quality of the blanking process mainly lies in the shape of the cut edge (see Figure 1). Four characteristic zones are distinguished, which are linked to the deformation mechanisms activated during the different stages of the process [GOD 55], [GOI 99]. The roll-over is due to the bending effect, which dominates at the beginning of the process. After some punch travel, shearing becomes the main mode of deformation and the sheared edge of the product is formed, while the roll-over keeps increasing as long as the punch force does. Finally, a fracture initiates at the surface, in the vicinity of the cutting edges of the tools, and propagates through the sheet, resulting in the fracture zone and the burr. The sheared zone is a smooth area, approximately perpendicular to the plane of the sheet, which should be as large as possible to ensure a good geometry of the product. Therefore, the whole blanking process (mainly the clearance) is optimised in order to reduce the roll-over, delay the fracture initiation and minimise the burr while keeping acceptable the punch force.

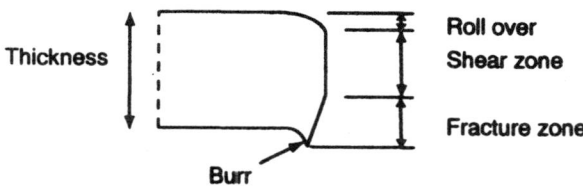

Figure 1. *Scheme of a blanked edge*

Modelling the blanking process remains a very difficult problem. One has to account for large deformations, friction and ductile fracture. Analytical approaches oversimplify the considered phenomena, whereas numerical analyses require the use of sophisticated numerical methods. In particular element eliminations or remeshing techniques governed by damage criteria are needed to model the crack propagation. Modelling this last phenomenon is still an open question. Despite many efforts and progress [MAI 91], [TAU 96], [HAM 01],[SAA 02], even complex finite element models fail to grasp all the details of the mechanisms involved. This is especially the case for the initiation and propagation of the ductile fracture through the sheet and many difficulties have to be solved before reaching acceptable results in terms of geometry of the cut edge. Hence, residual stresses are considered, at the time being, as a secondary phenomenon and very few results are available in the litterature.

We propose a simplified approach to overcome this difficulty and estimate the residual stresses induced by the blanking of thin metal sheets, without modelling the process itself. A phenomenological analysis of strain and stress mechanisms activated during the blanking process drives us to a simplified approach for the determination of effective plastic strain field to be taken into account when considering long range residual stresses [UED 86]. The model is applied to the case of punched disks.

3. Computation of residual stresses

Residual stresses over the structure are defined as the stress field respecting the equilibrium of the structure loaded by internal efforts induced by plastic strains incompatibilities [ROU 97]. For a given plastic strain field $\underline{\underline{\varepsilon}}^p$, the macroscopic residual stresses are defined by equation (1), in which $\underline{\underline{D}}$ denotes the elasticity operator and $\underline{\underline{\varepsilon}}(\underline{u})$ the tensor of total strains.

$$\underline{\underline{\sigma}}_{res} = \underline{\underline{D}}\left[\underline{\underline{\varepsilon}}(\underline{u}) - \underline{\underline{\varepsilon}}^p\right], \qquad [1]$$

These stresses are balanced when no external loading is applied if they satisfy the set of equations (2), where Ω denotes the considered domain, $\partial\Omega_F$ its boundary and $\underline{n}$ the normal vector to $\partial\Omega_F$.

$$\begin{cases} div\left(\underline{\underline{D}}\left[\underline{\underline{\varepsilon}}(\underline{u}) - \underline{\underline{\varepsilon}}^p\right]\right) = 0, \text{ on } \Omega; \\ \left(\underline{\underline{D}}\left[\underline{\underline{\varepsilon}}(\underline{u}) - \underline{\underline{\varepsilon}}^p\right]\right)\underline{n} = 0, \text{ on } \partial\Omega_F, \end{cases} \qquad [2]$$

This problem can be numerically solved for any given geometry and material properties, provided the plastic strain field is known.

In order to avoid the complex numerical simulation of the whole process, we use a phenomenological approach to reconstruct this field. The proposed model is based on the following experimental facts.

i) **Microhardness measurements** give an estimate of the equivalent plastic strain as a function of the distance to the cut edge (Figure 2). Two characteristics are noticed: very high values localised close to the edge and much smaller strain slowly vanishing as the distance to the cut edge increases.

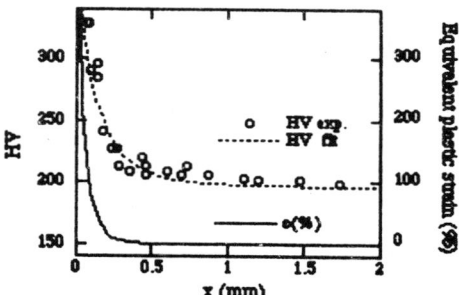

Figure 2. *Microhardness vs distance to the cut edge*

ii) **Micrographies** through the thickness of the sheet bring additional qualitative information (see Figure 3). The shape of the grains shows that shearing is localised in a narrow band along the cut edge. Traction deformations of much smaller magnitude are also visible and the global shape of the profile indicates bending effects.

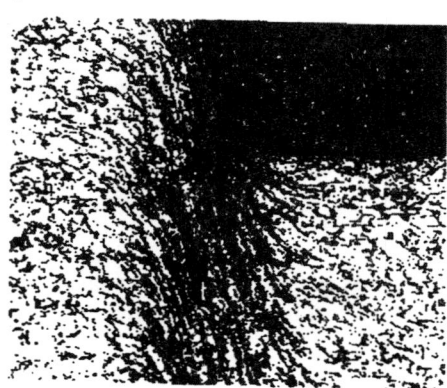

Figure 3. *Micrography of a blanked edge before fracture [MAI 91]*

iii) Surface displacement fields measured by image intercorrelation techniques confirm the existence of bending and tension and allow one to quantify those effects. The displacement field measured on the upper face of a punched disk corresponds to tension, whereas the one measured on the lower face corresponds to compression with a lower absolute value: $\varepsilon_{rr}^{top} \geq 6.5\%$, $|\varepsilon_{rr}^{bottom}| \leq 3\%$. Figure 4 shows an example of strains measured on the top surface of the sheet, in the case of a cylindrical hole.

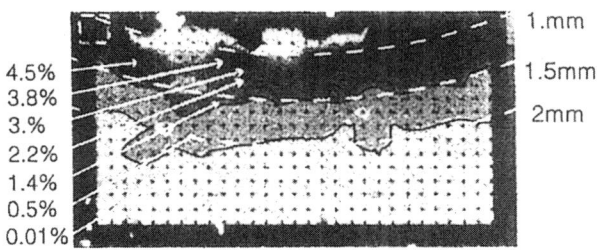

Figure 4. *Radial strain measured for a 5mm diameter punch. Top view of the sheet. Dashed circle : distance to the edge*

These observations lead us to split the plastic strain tensor into three contributions according to formula [3], in which $\underline{\underline{\varepsilon}}^p_s$, $\underline{\underline{\varepsilon}}^p_t$ and $\underline{\underline{\varepsilon}}^p_b$ respectively denote the strain parts due to shearing, tension and bending. The high level of plasticity and hardness close to the cut edge is due to shearing, whereas the slow decrease can be attributed to tension and bending effects. Residual stresses being the result of a linear problem, the contribution of each type of deformation can be analysed separately.

$$\underline{\underline{\varepsilon}}^p = \underline{\underline{\varepsilon}}^p_s + \underline{\underline{\varepsilon}}^p_t + \underline{\underline{\varepsilon}}^p_b \qquad [3]$$

Our main goal is to determine if long range residual stresses can be induced by the blanking process. Let us first consider shearing, the main deformation mode. If one assumes that $\underline{\underline{\varepsilon}}^p_s$ is uniform throughout the sheet thickness, there is rigourously no strain incompatibility, and hence no residual stresses, whatever the plastic strain level and the dependence on the distance to the cut edge. In order to examine the effect of non uniformity through the thickness, some finite element analyses were carried out, assuming $\underline{\underline{\varepsilon}}^p_s(x, z)$ profiles consistent with experimental observations. The main result of those simulations is that the residual stresses induced in such cases are very localised and do not spread far from the cut edge. Long range stresses can only be induced by tension and bending. In the present paper, we focus on tension. The "tension model" proposed in [ZHO 96] allows us to quantify this second order phenomenon, which is difficult to extract from available measurements.

4. Zhou and Wierzbicki tension zone model for the analysis of blanking

The analytical model proposed by Zhou and Wierzbicki can predict the major characteristics of the blanking process [ZHO 96]. It is based on numerous experimental observations and includes three major features of the blanking process. The first one is the shape of the deformed zone, wider at the center than at the surface of the sheet. The second one is the fracture process, with cracks initiating at the top and bottom surfaces and progressively growing inwards. The last one is the main mode of deformation before fracture: the flow pattern is not simple shear, but also includes a tensile component oblique to the parted surface. This last element is a keypoint of the model and of our discussion.

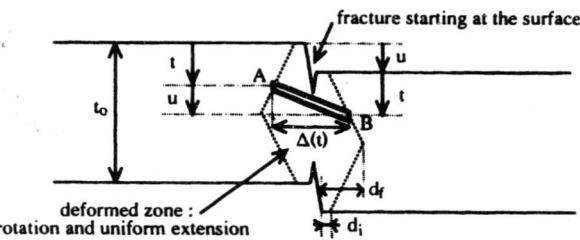

Figure 5. *Principle of the tension model [ZHO 96]*

Figure 5 summarises the principles of the model. The postulated shape of the deformed zone is a parallelogram, characterised by the distances d_i and d_f. The main deformation mechanisms of a representative element AB are rotation and tension. Failure and fracture of the element occur when the shear strain γ reaches a critical value γ_f. Elements being shorter at the surface, cracks will first appear at the top and bottom surfaces and propagate towards the center of the sheet. The evolution of the fracture is directly correlated to the shape of the deformed zone. These assumptions allow one to determine the punch force as a function of the punch displacement u. This point has been extensively discussed in [ZHO 96]. These assumptions also give the evolution of the tensile force at the interface between the sheared zone and the rest of the sheet, as developed hereafter.

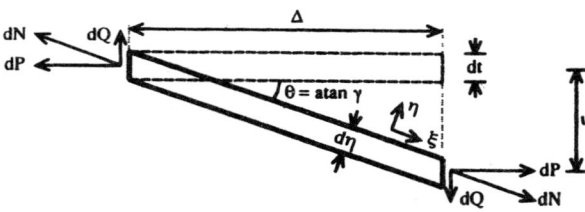

Figure 6. *Representative element of the tension model [ZHO 96]*

Figure 6 shows a representative element in the deformed zone. Following classical assumptions of limit analysis in plasticity, the shearing resistance for large localised deformation is neglected with respect to tension-compression resistance, so that the deformed zone is modelled as an assembly of non-interacting beams subjected to uniform extension in their current axial direction. Interaction with adjacent elements was shown to act only as a secondary effect, so that the top and bottom surfaces of the element are traction-free. Considering a plane-strain problem, the deformation components in the (ξ,η,z) coordinate system are given by equations [4] and [5].

$$\varepsilon_\xi = ln\frac{\Delta/\cos\theta}{\Delta} = ln\sqrt{1+\gamma^2} \qquad [4]$$

$$\varepsilon_\eta = ln\frac{d\eta}{dt} = ln\frac{1}{\sqrt{1+\gamma^2}} = -\varepsilon_\xi \qquad [5]$$

The equivalent stress and strain for the Von Mises criterion are defined by [6] and [7] and the material behaviour is modelled by the power-law [8].

$$\bar{\sigma} = \frac{\sqrt{3}}{2}|\sigma_\xi| \qquad [6]$$

$$\bar{\varepsilon} = \frac{2}{\sqrt{3}}|\varepsilon_\xi| \qquad [7]$$

$$\bar{\sigma} = \sigma_1 \bar{\varepsilon}^n \qquad [8]$$

The resulting shear force dQ and tensile force dP are given by formulae [9] and [10]. When the angle θ reaches the failure value θ_f, fracture occurs and the local values of dQ and dP drop to zero. The integration of dQ over the thickness of the sheet gives the resulting punching force. Results reported in [ZHO 96] show a very good agreement between the model and experimental data.

$$dQ = dN\sin\theta = \left(\frac{2}{\sqrt{3}}\right)^{n+1}\sigma_1\left(ln\sqrt{1+\gamma^2}\right)^n\frac{\gamma}{1+\gamma^2}dt \qquad [9]$$

$$dP = dN\cos\theta = \left(\frac{2}{\sqrt{3}}\right)^{n+1}\sigma_1\left(ln\sqrt{1+\gamma^2}\right)^n\frac{1}{1+\gamma^2}dt \qquad [10]$$

For our purpose, we focus on the tensile force distribution through the sheet thickness. The profile dP(t) was calculated using the tension model with the following parameters. The thickness t_o of the considered electrical steel is 0.5 mm. The model parameters are σ_1 = 795 MPa, n = 0.144, d_i = 0.2t_o and d_f = 0.6t_o and γ_f = 2.5.

Figure 7 shows the evolution of the tensile force distribution dP/dt(t) for increasing values of the punch displacement u. These curves clearly indicate that, at the beginning of the process, tensile forces reach values higher than the yield stress, which is 370 MPa. Tensile plastic strains will appear beyond the sheared area. The finite element analysis of this mechanism is presented in the next section.

246 Prediction of Defects in Material Processing

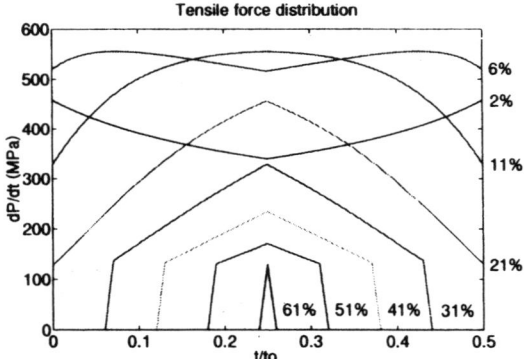

Figure 7. *Tensile force distribution for increasing punch displacements u*

5. Analysis of tension effects during the blanking process

We study here a simple case: a circular hole (ϕ=5mm) is punched in a disk (ϕ=20mm) of the considered 0.5 mm thick Fe-3%Si alloy. The elasto-plastic analysis of tension was carried out for the axisymmetric problem shown in Figure 8. Following the analytical results depicted in Figure 7, uniform traction (500 Mpa) is applied along the cut edge. The friction applied by the die and blankholder and their stiffness are modelled by attributing non uniform anisotropic elastic parameters in these regions. The elastic moduli E_{rr} and $E_{\theta\theta}$ are set to zero at the cut edge of the material and linearly increase to progressively prevent radial and orthoradial displacements at the top and bottom surfaces of the sheet. The E_{zz} component is very large in the die to forbid vertical displacements of the bottom surface of the sheet, and on the contrary very small in the blankholder, to allow free vertical displacement at the top surface. Vertical and in-plane strains are decoupled in the die and the blankholder by setting ν_{rz} and $\nu_{\theta z}$ to zero.

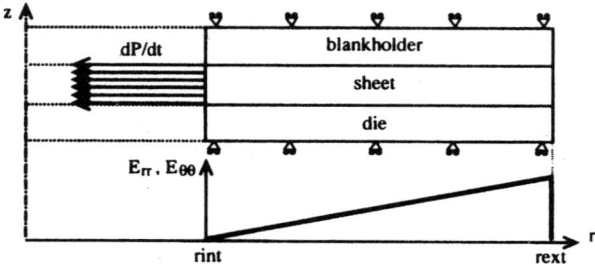

Figure 8. *Applied boundary conditions to analyse tension effects during the punching of a circular hole in a disk*

Figures 9 shows the components of the calculated plastic strain as a function of the distance to the cut edge. Their range of amplitude is very small (about 0.1%) and they disappear at 0.8 mm from the cut edge. This profile is consistent with Figure 2, although microhardness measurements are too coarse to predict such small strains.

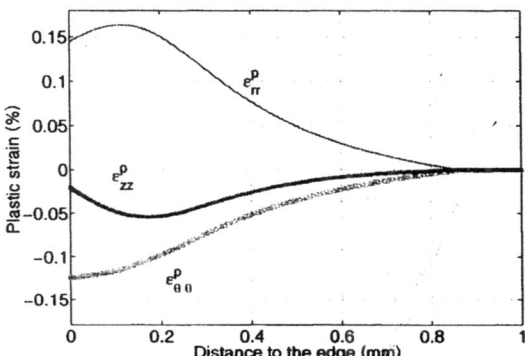

Figure 9. *Plastic strain as a function of the distance to the cut edge*

The resulting residual stresses are plotted in Figure 10. It is worth noticing that the components σ_{rr} and σ_{zz} are small. A different behaviour is observed for $\sigma_{\theta\theta}$ because of the axisymmetric configuration, which opposes orthoradial displacements of the sheet: $\sigma_{\theta\theta}$ reaches 150 MPa at the surface of the sheet and slowly evolves from -20MPa at the limit of the plastified zone to -4MPa at the outer surface of the sheet.

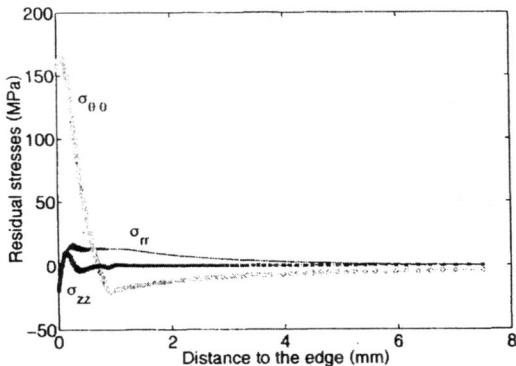

Figure 10. *Residual stresses as a function of the distance to the cut edge*

6. Conclusion

Magnetic properties are very sensitive to stresses and especially to compression in the case of Fe-3%Si alloys. The proposed analysis is a very simplified one, but it exhibits a possible mechanism, which creates long range small residual stresses. Such stresses could explain why in some cases, accounting only for plasticity does not allow to accurately model the degradation of the magnetic properties of punched electrical steels. More experimental analyses are underway to more qualitatively confirm these effects.

7. References

[BIL 99] BILLARDON R., HIRSINGER L., OSSART F., "Magneto-elastic coupled finite element analyses", *Revue Européenne des Eléments Finis*, vol. 8 n° 5-6, 1999, p. 525-551.

[GOD 55] GODTSCHALCK L., "Contribution à l'étude du poinçonnage des métaux", *Corrosion Industries – Métaux*, vol. 1, 1955.

[GOI 99] GOIJAERTS A., Prediction of ductile fracture in metal blanking, PhD thesis, Technische Universiteit Eindhoven, 1999.

[HAM 01] HAMBLI R., "Comparison between Lemaitre and Gurson damage models in crack growth simulation during blanking process", *Int. J. of Mech. Sci.*, vol. 43, 2001, p. 2769-2790.

[MAI 91] MAILLARD A., Etude expérimentale et numérique du découpage, PhD thesis, UTC, 1991.

[OSS 00] OSSART F., HUG E., HUBERT O., BUVAT C., BILLARDON R., "Effect of Punching on Electrical Steels: Experimental and Numerical Coupled Analyses", *IEEE Transactions on Magnetics*, vol. 36 n° 5, 2000, p. 3137-3140.

[ROU 97] ROUHAUD E., MILLEY A., LU J., "Introduction of residual stress fields in finite elements three-dimensional structures", *ICRS-5*, vol. 1, 1997, p. 386-391.

[SAA 02] SAANOUNI K., MARIAGE J., CHEROUAT A., "Sheet metal cuting simulation by continuum damage mechanics", 5th *Esaform Conference on Material Forming*, , Krakov Poland, 2002, p. 575-578.

[TAU 96] TAUPIN E., BREITLING J., WU W.-T., ALTAN T., "Material Fracture and Burr Formation in Blanking Results of FEM Simulations and Comparison With Experiments", *J. of Mat. Proc. Tech.*, vol. 59 n° 1-2, 1996, p. 68-78.

[UED 86] UEDA Y., FUKUDA K., KIM Y., "New measuring method of axisymmetric three-dimensional strains as parameters", *Transactions of the ASME*, vol. 108, 1986, p. 328-334.

[ZHO 96] ZHOU Q., WIERZBICKI T., "A tension zone model of blanking and tearing of ductile metal plates", *Int. J. of Mech. Sci.*, vol. 38 n° 3, 1996, p. 303-324.

Chapter 21
Wrinkling and Necking Instabilities for Tube Hydroforming

Germaine Nefussi and Noel Dahan
LMT-Cachan, CNRS, Université Pierre et Marie Curie, Cachan, France

Alain Combescure
INSA – LMC, Villeurbanne, France

1. Introduction

Tube hydroforming is the object of considerable attention, especially in the automotive industry, because of the progress in high-pressure technology and in computer control. The advantages of this process compared with traditional manufacturing are considerable and include cost and material savings, weight reduction, quality improvement and better accuracy. The main drawbacks of hydroforming are the equipment costs and the lack of theoretical knowledge.

In hydroforming operations, pressure is applied in order to shape the tube in the form of dies. No direct contact with the tool is needed at the beginning of the process; therefore, no friction occurs on the part until it hits the die. At the same time, axial forces are applied to the ends of the blank to prevent premature necking due to wall thinning, but if these axial forces are too large, buckling of the tube (which produces wrinkling) may occur. Thus, an appropriate combination of internal pressure and axial compression force is crucial to prevent necking as well as buckling.

A criterion for general instability and non-uniqueness was established by Hill [HIL 58] (developed in [HIL 91]) and Tvergaard [TVE 83]. Simplified methods for rigid-plastic materials in plane stress are also available for tube hydroforming. The well-known Swift's criterion [SWI 52] can be deduced from Hill's uniqueness principle, as done in Yamada and Aoki [YAM 66]. Xing and Makinouchi [XIN 01] applied this criterion to the hydroforming process. But this type of method does not take into account wrinkling of the tube under the axial load.

On the other hand, Tvergaard [TVE 83], for instance, analyzed buckling of elastic-plastic cylindrical shells under compressive axial force alone. Very few articles take into account necking and buckling instabilities at the same time. One can nevertheless quote Tirosh et al. [TIR 96] and Xia [XIA 01] who have pointed the two instabilities together. In both references the problem is solved with a simplified approach.

Finally, few Finite Element Codes (INDEED, LS-DYNA and PAM-STAMP) propose some kind of hydroforming simulation. But, in general, buckling and necking computations are not included in terms of stability. Hence this is still an open field.

In this paper, the hydroforming limit of isotropic tubes subjected to internal hydraulic pressure and independent axial load is discussed.

Swift's criterion is often used in this case for the prediction of diffuse plastic instability. For sheet-forming limit, the Cordebois-Ladeveze's criterion [COR 86] is well adapted and may be adapted to hydroforming process, by introducing the pressure.

We have already highlighted [NEF 02] the existence of two different Swift's criteria (for sheets and for tubes). We recalled that these types of approaches do not take into account buckling induced by axial loading. In fact, buckling may obviously occur before plastic instability; consequently, Swift's criterion must not be used alone to predict instability in the case of tube hydroforming. Numerical simulation is used here to confirm these points and to analyze both the buckling and bursting phenomena together. This is performed in the first part of this paper, in which we also confirm numerically that the Swift's criterion for tube hydroforming is a good criterion for the prediction of bursting. The J2-flow theory is used, with Jaumann's derivative. The material is supposed to follow the Prandtl-Reuss equations with von Mises' yield criterion and the associated flow rule. Isotropic hardening was taken into account.

An experimental setup is then presented in the second part. The first experimental results confirm the numerical analysis and we outlined, by this way, a method to obtain hydroforming limit diagrams.

2. Mathematical analysis

2.1. *Analytical results*

2.1.1. *Plastic instability of tubes and sheets*

It is shown in a precedent work [NEF 02] that the Swift's criterion is a good tool for the prediction of necking (diffuse instability) during a hydroforming process.

A thin, closed-end tube under internal pressure p and axial load P is considered. r and t are the current values of the cylinder's radius and wall thickness ($t \ll r$). The plastic material is isotropic and obeys von Mises' yield criterion and the hardening law is expressed as:

$$\bar{\sigma} = g(\bar{\varepsilon}) \qquad [1]$$

In [1], $\bar{\sigma}$ is the equivalent stress $\bar{\varepsilon}$ the equivalent strain. The cylinder is assumed to be thin enough for the plane stress hypothesis to be valid: σ_1 and σ_2 are the principal (respectively axial and hoop) Cauchy stresses. The tube is long enough ($l \gg r$) for the stresses to be assumed to be uniformly distributed (except, of course, near the ends). From the equilibrium equations, we can deduce:

$$\sigma_1 = \left(\bar{P} + p\pi r^2\right)/2\pi rt \quad \text{and} \quad \sigma_2 = pr/t \qquad [2]$$

In [2], $\bar{P}$ is negative when the axial force is compressive. Let $\sigma_1 = x\,\sigma_2$; then, plastic instability is expressed, as usual, as:

$$\frac{1}{z} = \frac{d\bar{\sigma}}{\bar{\sigma}d\bar{\varepsilon}} = \frac{g'(\bar{\varepsilon})}{\bar{\sigma}} = A(x).$$

Two choices, at least, are possible in order to obtain $A(x)$. One option is to assume that instability occurs when $dp = 0$ and $dP = 0$ simultaneously. In this case, after some calculations, the following instability criterion is obtained [SWI 52]:

$$\frac{1}{z_H} = \frac{d\bar{\sigma}}{\bar{\sigma} d\varepsilon} = \frac{4x^3 - 6x^2 + 3x + 4}{4(x^2 - x + 1)^{3/2}} \quad [3]$$

A second option is to assume, as in Hillier [HIL 63], that the pressure and the independent axial load are such that $dF_1 = 0$ and $dF_2 = 0$, with $F_1 = P + \pi r^2 p$ and $F_2 = prl$. One can note that these conditions are similar to those used in sheet forming. The first option leads to "the hydroforming criterion", or "tube Swift criterion, and the second one to the well-known sheet-forming criterion.

To obtain limit curves in terms of strains, we introduce the principal critical strains ε_{1c} and ε_{2c}. For instance, with a power hardening law: $\bar{\sigma} = g(\bar{\varepsilon}) = K\bar{\varepsilon}^n$, Swift's hydroforming criterion [3] yields:

$$\varepsilon_{1c}^H = (2x - 1)\frac{2n(x^2 - x + 1)}{4x^3 - 6x^2 + 3x + 4}; \varepsilon_{2c}^H = (2 - x)\frac{2n(x^2 - x + 1)}{4x^3 - 6x^2 + 3x + 4} \quad [4]$$

The corresponding curve is represented in Figure 1 for $n = 0.1$. It is similar to the curve of Xing and Makinouchi [XIN 01], when no torque is applied.

2.1.2. Plastic instability and buckling

The buckling value for such a tube was analytically calculated in [NEF 02]. It was found, of course, that the buckling value is much smaller than the necking value. The buckling value does not occur in the instability zone predicted by Swift's criteria. This point is very important, but seems to have been neglected in the hydroforming literature. Nevertheless, as said before, experimental results (guided by a limit analysis formulation) are given in Tirosh et al. [TIR 96] and an analytical method is developed in Xia [XIA 01].

As it is not easy to consider analytically the two failure modes together, we introduce in the next section a complete numerical simulation.

2.2. Numerical simulation

2.2.1. Problem formulation and basic equations

For cylindrical shells under internal pressure and axial load, the bifurcation analysis can be done entirely numerically. The simulation of the process can be performed in two steps: first, the equilibrium states must be predicted; then, their stability must be verified. Durand and Combescure [DUR 99] have developed large-strain elastic-plastic stability analysis. In their work, the principle of virtual work is used with Hill's criterion as plastic buckling criterion.

The equilibrium is written using the rate form of the principle of virtual work (with an internal pressure p and an axial load P):

$$\int_{\Omega_0} \left\{ \dot{\tau}^{ij}\delta\eta_{ij} + \tau^{ij}\dot{u}^k_{,i}\delta u_{k,j} \right\} d\Omega_0$$
$$= -\dot{p}\int_{S_0} g^{ir} n_r \delta u_i dS_0 - p\int_{S_0} \left[g^{ir}\dot{u}^j_{,j} - g^{ji}\dot{u}^r_{,j} \right] n_r \delta u_i dS_0 - \dot{P}\delta u_{\text{sup}} \quad [5]$$

In [5], η_{ij} is the Lagrangian strain tensor, u_i are the displacement components in the reference basis, and $()_{,i}$ designates covariant differentiation in the reference frame. The contra variant components of the Kirchhoff stress tensor are related to the corresponding components of the Cauchy stress tensor by: $\tau = J\sigma = \sqrt{g/G}\,\sigma\,g$ and G are the determinants of the metric tensors g_{ij} and G_{ij} in the reference state and in the current state respectively. u_{sup} designates the axial displacement of the upper end where P is applied.

With the J2-flow theory:

$$\dot{\tau}^{ij} = L^{ijkl} D_{kl} \quad [6]$$

where

$$L^{ijkl} = \frac{E}{1+\nu}\left\{ \frac{1}{2}\left(g^{ik}g^{jl} + g^{il}g^{jk}\right) + \frac{\nu}{1-2\nu}g^{ij}g^{kl} - \beta\frac{3}{2}\frac{E/E_t - 1}{E/E_t - (1-2\nu)/3}\frac{s^{ij}s^{kl}}{\sigma_e^2} \right\}$$
$$-\frac{1}{2}(g^{ik}\tau^{jl} + g^{jk}\tau^{il} + g^{il}\tau^{jk} + g^{jl}\tau^{ik}) \quad [7]$$

D_{kl} are the covariant components of the strain rate tensor D, E is the Young's modulus and ν the Poisson's ratio, $\sigma_e = (3/2 s_{ij}s^{ij})^{1/2}$ is the effective von Mises' stress, $s^{ij} = \tau^{ij} - 1/3 g^{ij}\tau^k_k$ is the stress deviator and β is 1 for plastic loading or 0 for elastic unloading.

The bifurcation analysis is based on Hill's theory [HIL 58] of uniqueness and bifurcation in elastic-plastic solids. The difference between two solution increments $\dot{u}$ corresponding to the same increment of the prescribed quantity is designated by $\tilde{u}$. A nonzero value of this difference results in non-uniqueness of the solution. Then, the bifurcation condition is a nonzero solution of:

$$\delta I = 0$$
$$I = \int_{\Omega_0}\left\{ L^{ijkl}\tilde{\eta}_{ij}\tilde{\eta}_{kl} + \tau^{ij}\tilde{u}^k_{,i}\tilde{u}_{k,j} \right\}d\Omega_0 + p\int_{S_0}\left[g^{ir}\tilde{u}^j_{,j} - g^{ji}\tilde{u}^r_{,j}\right]n_r\tilde{u}_i dS_0 \quad [8]$$

Finally, using the finite element method, this leads to the resolution of an eigenvalue problem defined by the following equation:

$$[\mathbf{K}_T + \mathbf{K}_\sigma + \mathbf{K}_p]\tilde{U} = 0 \qquad [9]$$

In equation [9], $\mathbf{K}_T$ is the tangent structural stiffness matrix; $\mathbf{K}_\sigma$ is the geometric stiffness matrix and $\mathbf{K}_p$ is the pressure matrix (symmetric in the case of an external energy potential); $\tilde{U}$ denotes the instability mode.

For axisymmetric structures loaded axisymmetrically (as is the case here) one can show that the eigenmode $\tilde{U}$ is a linear combination of Fourier modes and that the stability problem can be solved independently for each Fourier harmonic. Finally, we must solve one set of stability problems for each Fourier mode. The Fourier mode for which the instability first occurs defines the most probable instability shape. One of the advantages of the method is that it also allows the prediction of successive bifurcation modes in elastic-plastic situations. One can also predict non-axisymmetric bifurcation modes using only axisymmetric equilibrium states.

This method has been implemented into the INCA finite element program [COM 96] with 6- or 8-node isoparametric elements and 3 or 4 integration points respectively.

2.2.2. *Numerical results*

The numerical results presented here are applied to a tube fixed at both ends. The tube has an external radius of 25 mm, a thickness of 1.5 mm and a length of 150 mm. The radial displacement is fixed at both ends of the tube. The vertical displacement is fixed at the base of the tube. The material is stainless steel with Young's modulus 200,000 Mpa and Poisson's ratio 0.3. The stress/strain curve is represented by an Hollomon law where $K = 740$ Mpa and $n = 0.1$. The mesh is regular with two 8-node isoparametric elements through the thickness and 48 elements along the height of the tube.

The loads are a nominal axial stress σ_1 and a nominal hoop stress σ_2. They are assumed to increase proportionally. Some loading cases were studied, corresponding to the following situations:

Table 1. *Loading cases*

Case	σ_1 (Mpa)	σ_2 (Mpa)	x
1	− 50	100	− 0.5
2	− 33	100	− 0.33
3	− 20	100	− 0.20

4	0	100	0
5	50	100	0.5

The critical values of the pressure, computed with [9], and the corresponding critical strains are given in Table 2:

Table 2. *Critical pressure and strains*

Loading cases	p_c (MPa)	ε_{1c}	ε_{2c}	Instability type
1	21.3	−0.05	0.07	**buckling**
2	23.7	−0.10	0.14	**Necking/buckling**
3	25.1	−0.079	0.13	**Necking**
4	27.7	−0.041	0.11	**Necking**
5	30.6	0.004	0.065	**Necking**

The critical strains are plotted in Figure 1 (together with experimental results presented in the next section).

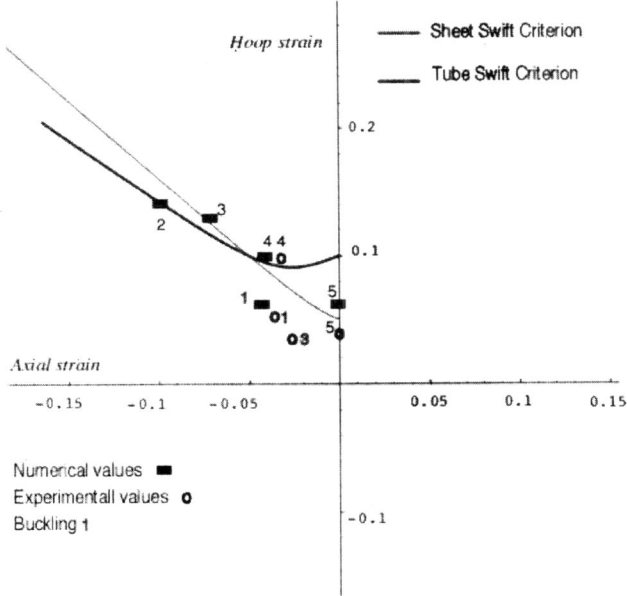

Figure 1. *Numerical and experimental critical strains*

It is obvious that the point corresponding to case 1, which is dominated by buckling, is in the stability area for the Swift criterion. The cases 3, 4 and 5, which are dominated by necking, are near the Swift curves. The case 2 is a very special: indeed it gave necking or buckling if the hardening law is slightly modified. In a further work, the sensibility of the simulation to the hardening law will be highlighted (the phenomenon is, of course, the same as for sheet-forming). Thus, we now have at our disposal a useful tool to predict any type of elastic-plastic instability associated with the simulation of hydroforming processes.

3. Experimental analysis

3.1 *Experimental setup*

In Figure 2, the complete experimental setup is presented; a desktop computer data acquisition system (LABVIEW) was used to monitor the displacement of the jack and the pressure during the whole test. Three double strain gages (axial and hoop) were mounted on each specimen.

Through the LABVIEW system, we have nine acquired data: the applied axial force, the displacement of the hydraulic jack, the pressure and the six strain-gages.

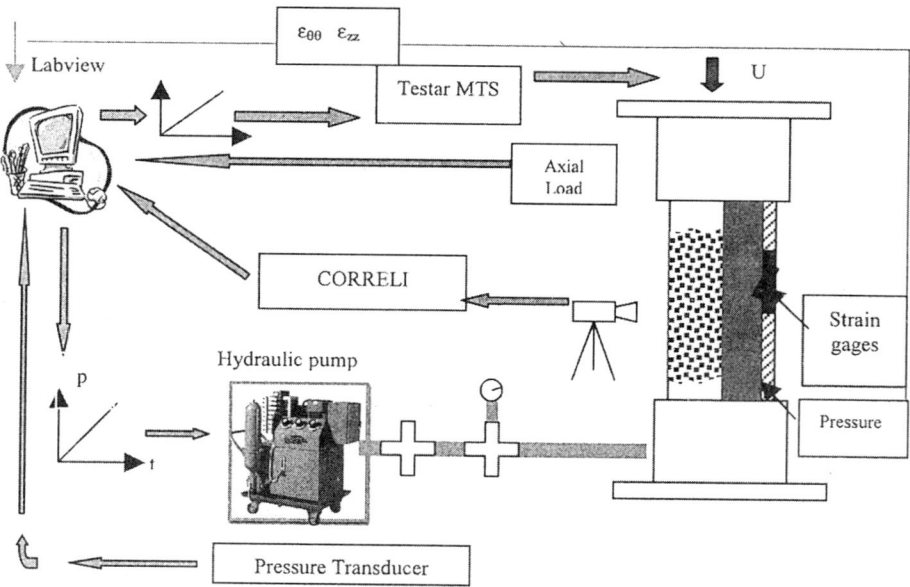

Figure 2. *Experimental setup*

3.2. Description of the test specimen

The test specimen are made of common commercial steel tubes (MATTHEY France 50X1, 5,430 TI, 1.4510), cut at the needed length (150 mm), and of two glued tips (ARALDITE 420 bi-components). These tips are connected to the test machine (MTS 250T) and allow the pressure to arrive from the hydraulic pump. The tubes have an external radius of 25 mm, and a thickness of 1.5mm. First, the material is supposed the same power law as in the numerical simulation.

3.3. Experimental results

First, determining material properties from a tension test, a true stress-strain curve has been drawn. Numerical simulations will be made in a further work with the proposed Hollomon law and with this true stress-strain curve. It will be shown, at this occasion, that the numerical results are strongly dependent of the hardening law.

Then, the same five cases as for the numerical simulation ($x = -0.5$, $x = -0.33$, $x = -0.20$, $x = 0$ and $x = 0.5$) were prepared and tested. To obtain a constant relation between the axial and hoop stresses, a preliminary simulation must be performed in each case, to obtain the curves (p, $u_{\sup}$), necessary to **drive** the pump and the jack. The tests were realized each twice, in order to test the fiability of the setup.

The first three cases lead to buckling. The cases 4 and 5 are necking test. These two different instability modes are shown in Figures 3–4.

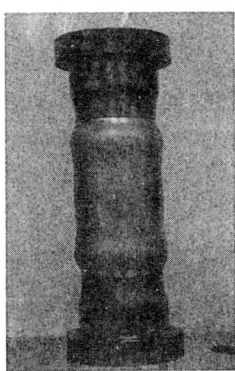
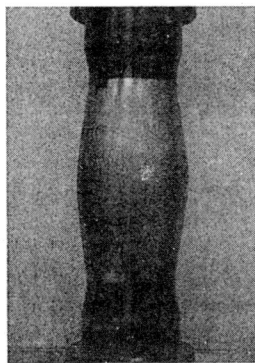

Figure 3. *Buckling*   **Figure 4.** *Necking*

The critical strains (given by the strain gages) are noted in Figure 1, together with the numerical results. Whenever the simulated and calculated modes are of the same kind, the critical strains are almost the same. Except for the case 3 ($x = -0.20$), where buckling was obtained instead of bursting), the results are in good agreement with the numerical simulation.

A possible explanation of the case 3, (buckling instead of necking), may be the following: the condition $\sigma_1 = x\, \sigma_2$ was not exactly obtained everywhere in this tube (but only near the ends), because of the bulging. Moreover, this case has been performed only once, and will have to be done again.

The experimental critical pressures are respectively $p =$ 23, 25.6, 26, 29 and 32 Mpa. This is in good agreement with the numerical simulation (see Table 2).

4. Conclusion

The so-called hydroforming Swift criterion is available to predict necking of tubes under inner pressure. But a special attention has to be paid to plastic-buckling, even though it is often disregarded for hydroforming. The axial load is generally added in order to force the material into the die, as well as to increase the thickness and delay necking. However, if this axial load is applied while necking is still a long way off, there is a risk of buckling. As shown in this paper, the critical strains corresponding to buckling are much smaller that the critical strains predicted by the necking criterion. The two failure modes are to be taken together, and a numerical simulation is a good tool to perform this.

With the complete simulation of the process, it is possible to plot hydroforming limit curves taking into accounts bursting as well as buckling. An advantage of this method is that it also enables us to differentiate between buckling and necking instabilities.

An experimental setup has been developed in order to verify these results. Most of the tests were performed twice. The experiments seem reproducible.

The first obtained results (except the case 3) are in good agreement with the numerical simulation, as well as with the analytical analysis. We have good tools to perform now "hydroforming limit curves".

Acknowledgements

The experimental part of this study was achieved with the financial support of Renault S.A. The authors are grateful to S. Godereaux for his interest in this work.

5. References

[COM 96] COMBESCURE A., "Static and dynamic buckling of large thin shells", *Nuclear engineering and design,* 1996, 92:339–354.

[COR 86] CORDEBOIS JP, LADEVÈZE P., "Sur la prévision des courbes limites d'emboutissage", *J. Mecanique théorique et appliquée*, 1986, **5**, No. 5: 341–370.

[DUR 99] DURAND S., COMBESCURE A., "Analytical and numerical study of the bifurcation of a cylindrical bar under uniaxial tension", *Revue européenne des éléments finis*, 1999, **8**, No. 7: 725–745.

[HIL 58] HILL R., "A general theory of uniqueness and stability in elastic-plastic solids", *J. Mech. Phys. Solids*, 1958, 6: 236–249.

[HIL 91] HILL R., "A theoretical perspective on in-plane forming of sheet metal", *J. Mech. Phys. Solids*, 1991, 39: 295–307.

[HIL 63] HILLIER MJ., "Tensile plastic instability under complex stress", *Int. J. Mech. Sci.* 1963, **5**: 57–67.

[NEF 02] NEFUSSI G., COMBESCURE A., "Coupled Buckling and Plastic Instability for Tube Hydroforming", *Int. J. Mech. Sci.* Under press.

[SWI 52] SWIFT HW, "Plastic instability under plane stress", *J. Mech. Phys. Solids 1952*; 1: 1–18.

[TIR 96] TIROSH J., NEUBERGER A., SHIRIZLY A., "On tube expansion by internal fluid pressure with additional compressive stress", *Int. J. Mech. Sci.*, 1996, 38: 839–851.

[TVE 89] TVERGAARD V., "Plasticity and creep at finite strains", *Theoretical and Appl. Mech, IUTAM*, 1989, 349–368.

[XIA 01] XIA Z.C., "Failure analysis of tubular hydroforming", *J. of Eng. Mat. and Tech.* ASME, 2001, **123**: 423–429.

[XIN 01] XING HL, MAKINOUCHI A., "Numerical analysis and design for tubular hydroforming", *Int. J. Mech. Sci.*, 2001, 43: 1009–1026.

[YAM 66] YAMADA Y., AOKI I., On the tensile plastic instability in axi-symmetric deformation of sheet metals. *Journal of the JSTP*, 1966, 7: 393–406.

Chapter 22

Forming Limit Curves in Blow Molding for a Polymer Exhibiting Deformation Induced Crystallization

Arnaud Poitou, Amine Ammar and Germaine Nefussi
LMT, ENS de Cachan, Cachan, France

1. Notations

$\underline{M}$	tensorial notation (M_{ij})
$\underline{M}:\underline{N}$	$Tr(\underline{M}.\underline{N}) = M_{ij} N_{ij}$
t	time
d/dt	time material derivative
$\partial/\partial t$	time partial derivative
$\delta/\delta t$	time upper convective derivative
$\widetilde{D}/\widetilde{D}t$	general notation for a convective derivative
X	thermodynamic internal variable
Y	thermodynamic dual variable
ϕ	intrinsic dissipation
ψ	free energy
ρ_0	density
$\underline{\tau}$	extra stress tensor
$\underline{D}$	strain rate tensor ($D_{ij} = (v_{i,j} + v_{j,i})/2$)
n, χ	kinetic parameters
T	temperature
α	relative degree of crystallinity
f	intrinsic degree of crystallinity
$\underline{Y}$	dual variable of elastic strain tensor
η	viscosity
$\underline{v}$	velocity field
G_0, ϕ_b	material parameter of the pompom model
λ	chain stretch
$\underline{S}$	chain orientation tensor
μ	dual variable of chain stretch
$\underline{y}$	dual variable of chain orientation tensor
θ_s	stretch relaxation time
θ_b	backbone relaxation time

2. Introduction

Blow molding is a process enabling to manufacture bottles. This process involves two steps. In a first step, a preform is injected molded and thus cooled quickly. For PET bottles this cooling time is too short to allow the material to crystallize. The preform is thus quite completely amorphous. In a second step the preform is heated, usually with infrared lights, at a temperature just above the glass temperature transition. The final shape of the bottle is then obtained with a combination of axial deformation and air blowing (Figure 1). During this final stage, the material is subjected to very high deformation (Chevalier *et al.*, 1999) and this deformation induces a crystallization, which varies form point to point with respect of the materials mechanical history (strain, strain rate, temperature, etc.).

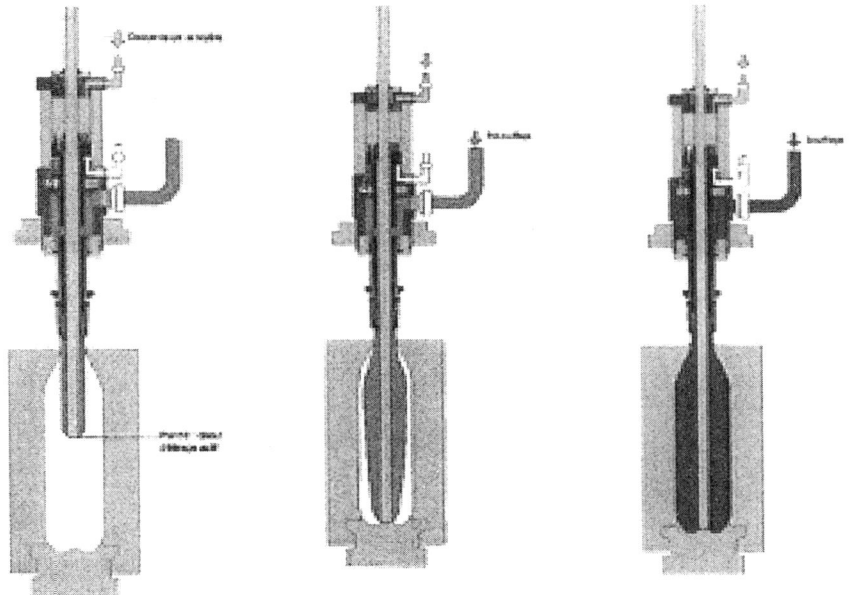

Figure 1. *Blow molding process*

The aim of this paper is not to propose a complete modelling of blow molding. We focus here on the flow induced crystallization phenomenon. In a first part we derive a framework for this modelling, which is illustrated here for a material whose constitutive relation is that of the Pom Pom model parameterized by the degree of crystallinity. In a second part, making a parallel between what is classically done for deep drawing, we analyze the occurrence of necking in terms of Considere and Swift Criterion.

3. Equations

Details of our model can be found in (Ammar, 2001, Poitou *et al.*, 2001, 2002). We give in the following the mathematical formulation of this model.

3.1. Rheology

The constitutive relation is chosen to be the Pom-Pom model, which exhibits two relaxation mechanisms: tube orientation $\underline{\underline{S}}$ and backbone stretch λ Figure 2). In its differential form, it can be written:

$$\underline{\underline{\tau}} = \frac{15}{4} G_0 \phi_b^2 \lambda^2 \underline{\underline{S}} \qquad [1]$$

$$\frac{d\underline{\underline{S}}}{dt} = \operatorname{grad} \underline{v} \cdot \underline{\underline{S}} + \underline{\underline{S}} \cdot \operatorname{grad} \underline{v}^T - 2(\underline{\underline{S}} : \underline{\underline{D}})\underline{\underline{S}} - \frac{1}{\theta_b}(\underline{\underline{S}} - \frac{1}{3}\underline{\underline{1}}) \qquad [2]$$

$$\frac{d\lambda}{dt} = \lambda(\underline{\underline{S}} : \underline{\underline{D}}) - \frac{1}{\theta_s}(\lambda - 1) \qquad [3]$$

Figure 2. *Branched polymer in a tube*

3.2. Quiescent crystallization

Following Avrami model written in a differential form for non isothermal conditions, we assume that the relative degree of crystallinity α (ratio between the actual degree of crystallization and the ultimate or maximum degree of crystallization allowed by the polymer structure), in quiescent conditions (*i.e.* without any deformation of the material), increases with respect to time as:

$$\begin{cases} \alpha = 1 - Exp(-f) \\ \dfrac{df}{dt} = \dfrac{n.f^{1-1/n}}{\lambda(T)} \end{cases} \qquad [4]$$

3.3. Deformation induced crystallization

In our formalism, strain (or flow) induced crystallization is directly deduced from equations [1–4]. If we assume indeed that the rheological parameters depend on the degree of crystallinity, a model of flow-induced crystallization is "naturally" associated to the rheology as soon as one writes the rheological model into a thermodynamical form. In our formalism, for the Pom Pom model, this strain induced crystallization model writes:

$$\dot{f} = -\frac{\partial}{\partial \Lambda}\left(\frac{1}{\theta_b(\Lambda)}\right)\left(-\frac{H}{6}\ln\det\frac{-2\underline{\underline{y}}}{H} - \frac{1}{3}\text{Tr}(\underline{\underline{y}}) - \frac{H}{2}\right)$$

$$- \frac{\partial}{\partial \Lambda}\left(\frac{1}{\theta_s(\Lambda)}\right)\int_0^\mu \left(\frac{m}{H} - 1 + \sqrt{1 + (\frac{m}{H})^2}\right) dm \qquad [5]$$

$$+ \frac{n}{\lambda(T)} f^{1-\frac{1}{n}}$$

$$H = \frac{15}{4} G_0\, \phi_b^2 \qquad [6]$$

$$\underline{\underline{y}} = \frac{\partial \psi}{\partial \underline{\underline{S}}} = -\frac{1}{6} H\, \underline{\underline{S}}^{-1} \qquad [7]$$

$$\mu = \frac{\partial \psi}{\partial \lambda} = H\left(\lambda - \frac{1}{\lambda}\right) \qquad [8]$$

$$\Lambda = \frac{\partial \psi}{\partial f} \approx L\, x_\infty\, (1 - \frac{T}{T_m})\exp(-f) \qquad [9]$$

Equation [5] means that the crystallization is induced by three different sources. The last one is the classical thermal activation. The first one means that crystallization occurs as soon as (i) the molecules are oriented and (ii) the orientation relaxation time increases with respect to the degree of crystallinity. In a very similar way, the second term means that crystallization occurs as soon as (i) the molecules are stretched and (ii) the stretch relaxation time increases with respect to the degree of crystallinity.

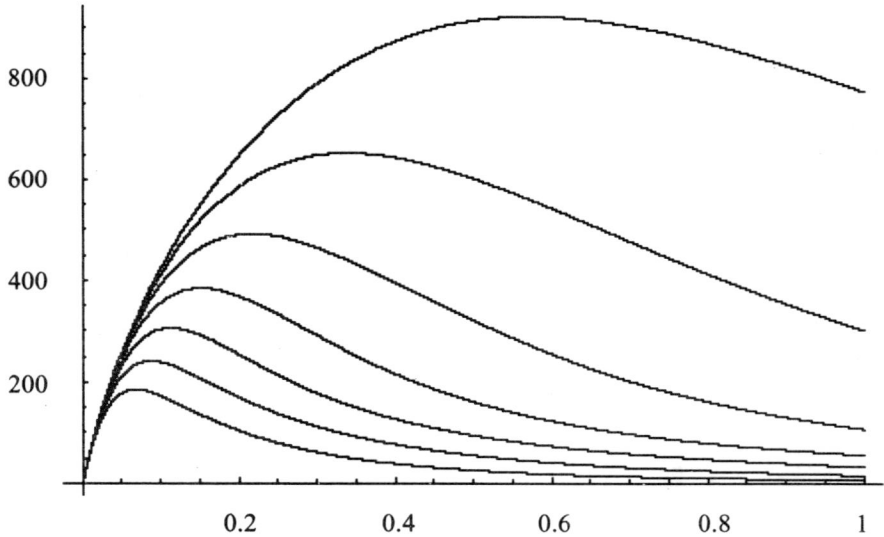

Figure 3. *Tensile test for increasing tensile velocities, without strain induced crystallization (Relaxation times $\theta_b=911$ s, $\theta_s=150$ s, initial length $L = 0.1$ m, tensile velocity $V=10^{-4}$, $2\ 10^{-4}$, $4\ 10^{-4}$, $8\ 10^{-4}$, $1.6\ 10^{-3}$, $3.2\ 10^{-3}$, $6.4\ 10^{-3}$ m/s)*

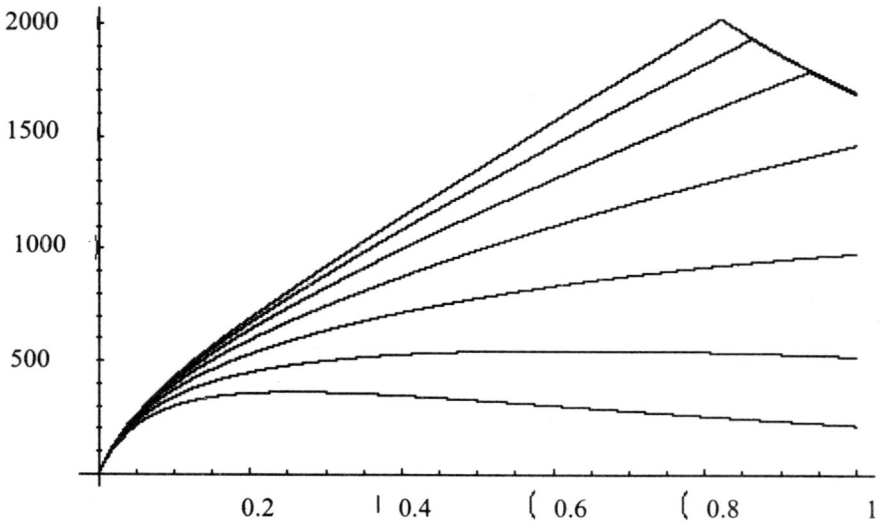

Figure 4. *Tensile test for increasing tensile velocities, for strong strain induced crystallization (Relaxation times $\theta_{b0}=911$ s, $\theta_{s0}=150$ s, $\theta_b = \theta_{b0}(1+A\ \alpha)$, $\theta_s = \theta_{s0}(1+A\ \alpha)$, $A= 100$, initial length $L = 0.1$ m, tensile velocity $V=10^{-4}$, $2\ 10^{-4}$, $4\ 10^{-4}$, $8\ 10^{-4}$, $1.6\ 10^{-3}$, $3.2\ 10^{-3}$, $6.4\ 10^{-3}$ m/s)*

4. Results and discussions

4.1. Tensile test

A tensile test can be simulated in solving simultaneously equations [1–3] and [5–9] for a constant tensile velocity. Figure 3 shows a typical tensile curve without any strain-induced crystallization for increasing tensile velocities. The effect of velocity is classically to increase the force and to allow for a larger deformation without necking. Figure 4 shows the same tensile curves for the same rheological parameters but with strain induced crystallization. This situation evidences a strain hardening and a shift of the limit deformation before necking, that increase both with crystallization and strain rate. This strain hardening is effectively due to crystallization as can be seen on Figure 5, where are plotted the relative degree of crystallinity during the test.

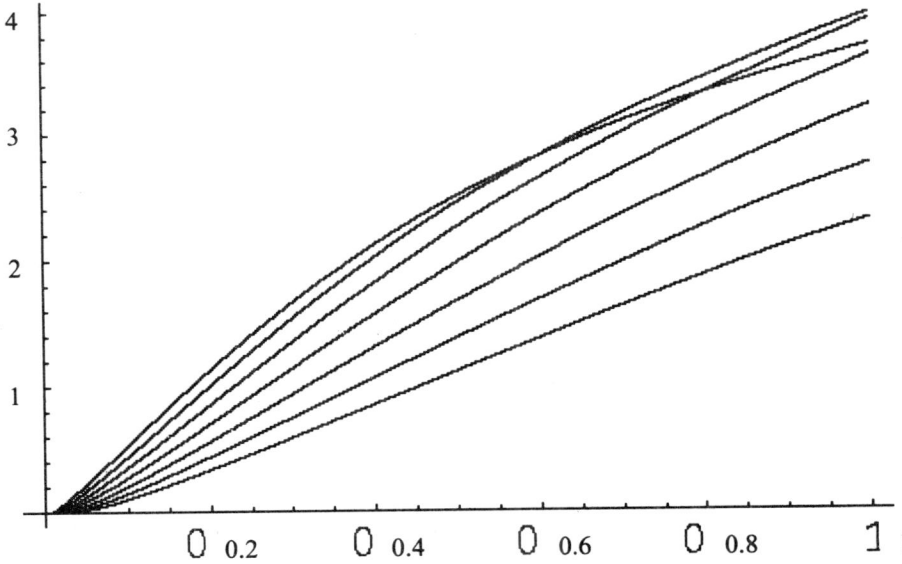

Figure 5. *Relative degree of crystallinity during the tests illustrated in Figure 4*

4.2. Bidimensional test

In order to see the influence of bidimensional deformations, we have solved the equations for constant tensile velocities along two directions. We have then plotted the "limit curve" associated to the maximum of the force along one direction. Figures 6 and 7 show these limit curves for different velocities, without and with strain induced crystallization. In comparison with the Swift criterion, if we associate these limit curves to a forming limit curve, we see the positive role of crystallization, which acts as a necking stabilizer.

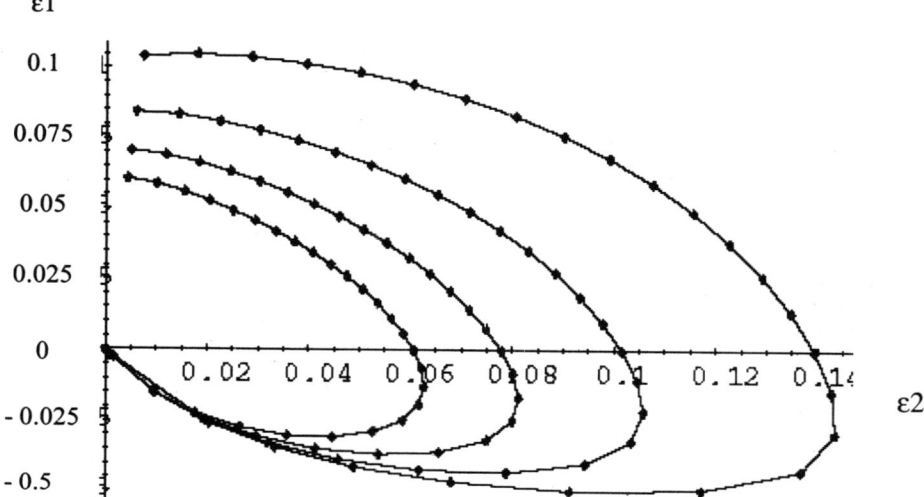

Figure 6. *"Limit" curve without crystallization*

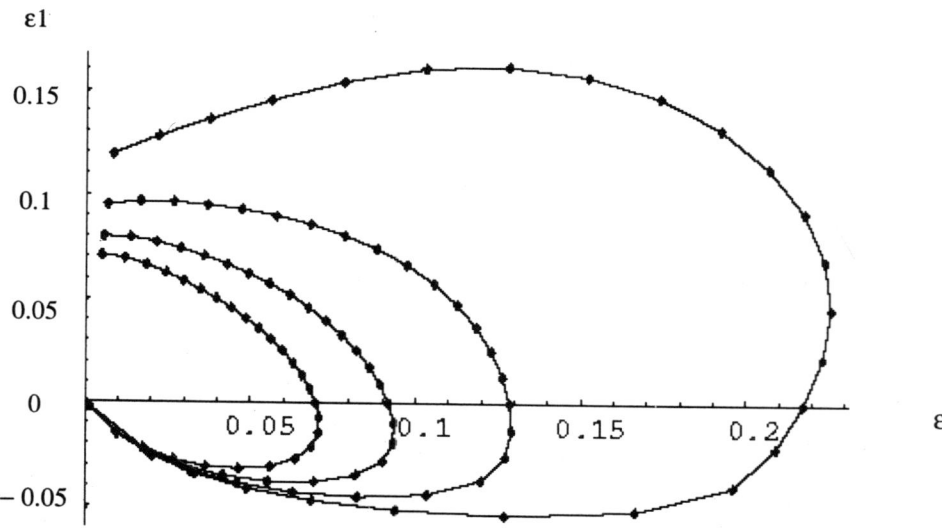

Figure 7. *"Limit"curve with crystallization, (Relaxation times $\theta_{b0} = 911s$, $\theta_{s0}=150s$, $\theta_b = \theta_{b0}(1+A\alpha)$, $\theta_s = \theta_{s0}(1+A\alpha)$, $A= 10$, initial length $L = 0.1$ m, tensile velocity $V=10^{-4}, 2\ 10^{-4}, 4\ 10^{-4}, 8\ 10^{-4}, 1.6\ 10^{-3}, 3.2\ 10^{-3}, 6.4\ 10^{-3}\ ms^{-1})$*

5. Conclusion

We have shown in this paper a model for flow-induced crystallization associated to a molecular model of the viscoelastic constitutive equation. This model is to be compared to McHugh one (Doufas *et al.*, 1999). It allows us to describe both a strain hardening during plane deformation and qualitative forming limit curves in two dimensions that outline the stabilizing role of strain-induced crystallization. This model remains qualitative for blow molding because (i) it does not contain any yield stress (ii) a precise stability analysis has not been done yet. However these new development are not out of reach.

6. References

Ammar A., Modélisation numérique de la cristallisation induite par l'écoulement d'un thermoplastique. Application à l'injection, PhD Thesis, ENS de Cachan France 2001.

Chevalier L., Linhone C., Regnier G., "Induced crystallinity during stretch-blow molding process and its influence on mechanical strength of poly(ethylene terephthalate) bottles", *Plastic, Rubber and Composite,* 1999, 28(8):385–392.

Doufas A.K., Dairanieh I.S., McHugh A.J., "A continuum model for flow induced crystallization of polymer melts", *J. Rheol,* 1999, 43(1):85–109.

Poitou A., Ammar A., "Polymer crystallization induced by flow: a thermodynamic approach", *C. R. Acad. Sci.,* 2001, 329(II b):5–11.

Poitou A., Ammar A, "A molecular model for flow induced crystallization of polymers", to appear in *Macromolecular symposia,* 2002.

Chapter 23

Modelling of Thin Sheet Blanking with a Micromechanical Approach

Application of the MTS Model

Christophe Poizat, Said Ahzi and Nadia Bahlouli
Institut de Mécanique des Fluides et des Solides, Université Louis Pasteur, Strasbourg, France

Laurent Merle
FCI, La Ferté Bernard, France

Christophe Husson
Institut de Mécanique des Fluides et des Solides, Université Louis Pasteur, Strasbourg, France, and FCI, La Ferté Bernard, France

1. Introduction

Blanking is a widely used sheet metal forming operation. In this process, a blank is separated from the sheet by a punch as shown in Figure 1. In engineering practice, the resulting shape of the product and the internal stresses in the damaged area near the cut edge are important properties towards quality. From a geometrical point of view, four characteristic zones can be distinguished (Figure 1):

– the roll-over zone: part drawn into the sheet by the punch,

– the sheared zone: zone that appears before ductile fracture,

– the fracture zone: this zone results from the propagation of a ductile crack. The sheared zone is quite smooth, whereas the fracture zone is rough,

– a burr: it depends mainly on the locus of the crack initiation. Burr can lead to cracks in subsequent forming operations like flanging (Kalpakjian, 1997).

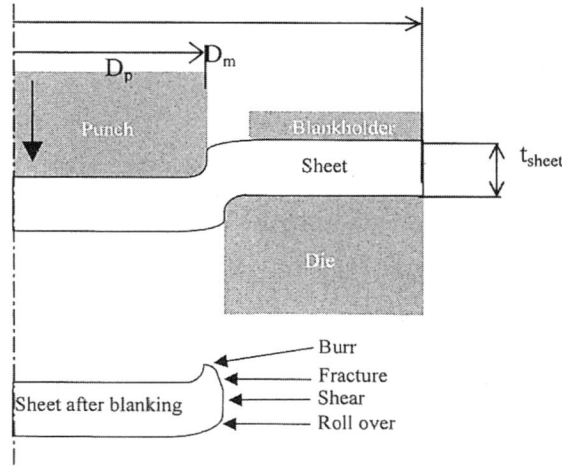

Figure 1. *Blanking process and blank product*

For a given material, several process parameters are known to greatly influence the shape and quality of the product (Brokken *et al.*, 1998; Kalpakjian, 1997):

– the blanking clearance, defined as $c = 100\,(D_m - D_p)/t_{sheet}$ in % where D_m, D_p and t_{sheet} are respectively the die diameter, the punch diameter and the sheet thickness. In practice, clearances usually range between two and ten percent of the sheet thickness,

– the edge condition,

– the wear state of the tool, which depends on lubrication,

– the thickness of the sheet, t_{sheet}. In this contribution, we consider thin sheets, (thickness between 0.1 mm and 1 mm),

– the punch force,

– the punch displacement is more or less sinusoidal, with an amplitude between 10 and 20 mm. The stroke rates typically lie between 50 and 1500 blanking operations/min, with a current manufacturing trend to higher stroke rates (up to 3000 operations/min),

– the resulting speed of the punch in the cutting phase, V_{punch}.

For the above mentioned stroke rates, the resulting strain rate range can be roughly evaluated with the following analytical equation (Kalpakjian, 1997):

$$\dot{\varepsilon} = V_{punch}/t_{sheet} \qquad [1]$$

where the punch speed is supposed to be constant during the cutting phase. It shows that, the higher the punch speed or the smaller the sheet thickness are, the higher the strain rates. Usually (Kalpakjian, 1997), typical strain rates for sheet metal forming lie between 1 and 100 s^{-1}. However, since the trends are to higher stroke rates and small sheet thickness for industry branches like the electronic industry, the strain rates can reach several 1000 s^{-1} in the shearing zone as detailed in section 3 (Figure 4). This requires advanced material models. It is namely well known that strain rate effects are involved in many metals (François *et al.*, 1991), and that many ductile metals display an enormous increase in yield stress for strain rates in excess of 10^3 s^{-1} (Follansbee and Kocks, 1988). Moreover, according to (Kalpakjian, 1997), the shearing zone is affected by the temperature rise, that may reach 300°C at slow speed and 600°C at high speed.

Adiabatic heating and temperature effects as well as strain rates effects have hence to be taken into account in the simulation of thin sheet blanking processes, over a wide range of temperatures and strain rates.

In this paper, we propose to use the Mechanical Threshold Stress (MTS) model (Follansbee and Kocks, 1988), with adiabatic heating, to take the viscous and temperature effects into account. With the help of this model and a Finite Element analysis, our aim is to get a better understanding of the blanking process of thin sheets.

In the next section, we describe the material model, that we have implemented in the FE code ABAQUS/Explicit (ABAQUS, 2001). In section 3, simulations of the blanking of a thin sheet of OFHC (Oxygen Free High Conductivity) copper are presented and discussed.

2. Material model

The Mechanical Threshold Stress model (Follansbee and Kocks, 1988) is based on the theory of dislocations motion. In other words, such a model attempts to establish a direct relation between the macroscopic behaviour and the mechanism of dislocations motion. This allows for a physically-based description of the material behaviour, over a large range of temperature and strain rates.

The MTS model consists more precisely of a description of the material flow stress at constant structure and a description of structure evolution during deformation. "Structure" is to be understood as "structure of the material at a microscopic level, accounting for the dislocations and grains structure". A reference threshold stress, $\hat{\sigma}$, is introduced as internal state variable. It represents a measure of the isotropic resistance to plastic flow of the material at its current dislocation structure (dislocation density, dislocation substructure, obstacles, etc.).

The mechanical threshold stress is separated into two components, an athermal stress $\hat{\sigma}_a$ and a thermal stress $\hat{\sigma}_t$. $\hat{\sigma}_a$ characterizes the rate independent interactions of dislocations with long range barriers such as grain boundaries. $\hat{\sigma}_t$ characterizes the rate dependent interactions with short range barriers. On the contrary to interactions with long range obstacles, the interactions with short range obstacles depend on temperature: the short range obstacles are easier to overcome by dislocations with an increase of the thermal energy. Finally, Follansbee and Kocks (1988) derived the following expression for the yield stress σ_y:

$$\sigma_y = \hat{\sigma}_a + (\hat{\sigma} - \hat{\sigma}_a)\left(1 - \left[\frac{kT\ln\left(\frac{\dot{\varepsilon}_0}{\dot{\varepsilon}_{eq}^p}\right)}{g_0 G b^3}\right]^{\frac{1}{q}}\right)^{\frac{1}{p}}$$ [2]

In equation [2], $\dot{\varepsilon}_{eq}^p$ is the equivalent plastic strain rate. g_0 is the normalized activation energy. G is the shear modulus, b is the magnitude of the Burgers vector. p and q are constants, and characterize the statistically averaged shape of the obstacle profile. The hardening law is given by equations [3], [4] and [5].

$$\theta = \frac{d\hat{\sigma}}{d\varepsilon_{eq}^p} = \theta_0 - \theta_r(T, \dot{\varepsilon}, \hat{\sigma})$$ [3]

where $\theta_r = \theta_0 F\left(\frac{\hat{\sigma} - \hat{\sigma}_a}{\hat{\sigma}_S(T,\dot{\varepsilon}) - \hat{\sigma}_a}\right)$ [4]

and F(x) = tanh(2X)/tanh2 [5]

This structure evolution is expressed as a change of the Mechanical Threshold Stress with strain. The hardening θ is the result of two competing processes: dislocation accumulation (θ_0) and dynamic recovery rate ($-\theta_r$). In equation [4], the function F is chosen to fit experimental data. It characterizes strain hardening for fcc materials with saturation-like behaviour (Follansbee and Kocks, 1988). The saturation level is described by the saturation stress $\hat{\sigma}_S$, which depends on temperature and strain rate, as shown in equation [6]. In this equation, $\dot{\varepsilon}_{S0}$, A and $\hat{\sigma}_{S0}$ are constants. $\hat{\sigma}_{S0}$ is the saturation threshold stress for deformation at 0 K.

$$\frac{\hat{\sigma}_S}{\hat{\sigma}_{S0}} = \left(\frac{\dot{\varepsilon}_{eq}^p}{\dot{\varepsilon}_{S0}}\right)^{\frac{kT}{Gb^3 A}} \qquad [6]$$

The MTS accounts for the softening due to adiabatic heating. In this case, the temperature rise ΔT is given by (see Table 2 for the definition of X, ρ, C_p):

$$\Delta T = \frac{X}{\rho C_p} \int \sigma_{ij} d\varepsilon \qquad [7]$$

where σ is the Cauchy stress tensor and $d\varepsilon^p$ is the plastic strain increment tensor.

Experimental studies are required to quantify the input parameters. In Table 1, we review two sets of parameters for OFHC copper (Follansbee and Kocks, 1988; Tanner et al., 1999). The material constants for OFHC Copper are listed in Table 2.

Table 1. *MTS model parameters for OFHC copper*

Parameters	(Follansbee and Kocks, 1988) MTS1	(Tanner et al., 1999) MTS2
Deformation conditions		
Strain rate $\dot{\varepsilon}$	10^{-4} to 10^4 s^{-1}	4×10^{-4} to 6×10^3 s^{-1}
Test temperature	25 °C	25 to 269 °C
Grain size	40 μm	62 μm
Normalized activation energy: g$_0$	1.6	2.98
p	2/3	0.318
q	1	1.2
$\dot{\varepsilon}_0$ [s^{-1}]	10^7	10^7
k/b^3 [MPa/K]	0.823	0.848
A	0.31	0.633
Stress at 0 K: σ$_{S0}$ [MPa]	900	1070
Saturation strain rate at 0 K: $\dot{\varepsilon}_{S0}$ [s^{-1}]	6.2 10^{10}	4.0 10^{23}
Initial strain hardening rate: θ$_0$ [MPa]	2315	2150+0.034 $\dot{\varepsilon}$
Athermal stress: $\hat{\sigma}_a$ [MPa]	40	40

Table 2. *Material constants for OFHC copper*

Constants	MTS1 and MTS2
Young's modulus: E [GPa]	124
Poisson's coefficient: ν	0.34
Initial yield stress: σ_0 [MPa]	90
Density: ρ[Kg/m³]	8960
Melting temperature: T_m [K]	1356
Test temperature: T_0 [K]	300
Inelastic heat fraction: X	0.9
Specific heat: C_p [J/kg K]	385

Numerous comparisons with the works of Follansbee and Kocks (Follansbee and Kocks, 1988) and Tanner et al. (Tanner *et al.*, 1999) with the help of a one element test show that the MTS subroutine in the FE code ABAQUS/Explicit is correctly implemented (Poizat and Ahzi, 2001). In Figure 2, we present one of these verification tests. A four nodes plain strain element with reduced integration (CPE4R element of ABAQUS) is stretched at several constant strain rates up to rupture. The parameter set MTS1 is used. Rupture is modelled by combining the MTS model with a strain based rupture criterion known as the Johnson-Cook criterion (Johnson and Cook, 1983). The Johnson-Cook rupture criterion is given by equations [8] and [9]:

$$\varepsilon_{cr} = \left(d_1 + d_2 \exp(d_3 \frac{\sigma_H}{\sigma_{eq}})\right)\left(1 + d_4 \ln\left(\frac{\dot{\varepsilon}_{eq}^p}{\dot{\varepsilon}_r}\right)\right)\left(1 + d_5 \frac{T - T_0}{T_m - T_0}\right) \quad [8]$$

$$w = \sum \frac{\Delta \varepsilon_{eq}^p}{\varepsilon_{cr}} \quad [9]$$

In [8], σ_H is the pressure stress, σ_{eq} is the Mises equivalent stress and $\dot{\varepsilon}_r = 1 s^{-1}$ is the reference strain rate. d_1, d_2, d_3, d_4 and d_5 are material constants and are given in (Johnson and Cook, 1983) for OFHC copper. In equation [9], $\Delta\varepsilon_{eq}^p$ is the increment of equivalent plastic strain. When w reaches 1, the stresses are set to zero, simulating rupture.

In Figure 2, the strain rate effect on our simulated results for the Mises equivalent stress is in good agreement with the experimental results of Follansbee and Kocks (1988). The slight increase of the strain at rupture with increasing strain rate for OFHC copper (Johnson and Cook, 1983) is modelled. Note that we impose w = 0.7 for rupture in order to model a smaller rupture strain than in the work of Johnson and Cook (1983). Our choice of w<1 is in accordance with several suggestions in damage mechanics literature (François *et al.*, 1991).

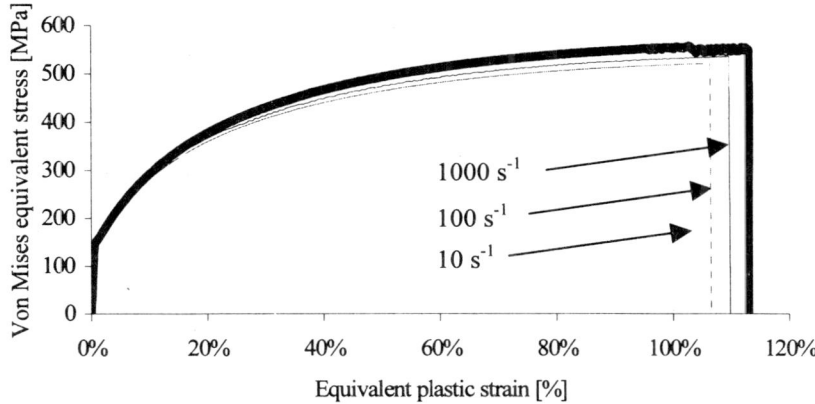

Figure 2. *One element test at $10\ s^{-1}$, $100\ s^{-1}$ and $1000\ s^{-1}$ constant strain rate for the MTS combined with the Johnson Cook rupture criterion*

3. 2D (plain strain) Finite Element simulation of the blanking process

In this section, we present and discuss results obtained from the Finite Element simulation of the blanking process of an OFHC thin sheet with the MTS model combined with the Johnson-Cook rupture criterion. The sheet thickness is of 0.58 mm and two punch velocities (0.05 and 0.5 m/s) are imposed. The punch and die radii are equal to 0.01 mm, which is representative of new tools (undamaged tools). The clearance is equal to 8.6%. Other geometrical data are standard data in the industry (see section 1) and are not necessary for our purposes. The simulation is two dimensional. The elements are CPE4R plain strain elements. Even though the experimentally observed deformation is not perfectly plain strain, this choice of elements is relevant for the simulation of blanking (Stegeman, 1999). The mesh remains relatively coarse in this study (50x20 in the blanking area, corresponding to an element size of 10 microns) to get reasonable CPU time but sufficiently small for good convergence. The tools are meshed with analytical rigid surface (Abaqus, 2001). The friction law is a Coulomb friction law with a coefficient of 0.02.

In Figure 3, we show the "damage" distribution at rupture in the case of a punch velocity of 0.5 m/s when the punch impacts the sheet. The element, where rupture occurred, are deleted. The roll over, shear and fracture zone are observed. No burr is obtained, which is in agreement with experimental works (Brokken *et al.*, 1998) with small clearances and undamaged tools.

These first results have shown that the values of the temperature due to adiabatic heating lies between 80 and 200 °C in the localization band. This is in accordance with experimental works (Kalpakjian, 1997). In Figure 4, we show the typical strain rate for a punch velocity of $0.5\ ms^{-1}$ (corresponding to 1000 strokes/min.) in the localization band (see the reference element shown in Figure 3). It's worth noting that the analytical approach (equation [1]) gives a good order of magnitude for the average strain rate in the blanking but underestimate the strain rate increase leading to fracture. Since ductile metals like copper change in their strain-stress behaviour around 1000 s^{-1}, this shows the necessity to use the MTS model for the blanking simulation at high stroke rates and/or for small sheet thickness.

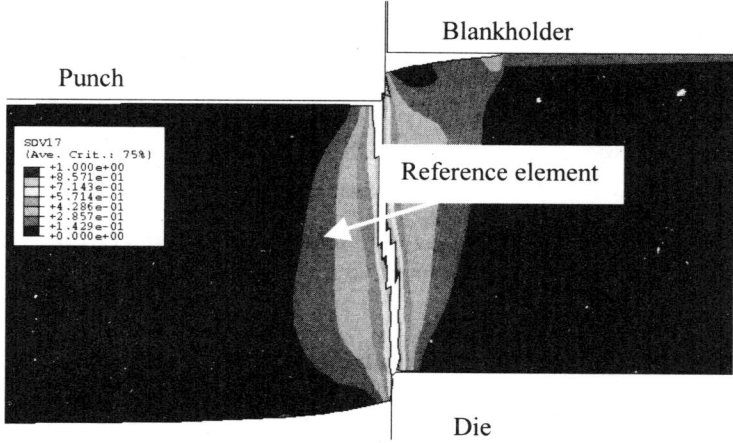

Figure 3. *Example of a typical "damage" distribution (w) at rupture*

The punch force-relative displacement curves for the two stroke rates are different as shown in Figure 5 (The relative punch displacement is defined as the ratio of the punch displacement and the sheet thickness). In this Figure, comparison with no adiabatic heating (isothermal case) is shown. These results show that adiabatic heating and strain rate effects have to be taken into account in a Finite element simulation of the blanking process. Moreover, depending of the sheet thickness and punch velocity, since the strain rates lie between quasistatic and very high strain rates, the MTS model is a pertinent model for the simulation of thin metal sheet blanking. This model is namely known to model strain rate effect over a wide range of strain rates as mentioned in section 2. The influences of the punch radii or clearances is established (Brokken *et al.*; 1998; Hambli, 1996). Using this model should help understand why the observed profiles of the cut edge are different when changing the stroke rate or the sheet thickness.

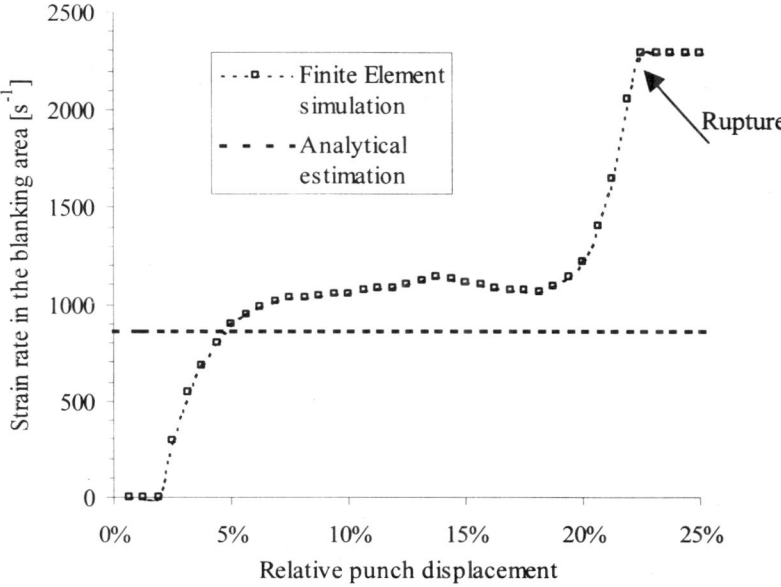

Figure 4. *Strain rate in the blanking zone (reference element in Figure 3) for a stroke rate of 1000 strokes/min. and a sheet thickness of 0.58 mm*

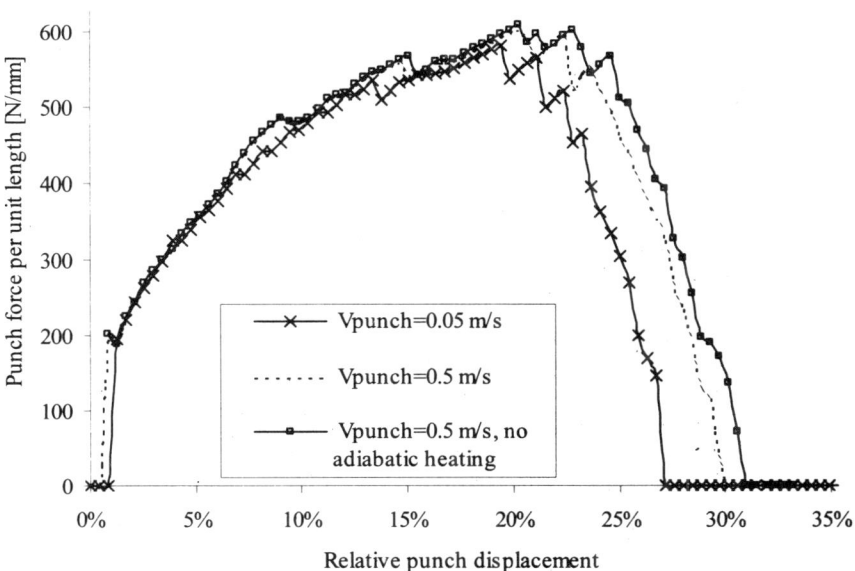

Figure 5. *Punch force per unit length versus displacement for two stroke rates*

4. Conclusion

In this contribution, we have shown that the large range of strain rates involved during the blanking process of thin sheet (thickness below 1 mm) requires the use of an advanced material model like the Mechanical Threshold Stress model. Combined with the Johnson-Cook strain based rupture criterion, this model is able to capture the influence of high strokes rates and small thicknesses involved in growing industry branches like the electronic industry. In a short future, parametric study with finer meshes will be undertaken to evaluate the influence of several parameters like the clearance, the punch radius, the stroke rates and the sheet thickness. In order to accurately model the damage, we also intend to couple the MTS material model with a physical based damage model. The aim is to model the transition between ductile failure at low strain rate and pure adiabatic shear band localization at high stroke rates.

5. References

Abaqus Manuals 5.8, Hibbitt, Karlsson and Sorensen, Inc., 2001.

Brokken D, Brekelmans W.A.M., Baaijens F.T.T., "Numerical modelling of the metal blanking process", *J. of Materials Processing Technology*, No. 83, 1998, pp. 192–199.

Follansbee P.S., Kocks U.F., "A constitutive description of the deformation of copper based on the use of the mechanical threshold stress as an internal state variable", Acta Metallurgica, 36 (1), 1988, 81–93.

François D., Pineau A., Zaoui A., *Comportement mécanique des matériaux*, tome 1 et 2, ed. Hermes, Paris, 2e édition, 1991.

Johnson G.R., Cook W.H., "A constitutive model and data for metals subjected to large strains, high strain rates and high temperatures", *Proc. 7th Int. Symp. on ballistics*, The Hague, Netherlands, 1983, pp. 541.

Hambli R., Etude expérimentale, numérique et théorique du découpage des tôles en vue de l'optimisation du procédé, Thèse de génie mécanique, ENSAM Angers (F), N°96.24 1996, p. 286.

Hammi Y., Simulation numérique de l'endommagement dans les procédés de mise en forme, Thèse, Université de Technologie de Troyes (UTT), Troyes, 2000, p. 241.

Kalpakjian S., "*Manufacturing processes for Engineering Materials*", Addison Welley Longman Inc., Menlo Park, CA (USA), 3rd ed., ISBN 0-201-82370-5, 1997, p. 950.

Poizat C., Ahzi S., Numerical simulation of cold metal forming: blanking, coining, bending, internal Report, Université Louis Pasteur, Strasbourg, France 2001, p. 48.

Stegeman Y.W. et al., "An experimental and numerical study of a planar blanking process", *J. of Mat. Proc. Tech.*, 87, 1999, pp. 266–276.

Tanner A.B., McGinty R.D., McDowell D.L., Modelling temperature and strain rate history effects in OFHC Cu, *Int. J. of Plasticity*, 15, 1999, pp. 575–603.

Chapter 24

Anisotropy in Thin, Canning Sheet Metals

David W.A. Rees
Department of Systems Engineering, Brunel University, Middlesex, UK

1. Introduction

Sheet metals used to produce metal cans possess a wide variation in strength and ductility depending upon the components produced from them. Body material requires ductility for ironing and final stiffness. Contoured ends and stiffened tabs require only moderate formability for shallow pressing. This paper examines the properties of two, nominal 3 mm thick materials under simple tension and bulge forming. Tensile strengths of heavily rolled canning steels lie in the range 450–550 MPa with approximately 5% elongation at fracture. Body material displays Hollomon hardening in tension ($n = 0.10$) while end material approximates more closely to elastic-perfectly behaviour. Bulging through elliptical apertures provides the hardening parameters in biaxial stress applications where the ductility of each material is extended. In addition to using r-values, the spread in flow curves, found from tension in different orientations, indicates the degree of in-plane anisotropy. Tension and bulge flow curves remain distinctly different. A satisfactory degree of equivalence between them is found from the Hill (1948) theory. However, to improve predicted flow behaviour under uniaxial and biaxial stress states, the Hollomon law should be based upon the acting stress state.

2. Theory

Referring to figure 1a, the rolling direction 1 is inclined at θ to the axis 1' of a tensile test piece, along which the stress is σ. Also, let axes I and II lie perpendicular and parallel respectively, to the direction of a local neck. The inclination of I to 1' is shown as α_θ indicating its dependence upon the rolling direction θ.

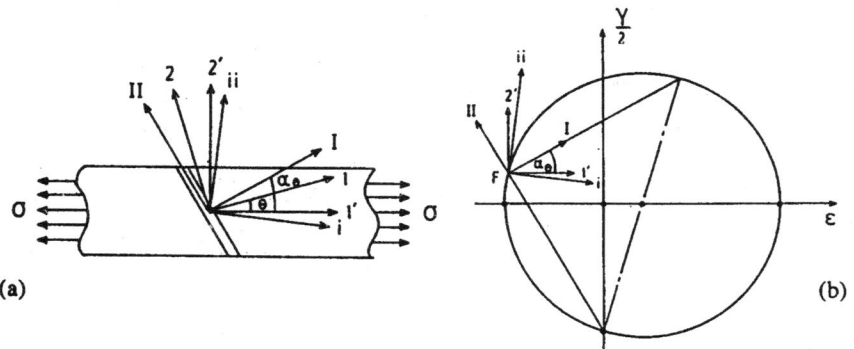

Figure 1. *Tensile neck showing (a) co-ordinates and (b) Mohr's circle*

The neck forms under plane strain, *i.e.* the direct strain along the II direction is zero. This strain is derived from a transformation of 3 plastic strain increments, $d\varepsilon_{1'1'}^{P}$, $d\varepsilon_{2'2'}^{P}$ and $d\gamma_{1'2'}^{P}$, lying along the test piece axes 1' and 2':

$$d\varepsilon_{1'1'}{}^P \sin^2\alpha_\theta + d\varepsilon_{2'2'}{}^P \cos^2\alpha_\theta - \tfrac{1}{2} d\gamma_{1'2'}{}^P \sin 2\alpha_\theta = 0 \qquad [1]$$

An incremental shear strain $d\gamma_{1'2'}{}^P$ will arise when axes 1 and 1' do not coinside. We may use trigonometric identities to write equation[1] as a quadratic in $\cos\alpha_\theta$:

$$(s^2 + q^2)\cos^2 2\alpha_\theta - 2pq \cos 2\alpha_\theta + (p^2 - s^2) = 0 \qquad [2]$$

where: $p = 1 + R_\theta$, $q = 1 - R_\theta$ and $s = T_\theta$. Now, $R_\theta = d\varepsilon_{2'2'}{}^P/d\varepsilon_{1'1'}{}^P$ and $T_\theta = d\gamma_{1'1'}{}^P/d\varepsilon_{1'1'}{}^P$ are test piece plastic strain increment ratios, which depend upon θ (Rees, 1995):

$$R_\theta = \{r_0 r_{90} - r_{45}(r_0 + r_{90})\sin^2 2\theta - 2r_0 r_{90}(\sin^4\theta + \cos^4\theta)\}$$
$$\div \{2r_{90}(1 + r_0)\cos^4\theta + 2r_0(1 + r_{90})\sin^4\theta - \tfrac{1}{2}[2r_0 r_{90} - (1 + 2r_{45})(r_0 + r_{90})]\sin^2 2\theta\}$$
$$[3a]$$

$$T_\theta = -\{[2r_0(1 + 2r_{90})\sin^2\theta - 2r_{90}(1 + 2r_0)\cos^2\theta + (1 + 2r_{45})(r_1 + r_2)\cos 2\theta]\sin 2\theta\}$$
$$\div \{2r_{90}(1 + r_0)\cos^4\theta + 2r_0(1 + r_{90})\sin^4\theta - \tfrac{1}{2}[2r_0 r_{90} - (1 + 2r_{45})(r_0 + r_{90})]\sin^2 2\theta\}$$

$$[3b]$$

in which r_0, r_{45} and r_{90} define width to thickness strain ratios for tension tests with $\theta = 0°$, $45°$ and $90°$ orientations respectively. Equations [3a] and [3b] combine an orthotropic, quadratic flow potential (Hill, 1948) with stress/strain transformation equations. It follows from equations [3a]–[3b], that the solution to the quadratic equation [2] enables these three r-values to be found from measured inclinations of the neck in failed test pieces. That is

$$\cos\alpha_\theta = \{pq \pm \sqrt{[s^2(q^2 - p^2 + s^2)]}\}/(q^2 + s^2) \qquad [4]$$

This method is suggested where, in heavily rolled materials with limited ductility, r-values cannot be found reliably by using extensometers. For example, setting $\theta = 0°$ and $90°$ in equations [3a] and [3b], the shear strain along tensile 1' axis disappears, i.e. $s = 0$, and simplified relationships apply:

$$r_0 = (1 - \cos 2\alpha_0)/(2\cos 2\alpha_0) \qquad [5a]$$

$$r_{90} = (1 - \cos 2\alpha_{90})/(2\cos 2\alpha_{90}) \qquad [5b]$$

Setting $\theta = 45°$ the solution to the quadratic equation [2] becomes

$$\cos 2\alpha_{45} = [AB^2 \pm C\sqrt{(A^2 B^2 - D)}]/(C^2 + A^2 B^2) \qquad [6]$$

where $A = 1 + 2r_{45}$, $B = r_0 + r_{90}$, $C = r_0 - r_{90}$ and $D = 4r_0 r_{90}$. Equation [6] may be solved for r_{45} knowing r_0 and r_{90} from equations [5a] and [5b], together with the measured α_{45} from a test piece with a $45°$ orientation. If $r_0 \approx r_{90}$, then the simplified solution to equation [6] will require only α_{45}. That is: $r_{45} = (1 - \cos 2\alpha_{45}) / (2 \cos 2\alpha_{45})$.

3. Test methods

Rolled steel sheets 0.23 mm and 0.19 mm thick were supplied by the canning industry in their tempered and lacquered condition prior to manufacture. Tension tests were performed on dumbbell shaped test pieces with parallel gauge dimensions 15 mm wide × 110 mm long. Their ends were enlarged to 25 mm wide with 5 mm transition radii to fit within the wedge grips of a screw-driven, tensile test machine. An extensometer was placed upon a 50 mm gauge length from which displacement was measured. A lateral extensometer was used where appropriate. Load was monitored continuously with a load cell as the cross-head moved the grips apart at a rate of 1 mm/min. The machine supplied a simultaneous recording of load versus displacement from which stress and strain were calculated.

Bulge testing was conducted upon each sheet using top dies with elliptical apertures for which the ratio of the major to minor axes lengths were: 1, 0.9, 0.78, 0.64 and 0.42. The length of the major axis was common at 180 mm. In all tests the rolling direction was aligned with the dies' minor axes. The top dies were interchanged and clamped with bolts to a common bottom die though which high pressure oil was regulated upon the sheet underside. This allowed the sheet to bulge through the aperture using a draw bead seal between top and bottom dies. The pressure was monitored with a pressure transducer and the bulge pole height measured with a displacement transducer. In addition, a pair of extensometers were positioned over the pole, lying at right angles to each other, in alignment with the ellipse axes. These were used to estimate principal pole strains from length changes to original gauge lengths of 50 and 70 mm astride the pole (Rees 2001).

4. Tensile test results

Figures 2 and 3 show the stress-strain behaviour of the can body and end material respectively. The two materials reveal contrasting strength, ductility and hardening behaviour. The thinner, 0.19 mm body material hardens to a strain approaching 2% but the thicker 0.23 mm material, used for large end manufacture, does not harden for an enhanced fracture strain between 5 and 10%.

Tensile testing aligned with orientations at θ = 0, 45° and 90° to the roll reveal the nature of anisotropy in each material. Ductility is at its greatest in the region of 0° and is at its least for 90°. Ultimate tensile strengths within the body vary from 485 MPa to 560 MPa with these orientations. The influence of orientation upon ultimate strength of the end reveals a lesser variation of 425–450 MPa. A clear region of elasticity exists in body and end material from which an elastic component of strain may be calculated from their respective moduli: $E = 170$ and 175 GPa.

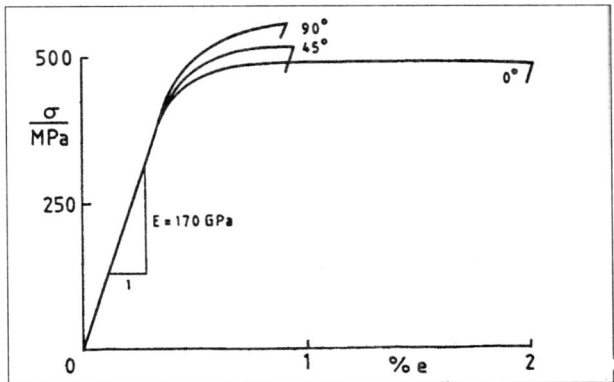

Figure 2. *Stress-strain curves for can body*

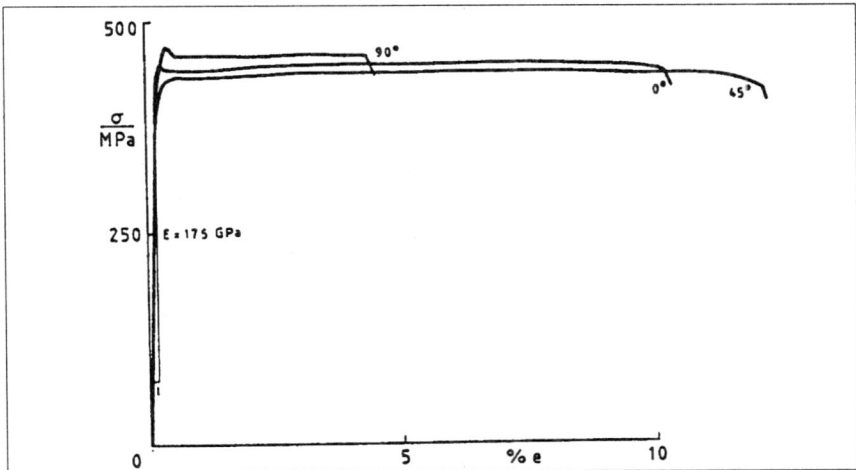

Figure 3. *Stress-strain curves for can end*

Such variations show no consistent trend except that strength is at its greatest and ductility is at its least in a direction transverse to the roll. This implies that a versatile plasticity model is required where coefficients, exponents etc, need to match the observed anisotropic behaviour. For example, an orthotropic theory (Hill, 1948), predicts an in-plane anisotropy within the body material as shown in Figure 4.

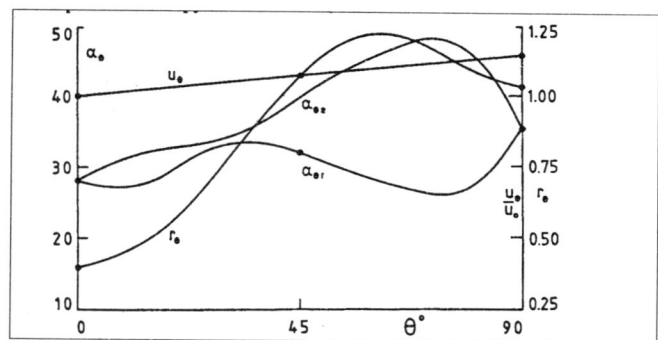

Figure 4. *In-plane anisotropy for can body material*

Because of its low ductility, it is not feasible to measure lateral displacement in a tension test on this material with the accuracy required for r-value determination. Instead, r-values were based upon the inclination of the neck by following the method outlined above. This shows the variations in the r_θ and α_θ − values based upon three measured local neck orientations, α_0, α_{45} and α_{90}, using equations [5a], [5b] and [6]. Intermediate r_θ-values are predicted from equation [3a] using the relationship

$$r_\theta = -R_\theta / (1 + R_\theta) \qquad [7]$$

from which intermediate α_θ then follow from equation [4]. Of the two α_θ variations shown in Figure 4, that which confirms the three measured values i.e. $\alpha_{\theta 1}$, is correct. Geometrically, the two solution to α_θ for a given θ, correspond to the two intersections points between a Mohr's circle (see Figure 1b) and its vertical axis. Both points give zero extension along the neck but only the true solution is shown in Figure 1a.

Compared to an isotropic sheet ($r_\theta = 1$ for all θ), the body material shows $r_\theta > 1$ for θ > 37.5°, consistent with an improved formability and $r_\theta < 1$ for θ < 37.5°, consistent with an impaired formability. Such mixed-mode behaviour can be tolerated for wall ironing and rolling of the can into its cylindrical shape. In applying the Hollomon hardening law: $\sigma = K\varepsilon^n$, to the can body material for 45° tension, it was found that $K = 870$ MPa and $n = 0.083$. The work hypothesis: $\overline{\sigma} \, \overline{\varepsilon}^P = \sigma \varepsilon^P$ shows that the corresponding Hollomon equivalent stress-strain relationship: $\overline{\sigma} = K'(\overline{\varepsilon}^P)^n$ employs a modified strength coefficient $K'' = K C^{n+1} = 1001$ where $C = 1.138$ is defined from the three r-values and the yield criterion as

$$C = \sqrt{\{(1+r_{45})(r_0 + r_{45})\{[r_{90}(1+r_0)]^{-1} + [r_0(1+r_{90})]^{-1} + (r_0 + r_{90})^{-1}\}/6\}} \qquad [8]$$

This equivalent flow law was able to correlate the remaining two orientations 0° and 90° in tension with flow stresses from bulge testing. Thus, predictions to the pole flow in a bulge test may be based upon a flow law found from a simple tension test.

A similar procedure was followed for the end material, where $K' = 500-550$ MPa from equation [3] with $r_0 = 0.9$, $r_{45} = 1.6$ and $r_{90} = 0.8$. This reflects the small spread observed when $n \approx 0$. The alternative method of r-determination, outlined above, was not used here. A local necking band was less well-defined and so the three r-values were based upon the measurement of test piece displacements with extensometers, for which ductility was adequate. This enabled r_θ to be found from equation [7] when R_θ is normally identified with the constant gradient to a plot of lateral versus axial plastic natural strains. Note, in the absence of hardening, no error will arise when r_θ-values are calculated from the total natural strains. That is, $R_\theta = d\varepsilon_{2'2'}^T / d\varepsilon_{1'1'}^T$ will apply to an off-axis tension test. Non-hardening is a feature of can-end material. The heavily rolled material is given a temper to promote the ductility required for a moderate amount of shallow forming and installation of an opening tab before its attachment to the body.

Previous studies on other canning sheets (Rees, 2002), with contrasting ductilities, showed that Hill's yield criterion was unable to provide an equivalence between tension and bulge test flow curves. The biaxial stress state at the pole of a bulge promoted a greater strain to fracture and a spread between curves that defied a unique equivalence relation based upon this criterion. The test-dependent coefficient K in the Hollomon law was reflected within a separation of linearised experimental stress-strain plots within double logarithmic axes (true stress versus natural plastic strain). Also, the slightly differing gradients of these lines altered the hardening exponent. The alternative approaches considered were either to seek a unique flow curve relation from another yield criterion or to abandon this need and simply employ the Hollomon flow law with Hill's theory, as found for each test. Having adopted the second approach, each elliptical bulge test was expressed with its own equivalent flow law for predictive purposes. This provided satisfactory predictions to pressure versus height curves during bulging testing. In contrast, here we shall employ an average equivalence relation, based upon the tensile tests, to predict bulge pole flow.

5. Bulge test results

The theory of ellipsoidal bulging of rolled, orthotropic sheet has been covered in previous papers (Rees, 2000). In summary, when 1 and 2 are aligned with the minor and major axes respectively, this theory provides a relation between normal pressure p and pole height h together with expressions for the principal pole strains ε_1^P and ε_2^P as

$$p = \frac{\bar{\sigma} t_o}{R X} \left[1 + \left(\frac{h}{b}\right)^2 \right]^q \qquad [9]$$

$$\varepsilon_1^P = \ln[1 + (h/b)^2] \qquad [10a]$$

and $\quad \varepsilon_2^P = \ln[1 + (h/a)^2] \qquad$ [10b]

Equations [9], [10a] and [10b] apply to bulging a sheet of initial thickness t_o through an elliptical die with semi-major and semi-minor axes lengths a and b respectively. The previous r_θ-values and the equivalence relation applies to equation[9] in which the parameter R contains the two, principal pole curvatures $R_1 = (h^2 + b^2)/2h$ and $R_2 = (h^2 + a^2)/2h$ and X is an equivalent stress coefficient as follows:

$$R = \frac{R_1 R_2}{QR + R_{21}} \quad \text{where} \quad Q = \frac{R_2 + (1 + 1/r_0)\ R_1}{R_1 + (1 + 1/r_{90})\ R_2} \qquad [11a]$$

$$X = \frac{\bar{\sigma}}{\sigma_1} = \sqrt{\frac{3[r_{90}(1 + r_0) - 2Q r_0 r_{90} + Q^2 r_0 (1 + r_{90})]}{2(r_0 + r_0 r_{90} + r_{90})}} \qquad [11b]$$

In equations [11a] and [11b] $Q = \sigma_2/\sigma_1 < 1$ is the pole stress ratio. In the experiment, the pressure and pole height are measured. The theoretical principal strains may be calculated from the bulge height and also measured directly with pole extensometers (Rees, 2001). A comparison between equations [9], [10a] and [10b] and experiment, using paired couplings of these quantities, are given for each material in Figures 5–7. For the end material, the band in the pressure versus height prediction (see Figure 5) is based upon the spread observed in equivalent flow stress data (Figure 3).

Pole instability failures are predicted to occur at maximum pressure on this curve. The experimentally observed behaviour (single line) reveals some evidence that the corresponding critical bulge height is attained in this more ductile material. The peak pressure is approached more accurately as the elliptical aperture sharpens into its least axes ratio of 0.42 for die 5. In contrast, the failures found from bulging body material all occur along the rising pressure curve (Figure 6).

That is, due to a limited ductility of body material, pole failures occur well before a condition of pole instability can be achieved. All errors in predicted pressure are likely due to the uniaxial Hollomon flow law used with equation [9], since this was based upon an even lower tensile strain range.

Anisotropy in Thin, Canning Sheet Metals   289

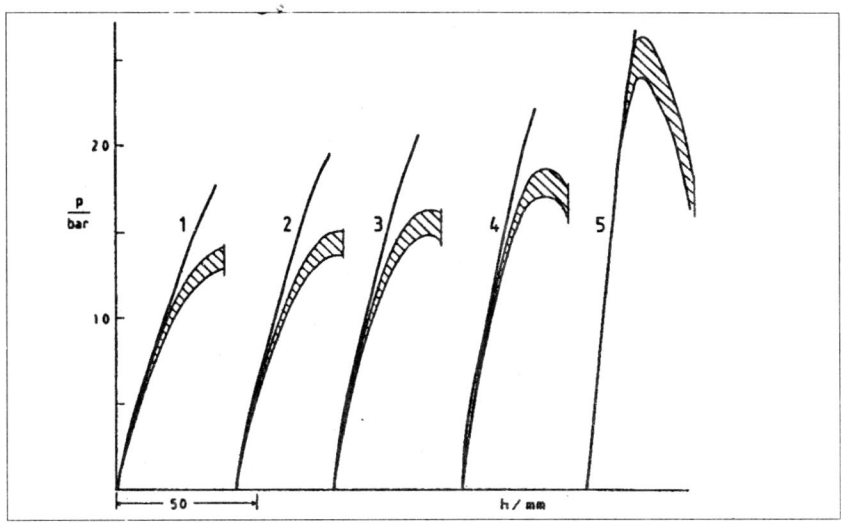

Figure 5. *Pressure versus bulge height for can-end material*

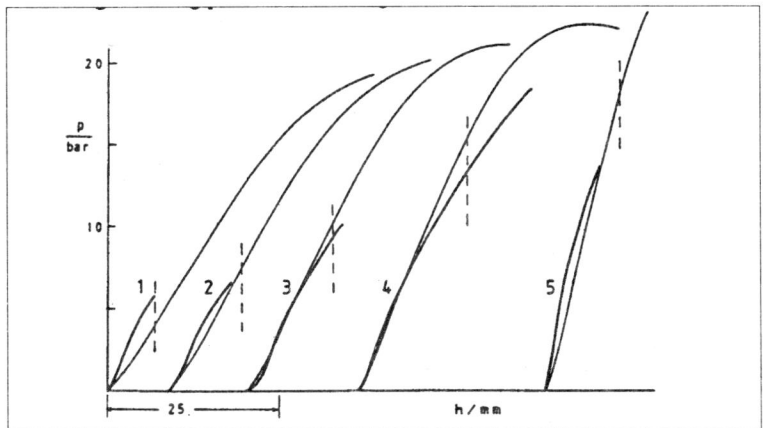

Figure 6. *Pressure versus bulge height for can-body material*

True pole instability failures at maximum pressure were observed in ductile steels (Rees, 1997). With the relatively low ductility of canning material, it appears that a different failure criterion is required. To estimate a failure point on a rising p versus h plot, it is proposed that either a limiting equivalent plastic strain, or a limiting major principal strain, be imposed upon the pole's principal strain state. Biaxial tension promotes a greater ductility compared to simple tension in these materials. Therefore, it is preferable to apply a strain limit from an equi-biaxial bulge test, *i.e.* bulging through a circular aperture. For an orthotropic sheet the respective criteria become:

$$(\varepsilon_{1f}^{P})_c = (\varepsilon_{1f}^{P})_e \qquad [12a]$$

$$\text{and} \quad (\overline{\varepsilon}_f^{P})_c = (\overline{\varepsilon}_f^{P})_e \qquad [12b]$$

where *c* and *e* refer to circle and ellipse apertures and $\overline{\varepsilon}^P$ for each test follows from:

$$\frac{\overline{\varepsilon}^P}{\varepsilon_1^P} = \frac{\sqrt{(2/3)(r_0 + r_0 r_{90} + r_{90})[Q^2 r_0(1+r_{90}) - 2Qr_0 r_{90} + r_{90}(1+r_0)]}}{r_{90}[1 + r_0(1-Q)]} \qquad [13]$$

When equation [12b] is combined with equations [10a] and [10b] and equation [13] the critical pressure and height at failure in elliptical bulge tests may be obtained with improved accuracy, as shown in Figure 6. Similarly, these limiting strains may be imposed upon equations [10a] and [10b] to provide the fracture point along each principal, engineering, pole strain path (*e.g.* see Figure 7a).

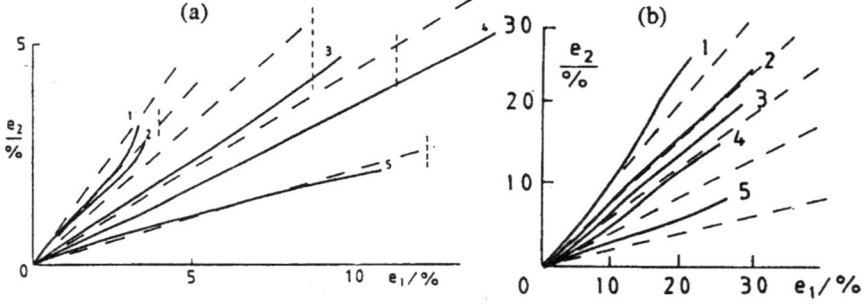

Figure 7. *Engineering pole strain paths for a) body and b) end material*

Figure 7 shows a greater strain range than can be achieved from simple tension (see Figures 2 and 3). Tension ($Q = 0$) is limited to small strain by the formation of a local neck. The latter is inhibited by increasing Q under a biaxial stress state, *i.e.* where the Mohr's circle (see Figure 1b) is shifted rightward, until there is no intersection point with its vertical axis. When $Q \geq 0.5$ for isotropic material, there is no zero strain direction along which a local necking band can form. As a result, bulge pole straining becomes more diffuse. Finally, it is noted in Figure 7 that the pole extensometers provide an average strain across the pole (Rees, 2001). Consequently, their paths differ slightly from the theoretical pole point strains when bulging through dies 1–5.

6. Conclusions

The quadratic yield function is particularly useful when an in-plane anisotropy of rolled sheet is characterised by its 3, r-values. It appears to be less successful as a basis for equivalence between flow curves for uniaxial and biaxial plasticity. Bulging of thin sheet produces tensile, principal pole stresses, where the greatest disparity between observed and predicted yielding has been reported for $r < 1$ materials (Mellor, 1982). Alternative, yield functions (Bassani, 1977; Hosford, 1979; Hill, 1990), claim to account for the anomaly within a single r-value: $\bar{r} = (r_0 + 2r_{45} + r_{90})/4$ and integer stress exponents > 2 but these offer little advantage for canning materials (Rees, 2002).

7. References

Bassani L., "Yield characterisation of metals with transversely isotropic plastic properties", *Int Jl Mech Sci*, Vol. 19, 1977, pp. 651–660.

Hill R, "A theory of the yielding and plastic flow of anisotropic metals", *Proc Roy Soc, London*, Vol. 193A, 1948, pp. 281–297.

Hill R, "Constitutive modelling of orthotropic plasticity in sheet metals", *Jl Mech Phys Solids*, Vol. 38, 1990, pp. 405–417.

Hosford W.F, "On yield loci of anisotropic cubic metals", *Proc: 7th North Amer Metal Working Res Conf, SME*, Dearbon Michigan, 1979, pp. 191–197.

Mellor P.B, *Mechanics of Solids*, Eds: Hopkins, H. G. And Sewell, M.J.. 1982, pp. 383–402, Pergamon.

Rees D.W.A, "Tensile flow and instability in galvanised, rolled steel sheet", *Jl Strain Analysis*, Vol. 30, No 4, 1995, pp. 305–315.

Rees D.W.A, Instability analysis for ellipsoidal bulging of sheet metal, *Proc. Advanced Methods in Materials Processing Defects*, Eds: Predeleanu, M and Gilormini P., CNRS Cachan 1997, pp. 235–244, Elsevier.

Rees D.W.A, "Rim and pole failures from elliptical bulging of oriented orthotropic sheet metal", *Jl Strain Analysis*, Vol. 35, No 2, 2000, pp. 109–123.

Rees D.W.A, "A bi-axial pole strain transducer device", *Meas Sci and Tech*, Vol. 12, 2001, 97–102.

Rees D.W.A, "Anisotropy in thin canning sheet metals", *Proc. 5th Int Conf on Mat Form., ESAFORM 2002*, Eds Pietrzyk M., Krakow, AGH Krakow 2002, pp. 487–490, Akapit.

Chapter 25
Progress in Microscopic Modelling of Damage in Steel at High Temperature

Marc Remy, Sylvie Castagne and Anne Marie Habraken
Institute of Mechanics and Civil Engineering, University of Liège, Belgium

1. Introduction

This research, performed in collaboration with industrial partners of the ARCELOR group (IRSID and the Technical Direction of Cockerill Sambre), aims to study steel at high temperature. A damage model using a numerical microscopic[1] approach is identified by experimental measurements obtained from the microscopic and macroscopic scales. The final goal is a parametrical study of various microscopic factors such as grain size or precipitation state in continuous casting process.

Constitutive laws have been implemented into a finite element code to study the initiation and propagation of cracks in the industrial process of continuous casting. A macroscopic finite element study of this process (see [PAS 00]) provides global stress and strain fields defining the loading of the microscopic cell during the process.

For the studied low carbon steel, transverse cracking between austenitic grains is recognized as a major problem in this process. Hot tensile tests have demonstrated that steel is more brittle in the temperature range from 1000°C to 600°C. The allotropic austenite/ferrite phase transformation occurs in this temperature range. This loss of ductility is responsible for the appearance of cracks during the bending and unbending operations of the strand. Other researches (see [GAM 01]) also tend to demonstrate that austenitic grain boundary is a favorable place for cracks to begin. They appear by strain concentration and micro-void coalescence at grain boundaries and by grain boundary sliding. The influence of creep, controlled by diffusion, is important in the studied temperature range. Castagne [CAS 01] proposes a literature review on this topic.

The first step in the development of the model is the building of a representative microscopic cell of the material. Diffusion of voids at grain boundary and sliding are introduced in the model. Its identification uses comparisons between simulations performed at the microscopic scale level and macroscopic experiments.

As the micro-structure influences damage and creep, it has to be introduced if we want to determine the sensibility factors for crack initiation. The microscopic cell consists in grains and grain boundaries, as shown in Figure 1. Three essential features require identification: the grain size and shape, the constitutive law inside the grain, and the constitutive laws defining the grain boundary behavior: sliding, pressure, cohesion ...

Metallographic and texture analysis combining optical microscopy (OM), scanning electron microscopy (SEM) and orientation image microscopy (OIM), as well as various chemical etching and visual observation on steel samples, have been performed at room temperature. These efforts to determine the previous austenitic grain size and morphology are presented. Hence, the geometry of the microscopic cell is defined taking into account the results of this micro-graphic study.

---

1. Throughout this article "microscopic" must be understood as "comparable in size with austenic grains".

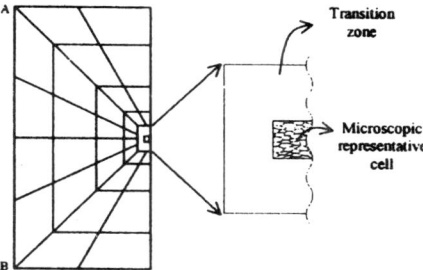

Figure 1. *Macroscopic section of the slab on which the thermo-mechanical state is computed by a macroscopic model (left picture). Microscopic cell (right picture) on which the loading from the macroscopic model will be applied*

The cell discretization is obtained either by automatic meshing of observed grains by OIM or on purely geometrical meshes (hexagon, square, ...).

A law of Norton-Hoff type has been used to quantify the visco-plastic behavior inside the grain for the studied steel. Its parameters have been identified using experimental results (see [WOL 01]). Compression tests of cylindrical samples, after a thermal treatment aiming to reproduce the thermal cycle in continuous casting have been performed. Various strain rates and temperatures have been tested and compared with simulations in order to identify this mechanical law.

Crack appearance depending on temperature, strain rate and strain level is identified thanks to acoustic detection during the tests [KOP 99]. Such a set of experiments is applied to the studied steel to identify the constitutive law used at grain boundaries.

This paper is subdivided in the following sections. Section 2 describes the microscopic cell. Section 3 explains the efforts to get information on the austenitic grain. Section 4.1 describes the elasto-visco-plastic law used at high temperature. The interface finite elements and the interface law are briefly described in section 4.2, details of the model are given in [CAS 02]. As a first application, encouraging results of crack propagation with a simple model are presented in section 5.

2. Microscopic cell

In order to represent inter-granular creep fracture, the developed model contains solid finite elements for the grains and interface elements for their boundaries (see also [ONC 99]). In the grains an elasto-visco-plastic law without damage is used (see section 4.1), and at the boundaries a law with damage is preferred (see section 4.2). The relevant damage mechanisms at the micro-scale are viscous grain boundary sliding, nucleation, growth and coalescence of cavities leading to micro-cracks. The linking-up process subsequently leads to the formation of a macroscopic crack.

296 Prediction of Defects in Material Processing

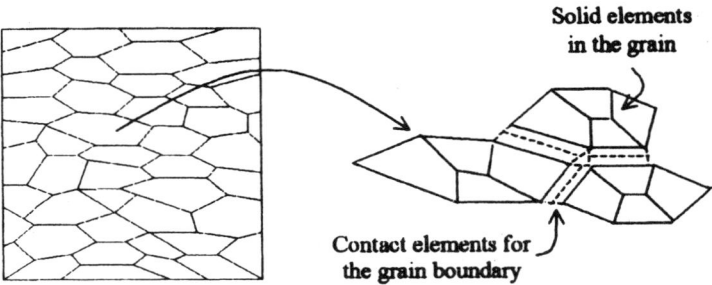

Figure 2. *Microscopic cell with solid and interface elements*

2.1. *Solid finite elements and grain representation*

The grains are modeled by thermo-mechanical 4-nodes quadrilateral solid elements BLZ2T of mixed type [ZHU 95]. Figure 2 shows a sketch of the microscopic representative cell. On the right, a zoom on three grains allows one to visualize the meshing of these grains and their interfaces.

The number of grains in the microscopic cell as well as the density of the mesh inside the grains have to be tested to define a cell which gives accurate results within an acceptable computing time. The shape and size of grains has been determined from the results of the microscopic analysis (see section 3).

3. Micro-structure of the studied steels

Microscopical studies have been performed on samples extracted near one corner of the slabs, using optical microscopy, scanning electron microscopy and orientation image microscopy. Of course, these observations at room temperature detect the microstrucuture of ferrite and not austenite. Indeed, austenite exists only in a temperature range with lower and upper limits of approximately 800°C and 1100°C. As we will see in this section, observations of micro-structures at room temperature can give a reasonable idea of the austenite grain size and of its morphology at high temperature.

3.1. *From austenite to ferrite*

During the allotropic phase transformation, each austenitic grain can give birth to many ferritic grains, depending on the number of nucleation sites of ferrite, which growth successfully inside the austenite, as temperature drops. The cooling rate is the key macroscopic parameter which controls this number of ferritic grains ; a larger cooling rate implies a larger number of ferritic grains and ferritic grains of smaller size. Decrease of the ferritic grain size tends to improve the mechanical properties of

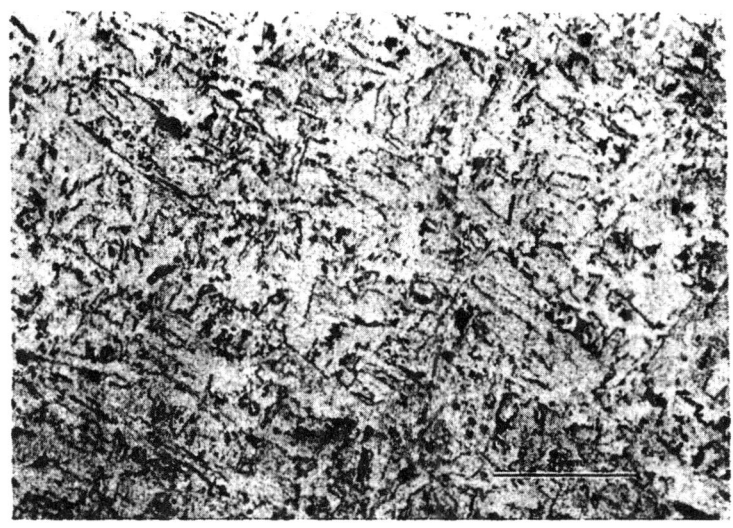

Figure 3. *Probable widmanstättite in the studied steel. OIM taken inside the displayed zone indicates a common orientation for the micro-crystals of ferrite. Rolling direction of the slab is horizontal and normal direction vertical*

the steel. In the most usual case which leads to a micro-structure of ferrite/pearlite, nucleation of ferrite starts at the boundary of austenitic grains.

3.2. *Optical microscopy*

From the microscopical optical investigations, complex structures appear in the studied steel, among which widmanstättite embedded in ferrite/pearlite. Widmanstättite is the micro-structure which is formed when the precipitation of ferrite grains starts inside the crystal of austenite. Micro-crystals of ferrite are oriented along the directions of the previous austenitic micro-crystal. Such a micro-structure is also called acidular ("needle-like") (see [COH 95] and Figure 3). The probable origin of this widmanstätten structure is a fast cooling of the slab. This hypothesis seems to be confirmed by the observation of small ferrite/pearlite grains in the first mm near the surface (not displayed).

Among these complex structures, we have identified 3 large widmanstättite structures (typical size of 3 mm) near the slab surface in a section of 1.5 cm x 1.0 cm perpendicular to the transverse direction of the slab.

The identification of 3 distinct widmanstättite zones result from the following observations: 1) inside each zone, elongated structures are parallel and acidular and 2) a clear texture is observed by OIM near the center of one zone (see next section, where more details are given). This implies a common micro-orientation for each zone. From this we deduce that these 3 zones are originated from 3 distinct austenitic grains. According to these considerations, the size of the austenitic grains would be of the order of 3 mm.

3.3. Orientation imaging microscopy

The concepts of texture of materials, orientation distributions function (ODF) and related topics are fully described in [BUN 82]).

The sample preparation consists in several mechanical polishing steps then a final electro-polishing step. For this analysis, etching was done with nitric acid and stopped by methanol.

The orientation image microscopy is a technique coupled with SEM aiming at a precise measurement of the three-dimensional orientation of micro-crystal lattice at many regularly spaced positions in the plane of the sample. This technique creates a rectangular map of the lattice orientations (Euler angles) of the grains in the sample. Post-processing technique of those images could include the reconstruction of grain based on the orientation of the scanned positions, pole figures, orientation distribution function, etc.

Four images with different resolutions and positions have been obtained of one of the 3 observed widmanstättite zones. The first two ones, of regular quality, were taken in the central part of the zone, the last ones, of excellent quality, in the interface between this zone and the embedding ferrite/pearlite.

3.3.1. *The post-processing analysis software*

From the measurements, a dedicated software "TSL" allows one to construct pole figures and texture maps (orientation distribution functions) and a pseudo-color image of the observed sample. This software groups scans together into grains, on the basis of the observed Euler Angles of all the scans. It takes into account the symmetry group of the studied phase in its regrouping computation.

As a conclusion to this study, texture was detected only for the observed zone of widmanstättite. The result for this sample is that a dominant orientation appears clearly in its center. Nevertheless, this texture does not appear near the boundary of the same zone.

3.3.2. *OIMesh*

Dedicated software "OIMesh" has been developed by our team to achieve a regrouping of ferrite grains of common austenitic origin [REM 01]. What has been tested with OIMesh is very simple. We have assumed that in the austenite to ferrite phase transformation, all the austenite atoms have a movement parallel to the crystal principal axes. (ie. the strain principal directions coincide with crystal lattice axis).

Damage in Steel at High Temperature 299

This kind of "minimal" movement is what is expected if widmanstättite is formed and nucleation of ferrite started inside the austenitic grain. As in the transformation, one of the axes is stretched and the others shrink, it can be seen that a given austenite grain gives three classes of possible ferrite grain. Each class is achieved when the corresponding axis (out of 3) becomes the stretching axis.

It can be shown with this hypothesis in mind that two given ferrite crystals coming from the same austenite grain will be rotated by a multiple of 90°, the axis of rotation being a line joining the centers of two opposite edges of the bcc lattice. This hypothesis has been introduced in the algorithm of grain regrouping of OIMesh.

Processing of the 4 images obtained by OIM with OIMesh does not result in any significant regrouping of ferrite grains, but for some very marginal low quality measurements. Despite this rather deceptive result, modification of the detection algorithm or application to other cases could possibly lead to better results.

3.4. Metallographic analysis

Picral an nital tests have also been performed to try to enhance the austenitic boundary. Using this technique, metallographists from industry were able to directly detect some part of the austenitic boundary by direct observation of the sample. These findings confirm the large value for the austenitic grain size.

With micro-structures as large as several mm at high temperature, one could possibly question the validity of continuous mechanics and of any isotropy or homogeneity hypothesis which could be made without taking care of the size of studied structures. This justifies the idea of using a representative cell in which the grains are modeled individually to study crack initiation and propagation.

4. Laws for the solid and interface elements

4.1. Norton-Hoff constitutive law

An isotropic model for the constitutive elasto-visco-plastic law inside the grain is used. This approximation is reasonable at the current stage of our study and could be replaced using an anisotropic law, if necessary.

The creep behavior inside the grain can be represented by an elasto-visco-plastic law of Norton-Hoff type, Equation 1. The Norton-Hoff law gives us the visco-plastic limit as:

$$\sigma_e = e^{-p_1 \varepsilon_e} \sqrt{3}\, p_2 \left(\sqrt{3}\dot{\varepsilon}_e\right)^{p_3} \varepsilon_e^{p_4} \quad , \qquad [1]$$

expressed in term of von Mises equivalent stress σ_e, strain rate $\dot{\varepsilon}_e$ and strain ε_e.

Its parameters (p_1, p_2, p_3, p_4) have been identified with help of results of compression tests (stress σ vs. strain ε) obtained by the IBF Aachen at different strain rates $\dot{\varepsilon}$

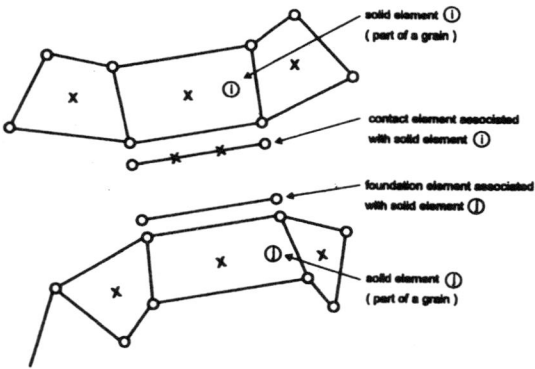

Figure 4. *Interface element: contact element, associated foundation, linked solid elements. The dots symbolize nodes and the crosses represent integration points. For clarity, a gap has between introduced between the contact element and the associated solid element, as also between the foundation and the solid element*

(0.0001s^{-1}, 0.001s^{-1}, 0.01s^{-1}) and temperatures T (800°C, 900°C, 1 000°C, 1 100°C), we have determined the parameters if the Norton-Hoff law independently at each temperature by means of a χ^2-fitting of the Norton-Hoff law to experimental points.

4.2. *Interface finite elements, grain boundary representation and acoustic tests*

The solid elements modeling the grains are connected by interface elements to account for cavitation and sliding at the grain boundary (see [ONC 99]). As the thickness of the grain boundary is small compared to the grain size, the grain boundary is represented by one-dimensional interface elements. These elements are associated with a constitutive law which includes parameters linked to the presence of precipitates, voids, etc. The damage variable explicitly appears in this law. The interface element and interface laws are described with full details in [CAS 02].

The interface element is composed of a modified contact element and a foundation element (see Figure 4).

For each integration point of the contact element, the program determines the associated foundation segment as well as the solid element to which the foundation is associated. The state variables for the interface element are computed using information of the two solids elements in contact (elements i and j in Figure 4). The state variables in the interface element are the algebraic average of the corresponding values at the integration points of elements i and j. This is obviously a rough approximation. Better schemes will be tested in the near future.

The original contact element has been modified from [HAB 98] to introduce a new interface law and a cohesion criterion with help of a penalty method.

The interface element is supposed to be in contact with the foundation (i.e. no crack) until the damage variable reaches a threshold value, following [ASH 72] for grain boundary sliding, [GIE 94] for nucleation of cavities, [TVE 84] for cavity growth under diffusion and creep deformations, [ONC 99] for void coalescence.

Identification of damage parameters

The goal of the acoustic tests is to provide a set of experiments to identify the parameters of the interface law (grain boundary behavior). They were developed at IBF in Aachen to predict the formability of metals at high temperature.

During forming, when the internal stresses at a material point exceed locally a stress limit, there is a sudden microscopic crack. This new state of equilibrium has a lower potential energy. The potential energy emitted in this way is dissipated in form of elastic waves and can be detected as sound pulses. Piezoelectric sensors are used to record these sound signals. The weak sound signal is pre-amplified, filtered to separate process-relevant and interfering signals and amplified again before being stored into data acquisition of the computer.

To recover casting state, the samples were first heated to 1375°C in a radiation furnace with an argon inert gas to prevent oxidation. Finally, the samples were compressed with a constant strain rate up to crack initiation, while upcoming acoustic emission events caused by material failure were recorded.

Sets of acoustic tests were obtained with different geometries of the samples to achieve different stress-state histories at their critical point. There are four geometries, two cylindrical ones characterized as the flat and the slim, and two non-cylindrical ones, called the concave and the flange. IBF has tested three temperatures (800°C, 900°C, 1000°C) and two strain rates (0.001 s-1, 0.0005 s-1). At least four samples from the surface of the slab have been tested for each combination and for each geometry to ensure statistically relevant results.

Macroscopic finite element simulations of the tests using constitutive law identified in section 4.1 give the loading curve as well as the stresses, the strains and temperature histories in the region around the crack. Then these histories will be imposed as boundary conditions to the representative microscopic cell to identify the damage law parameters by inverse modeling.

5. First application

The growth of a crack for the simple model of Figure 5a has been simulated as a first application.

With this first example, the stability of the code when a crack is propagating has been analyzed. The parameters for the interface law are issued from [ONC 99]. For the continuous zone we used the NH law indentified below, with an elastic law with an elastic modulus of 32400 MPa which correspond to the elastic modulus of steel at

302 Prediction of Defects in Material Processing

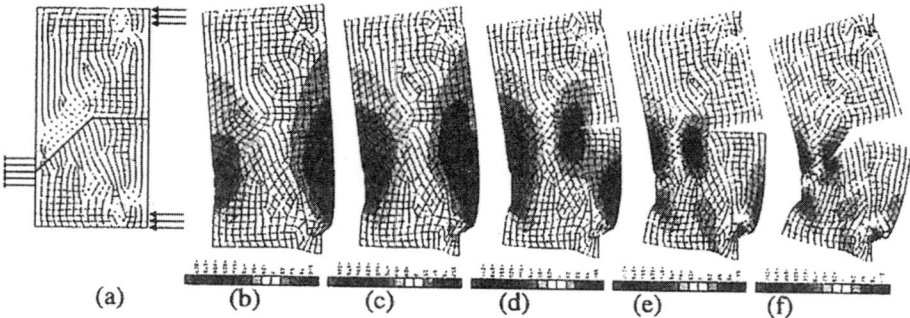

Figure 5. *a: Loads and fixations applied to the simple model. The arrows indicate the imposed displacements, the fixations are drawn on the left. The lines in bold represents the zone where interface elements have been put and where the crack can propagate. The size of the model is 1mm x 2mm. b-f: Stress maps (σ_y) at different steps of the crack propagation in the course of a simulation*

900°C. Figure 5b-f shows the evolution of the crack for a large value of the penalty coefficient. In each of them, the stress σ_y in the direction perpendicular to the interface zone is represented. Picture b) is a reference case before the initiation of the crack. From picture c) to f) we can follow the propagation of the crack. The stress concentrates around the crack tip and decreases in the rest of the specimen as the crack propagates.

6. Conclusions

The research is dedicated to a microscopic study of damage at high temperature of micro-alloyed steel for continuous casting. The development of the damage model is progressing in parallel with the analysis of the experimental tests.

The Norton-Hoff law has been identified and will be used in all the future simulations. Most of all, through the damage experimental study, we have already at our diposal a set of identification tests for the damage law as well as validation tests for this study.

We have succeeded in using our contact element to model crack propagation in a simple test and the results were very encouraging. The next step is to apply it to more complex cases.

Acknowledgements

S. Castagne and A.M. Habraken thank the National Fund for Scientific Research of Belgium (FNRS) for its support. Industrial partners ARCELOR and its research

Damage in Steel at High Temperature 303

teams IRSID and the Technical Direction of Cockerill Sambre are acknowledged as well as IBF and CRM who performed experimental studies. The authors thank Prof. J. Lecomte-Beckers (ULg) for her help in the interpretation of the micro-graphic analysis.

7. References

[ASH 72] ASHBY M. F., "Boundary defects and atomistic aspects of boundary sliding and diffusional creep", *Surface Sci.*, vol. 31, 1972, p. 498-542.

[BUN 82] BUNGE H. J., *Texture Analysis in Materials Science*, Butterworths Publishers, London, 1982.

[CAS 01] CASTAGNE S., Damage in continuous casting of steel: towards a microscopic representative cell, D.E.A. graduation work, University of Liège, 2001.

[CAS 02] CASTAGNE S., REMY M., HABRAKEN A. M., "Development of a mesoscopic cell modeling the damage process in steel at elevated temperature", *Proceeding of The 6th Asia-Pacific Symposium on Engineering Plasticity and Its Applications*, Sidney, 2002.

[COH 95] COHEUR J.-P., *Connaissance des matériaux métalliques*, University of Liege, 1995.

[GAM 01] GAMSJÄGER E., FISHER F. D., CHIMANI C. M., SVOBODA J., "Large strain concentration during continuous casting — a micromechanical study of the diffusional phase transformation", HABRAKEN A. M., Ed., *Proceedings of the 4th international ESAFORM conference on material forming*, Liege, 23-25 April 2001, University of Liege, p. 871-874.

[GIE 94] VAN DER GIESSEN E., TVERGAARD V., "Development of final creep failure in polycrystalline aggregates", *Acta Metall.*, vol. 42, 1994, p. 959.

[HAB 98] HABRAKEN A. M., CESCOTTO S., "Contact between deformable solids, the fully coupled approach", *Mathematical and Computer Modelling*, vol. 28, n° 4-8, 1998, p. 153-169.

[KOP 99] KOPP R., BERNARTH G., "The determination of formability for cold and hot forming conditions", *Steel Research*, vol. 70, n° 4-5, 1999, p. 147-153.

[ONC 99] ONCK P., VAN DER GIESSEN E., "Growth of an initially sharp crack by grain boundary cavitation", *J. Mech. Phys. Solids*, vol. 47, 1999, p. 99-139.

[PAS 00] PASCON F., Finite element modelling of contact between the strand and the mould in continuous casting, D.E.A. graduation work, University of Liege, 2000.

[REM 01] REMY M., "OIMesh (alpha release): Brief Status of the program and preliminary user's manual", technical report, Nov. 2001, University of Liege.

[TVE 84] TVERGAARD V., NEEDLEMAN V., "Analysis of the cup-cone fracture in a round tensile bar", *Acta Metall.*, vol. 32, 1984, p. 157-169.

[WOL 01] WOLSKE M., Investigations to the formability of micro-alloyed steel, Final report for R&D project ix/3/2000, April 2001, IBF and MSM.

[ZHU 95] ZHU Y., CESCOTTO S., "Unified and mixed formulation of the 4-node quadrilateral elements by assumed strain method: Application to thermomechanical problems", *Int. Journal for Numerical Method in Engineering*, vol. 38, 1995, p. 685-716.

Chapter 26

Use of Laser-Doppler Velocimetry and Flow Birefringence to Characterize Spurt Flow Instability during Extrusion of Molten HDPE

Bruno Vergnes
Centre de Mise en Forme des Matériaux, Sophia Antipolis, France

Laurent Robert and Yves Demay
Centre de Mise en Forme des Matériaux, Sophia Antipolis, France, and Institut du Non-Linéaire de Nice, Valbonne, France

1. Introduction

In many polymer processing operations, flow instabilities occur at output rates above critical values [LAR 92]. For linear long-chain polymers, such as linear low density polyethylenes (LLDPE) and high density polyethylenes (HDPE), the typical flow curve exhibits two stable branches (branch I and branch II), separated by the so-called spurt or stick-slip instability. In this area, the pressure oscillates between two extreme values, although the imposed flow rate is kept constant. During these oscillations, an hysteresis cycle is described between the two stable branches [DUR 96]. The origin of this instability seems to involve a cohesive failure i.e. macroscopic slip very close to the wall [MÜN 00, DEN 01]. A sudden chain disentanglement at the first molecular melt/wall interfacial layer (chains adsorbed on a high energy surface) is a possible molecular process [WAN 99].

The aim of this work is to give a contribution to a better understanding of this instability by performing reliable quantitative direct measurements of the flow of a linear polyethylene in stable and unstable flow conditions. Particularly, our efforts were to study simultaneously pressure oscillations, birefringence fringes and velocity pulsations in the die. Flow on fluoropolymer coated die, which is known to postpone the instability by promoting wall slip (see [PIA 95]), is also studied.

2. Experimental

The polymer used is a commercial HDPE provided by Elf-Atochem company, with reported molecular weight $M_w = 200\ kg/mol$ and a polydispersity index $M_w/M_n = 17$.

Experiments are performed with a single screw extruder equipped with a servo-controlled pump in order to ensure the flow rate to be as constant as possible. The polymer is extruded between two high energy polished steel movable plates (and two glass surfaces on each side for observations). For the modified die, two thin metal surfaces covered with an industrial PTFE coating (similar to those of cooking devices) are stuck on the two plates. The working temperature is 160°C.

Figure 1 shows the flow channel, which consists of a reservoir with a cross section $H_r \times W = 18 \times 10.2\ mm^2$ and a slit die of height $H = 1.02\ mm$ for the polished steel die (height $H = 1.0\ mm$ for the PTFE coated die) and length $L = 24.5\ mm$ (contraction ratio 18:1). The different axis are indicated in Figure 1. The velocity measurements are performed by a 1D Laser-Doppler Velocimetry (LDV) system from Dantec MT©, with high local (54 μm) and temporal (less than 5 ms) resolutions.

Birefringence fringe pulsations are recorded with a CCD video camera and treated by image analysis, in order to quantify the time dependant evolution of the Principal Stress Difference (PSD) along the symmetry axis X. We used a high speed CCD camera with a synchronized recording device for the pressure transducer signal.

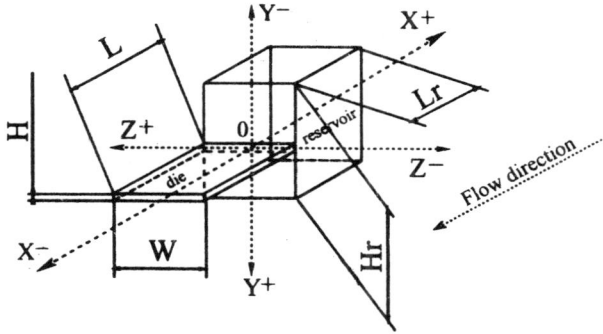

Figure 1. *Geometry of the flow channel*

3. Results and discussion

3.1. *Flow in the polished steel slit die*

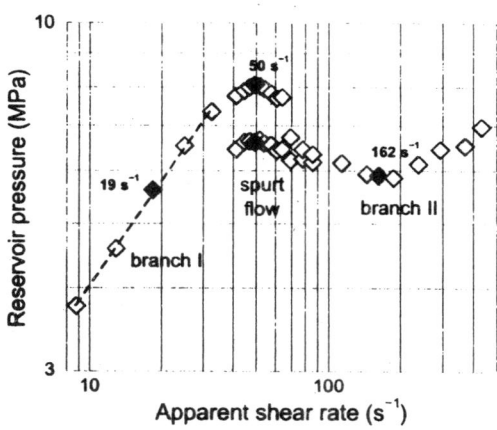

Figure 2. *Apparent flow curve for the polished steel die*

As expected, the **flow curve** plotted in Figure 2 presents two stable branches separated by the oscillating area : the branch I exhibits a power law behaviour whereas the branch II appears as particularly flat. Apparent flow rate values ($\dot{\gamma}_{app}$), deduced from the output rate, and corresponding to black dots indicated in Figure 2, are related to the different velocity profiles presented in Figures 3, 5 and 6.

Figure 3 shows the result of a typical velocity profile during **flow along branch I** $(0 < Z < 5.1\ mm,\ X = -12\ mm,\ \dot{\gamma}_{app} = 19\ s^{-1})$ measured by LDV. Experimental

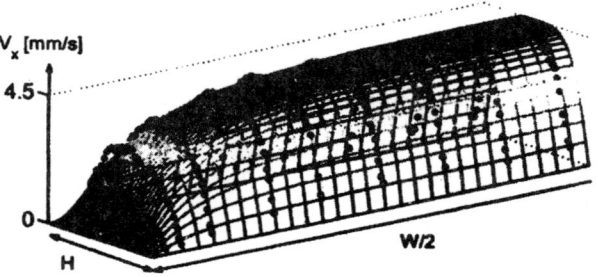

Figure 3. *Velocity profile on branch I ($\dot{\gamma}_{app} = 19\ s^{-1}$)*

values (black dots) are compared to the result of a numerical computation of the profile (finite element method) calculated with classical hydrodynamic boundary conditions (no slip) at the same input flow rate. In this case, a very good agreement is obtained.

During the **oscillating flow instability**, periodic oscillations of pressure and velocity are observed [MÜN 00]. As it is presented in Figure 4, pressure and velocity are in phase quadrature. Figure 5 presents velocity profiles obtained with the max-

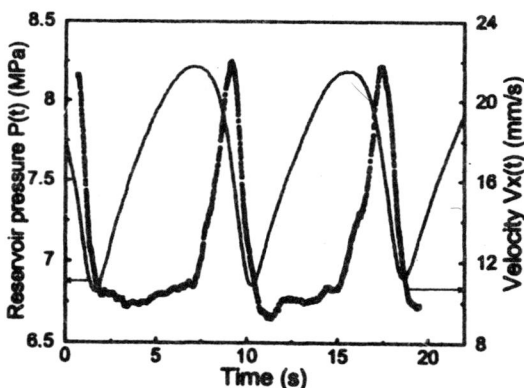

Figure 4. *Velocity and pressure oscillations during oscillating flow ($X = -10\ mm$, $Y = Z = 0,\ \dot{\gamma}_{app} = 50\ s^{-1}$)*

ima (Δ) and the minima (∇) of the velocity $V_x(t)$ during the spurt flow instability $(0 < Z < 5.1\ mm,\ X = -12\ mm,\ \dot{\gamma}_{app} = 50\ s^{-1})$. Profiles are deduced from each transient velocity signal $V_x(t)$, as it is shown in Figure 4. As a result, during the spurt instability, the velocity profile oscillates between two steady state profiles : profile of

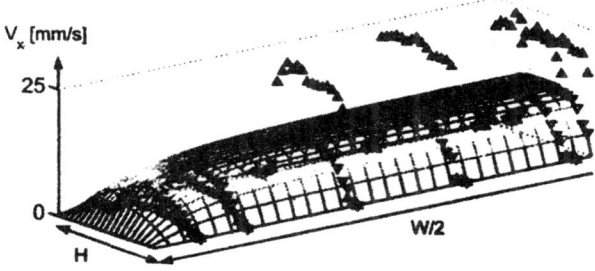

Figure 5. *Velocity profiles of maxima ($\triangle$) and minima ($\triangledown$) during spurt flow, based on Figure 4 ($\dot{\gamma}_{app} = 50\ s^{-1}$)*

the minima (classical viscous profile, e.g. branch I profile) and profile of the maxima, that shows a strong macroscopic wall slip. It is worth noting that for this profile, wall slip is not homogeneous across the die width.

Figure 6 shows a typical velocity profile during **flow along branch II** ($0 < Z < 5.1\ mm$, $X = -12\ mm$, $\dot{\gamma}_{app} = 162\ s^{-1}$). A steady state profile is obtained for

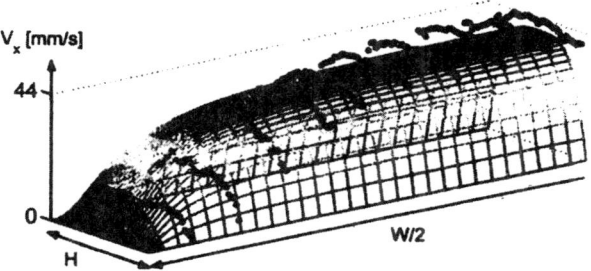

Figure 6. *Velocity profile on branch II at $\dot{\gamma}_{app} = 162\ s^{-1}$*

apparent flow rates up to $100\ s^{-1}$. The polymer melt slips at the wall in the middle part of the die. Experimental profiles are flat and above the theoretical one for $0 < Z < 3.5\ mm$ (wall slip), and are parabolic and below the theoretical one for $3.5 < Z < 5.1\ mm$ (wall stick). Flow conservation is obviously verified.

Figure 7 shows a velocity profile during flow along branch II for a high output rate ($0 < Z < 5.1\ mm$, $X = -12\ mm$, $\dot{\gamma}_{app} = 293\ s^{-1}$). Comparison with the Figure 6 indicates that the slip region across the die width increases as the flow rate in branch II increases. Complementary experiments have shown that wall slip is initiated at the contraction and occurs all along the die land [ROB 01]. Wall slip is not homogeneous across the die width for shear rates between 100 and $350\ s^{-1}$, that

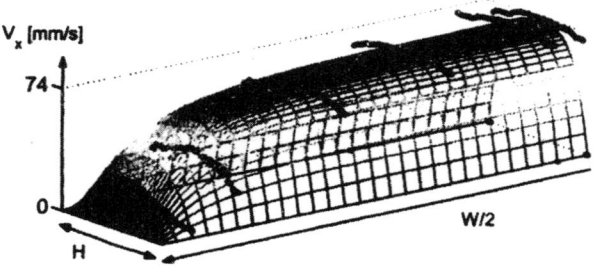

Figure 7. *Velocity profile on branch II at $\dot{\gamma}_{app} = 293\ s^{-1}$ (black dots at zero velocity near the wall are related to an artefact)*

explains the curvature of the branch II (Figure 2), and becomes homogeneous only for very high shear rates. Slip velocities have been deduced, and are of the same order of magnitude as those measured with a capillary rheometer [ROB 00].

During the instability, **birefringence fringes pulsations** are observed. Figure 8 depicts two periods of the time dependant synchronized evolution of the reservoir pressure (solid lines) and the PSD along the symmetry axis in the reservoir region (dotted lines) during the instability ($\dot{\gamma}_{app} = 50\ s^{-1}$). We observe for the first time that

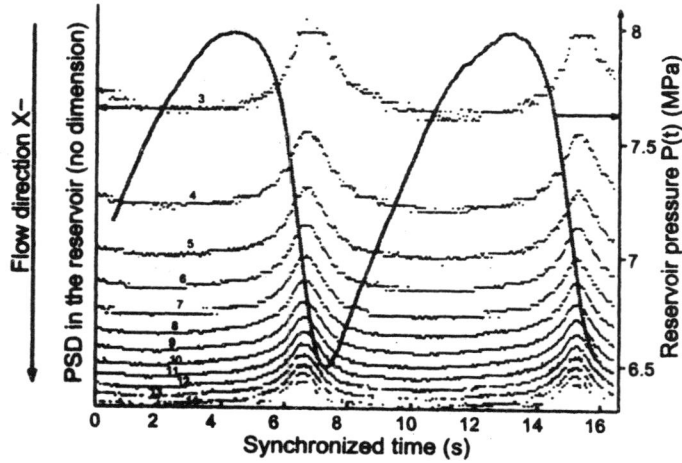

Figure 8. *Evolution of pressure and PSD along the symmetry axis in the reservoir during spurt flow*

periodic oscillations of the PSD are in phase quadrature with the pressure, and thus in phase with the velocity (see Figure 4). As a result, during the transition branch I →

branch II, the wall shear stress decreases (wall slip) but the PSD increases because the flow rate increases (elongational stresses).

3.2. Flow in the PTFE coated slit die

Figure 9 presents flow curves for three different dies : the polished steel die (◇), the PTFE coated die (●) and an abrupt contraction (△). The contraction die, with the same height ($H = 1\ mm$) and a length $L \approx 0\ mm$, represents the entry effects. The flow

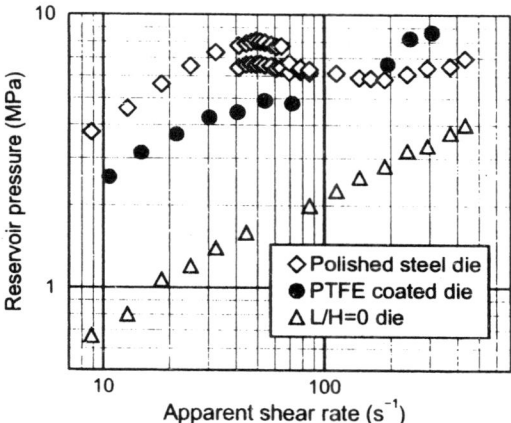

Figure 9. *Apparent flow curve for both steel and PTFE coated die*

curve for the polished steel die (◇) is the same as in Figure 2, with the two branches separated by the spurt instability. According to previous work [PIA 95, WAN 97], the flow curve obtained with the PTFE coated die is continuous and the extrusion is always stable. We observe that this flow curve is below the flow curve for the polished steel die for apparent shear rates below 150 s^{-1}, and above for higher apparent shear rates. In this case, wall slip is expected at all shear rates.

Figure 10 shows the result of a velocity profile in the PTFE coated die for $\dot{\gamma}_{app} = 19\ s^{-1}$ ($0 < Z < 5.1\ mm$, $X = -12\ mm$), corresponding to flow along branch I for the polished steel die. The polymer melt slips at the wall across the entire die width, except on each side close to the glasses, so side effects are reduced. All experimental profiles show wall slip, and consist of a parabolic profile plus a wall slip value. As there is always wall slip in the entire die width, profiles are below the theoretical one and flow conservation is obviously verified.

We present in Figure 11 a velocity profile for $\dot{\gamma}_{app} = 308\ s^{-1}$ in PTFE coated die ($0 < Z < 5.1\ mm$, $X = -12\ mm$). In comparison with the profile presented in the Figure 7 for flow in the polished steel die at the same apparent flow rate value, wall

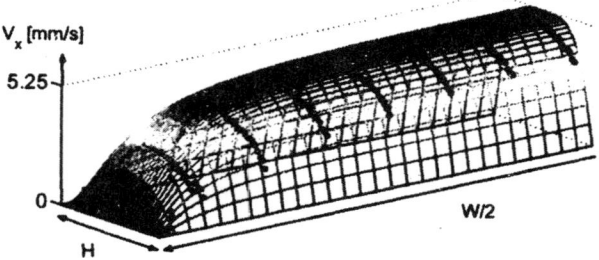

Figure 10. *Velocity profile for $\dot{\gamma}_{app} = 21.6\ s^{-1}$ in PTFE coated die*

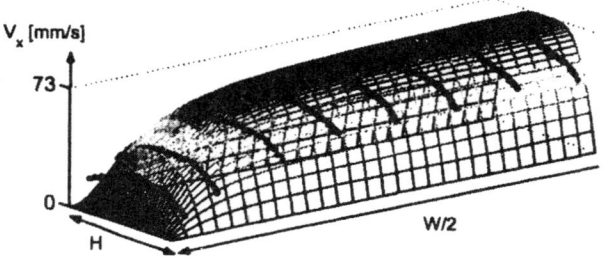

Figure 11. *Velocity profile for $\dot{\gamma}_{app} = 308\ s^{-1}$ in PTFE coated die*

slip is nearly homogeneous across the die width. Furthermore, the upstream measured pressure is higher for the PTFE coated die (8.5 MPa at $\dot{\gamma}_{app} = 308\ s^{-1}$) than for the polished steel die (6.35 MPa at $\dot{\gamma}_{app} = 293\ s^{-1}$). All that features suggest a different mechanism for slip on PTFE and on polished steel.

4. Conclusion

We have experimentally investigated the spurt or stick-slip defect, an extrusion instability of linear polymer melts, on a high density polyethylene with an extruder equipped with a transparent slit die, where transient measurements of pressure drop, local velocities by laser Doppler velocimetry and stress fields by flow birefringence were possible. Velocity measurements confirm that, during the instability, the velocity profile oscillates between two steady state profiles : a classical viscous profile on branch I, and a profile showing a strong macroscopic wall slip on branch II. We showed that oscillations of stresses and velocities are in phase, and in quadrature with the pressure. Wall slip seems to be initiated at the entry of the die land and occurs all along, but is not homogeneous across the die width (confined slip phenomenon).

Finally, we have shown that using fluoropolymer coated surfaces eliminates the instability by promoting a homogeneous wall slip, suggesting a different slip mechanism.

5. References

[DEN 01] DENN M., "Extrusion Instabilities and Wall Slip", *Ann. Rev. Fluid Mech.*, vol. 33, 2001, p. 265–287.

[DUR 96] DURAND V., VERGNES B., AGASSANT J.-F., BENOIT E., KOOPMANS R., "Experimental study and modeling of oscillating flow of high density polyethylenes", *J. Rheol.*, vol. 40, n° 3, 1996, p. 383–394.

[LAR 92] LARSON R. G., "Instabilities in viscoelastic flow", *Rheol. Acta*, vol. 31, 1992, p. 213–263.

[MÜN 00] MÜNSTEDT H., SCHMIDT M., WASSNER E., "Stick and slip phenomena during extrusion of polyethylene melts as investigated by laser-Doppler velocimetry", *J. Rheol*, vol. 44, n° 2, 2000, p. 245–255.

[PIA 95] PIAU J.-M., KISSI N. E., MEZGHANI A., "Slip flow of polybutadiene through fluorinated dies", *J. Non-Newt. Fluid Mech.*, vol. 59, 1995, p. 11–30.

[ROB 00] ROBERT L., VERGNES B., DEMAY Y., "Experimental investigations of the spurt instability in the flow of molten high density polyethylene", *Proc. XIIIth Int. Congr. on Rheology*, 2000.

[ROB 01] ROBERT L., Instabilité oscillante des polyéthylénes linéaires : observations et interprétations, PhD thesis, Université de Nice Sophia-Antipolis, 2001.

[WAN 97] WANG S., DRDA P., "Stick-slip transition in capillary flow of linear polyethylene: 3. Surface conditions", *Rheol. Acta*, vol. 36, 1997, p. 128–134.

[WAN 99] WANG S., "Molecular transitions and dynamics at polymer/wall interfaces : Origins of flow instabilities and wall slip", *Adv. Polym. Sci.*, vol. 138, 1999, p. 227–275.

Chapter 27

Induced Non Homogeneity in a Saturated Granular Media Submitted to Slow Shearing

Nicolas Roussel, Christophe Lanos and Yannick Mélinge
Groupe de Recherche en Génie Civil de Rennes, INSA, Rennes, France

1. Introduction

When an initially homogeneous saturated granular material is slowly sheared, the induced pressure field creates a pore pressure gradient in the sample that enables a non-homogeneous water flow to take place. If no water is brought to the studied material, this results in a lowering void ratio and a heterogeneous hardening of the sample. The behaviour relation changes through the drying. In this case, an eulerian formulation of this biphasic flow problem is proposed. These phenomena are studied in the case of a squeezing flow induced by a simple compression test (plastometer test). Theoretical and experimental values are compared.

2. The simple compression test

2.1. *Description of the test*

This study is based on a simple compression test (plastometer test) carried out on cylindrical samples with reduced slenderness.

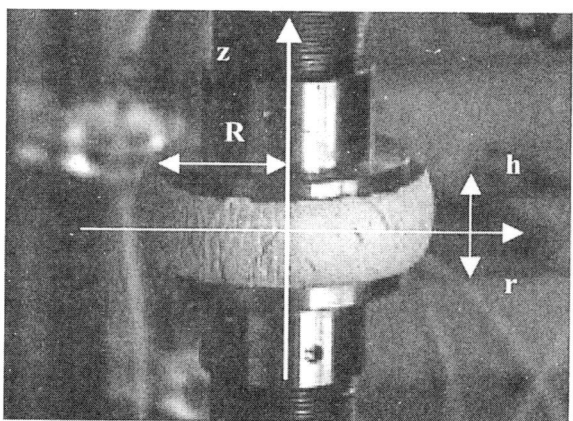

Figure 1. *Compression test geometry*

The plastometer consists in two coaxial circular and parallel plates, without any rotation. The squeezing of the sample between the two plates induces a radial flow. R is the radius of the plates, h the height of the sample, F the compression load applied on the plates and hs the compression speed (Figure 1). The recording of the compression load versus the height of the sample, the plate radius and the compression speed allows the identification of parameters linked to an average behaviour relation [LAN 00]. Testes are made with constant compression speed and the tested material fills permanently the area between the plates.

3. Stress and strain field in an homogeneous plastic material

This problem presents cylindrical symmetry and, for a perfect plastic material considered as homogeneous, it is assumed that the studied phenomena only depend on r (elongational flow).

The radial flowing speed is [COV 77]:

$$V_r(r,t) = \frac{r\dot{h}}{2h(t)}$$

The total radial stress is:

$$\sigma_r(r,t) = \sigma_r^{(d)}(r,t) + p(r,t)$$

where the deviatoric radial stress is

$$\sigma_r^{(d)}(r,t) = \frac{K_i}{\sqrt{3}}$$

The von Mises criterion is fulfilled everywhere in the sample.

The pressure gradient writes:

$$\frac{\partial p(r,t)}{\partial r} = -\frac{2}{h(t)} K_i \qquad [1]$$

where Ki is the plastic yield value.

The recording of the compression load versus the height of the sample allows the determination of Ki if the sample stays homogeneous [LAN 99].

All those statements are still true if there is a water migration but Ki would then depend on time and radius.

4. Biphasic model

4.1. *Mass balance*

The filling saturated material is composed with skeleton hs(r,t) and water hw(r,t) following:

$$h_s(r,t) + h_w(r,t) = h(t)$$

The local radial speed of the skeleton is Vs(r,t) and the radial speed of the fluid is Vw(r,t) following:

$$\frac{h_s(r,t)V_s(r,t)+h_w(r,t)V_w(r,t)}{h_s(r,t)+h_w(r,t)}=V_r(r,t) \qquad [2]$$

which gives, by substituting the void ratio in [2]:

$$e(r,t)=\frac{h_w(r,t)}{h_s(r,t)}$$

$$\frac{V_s(r,t)+e(r,t)V_w(r,t)}{1+e(r,t)}=V_r(r,t)$$

The mass balance for the skeleton is:

$$\frac{\partial h_s(r,t)}{\partial t}=-h_s(r,t)\left(\frac{\partial V_s(r,t)}{\partial r}+\frac{V_s(r,t)}{r}\right)-V_s(r,t)\frac{\partial h_s(r,t)}{\partial r}$$

And for the fluid:

$$\frac{\partial h_w(r,t)}{\partial t}=-h_w(r,t)\left(\frac{\partial V_w(r,t)}{\partial r}+\frac{V_w(r,t)}{r}\right)-V_w(r,t)\frac{\partial h_w(r,t)}{\partial r}$$

Noting that:

$$\frac{\partial e(r,t)}{\partial t}=\frac{1}{h_s(r,t)}\frac{\partial h_w(r,t)}{\partial t}-\frac{h_w(r,t)}{h_s^2(r,t)}\frac{\partial h_s(r,t)}{\partial t}$$

and

$$-\frac{\partial h(t)}{\partial t}=\dot{h}_s=-\frac{\partial h_w(r,t)}{\partial t}-\frac{\partial h_s(r,t)}{\partial t}$$

and, as the plates stay horizontal,

$$\frac{\partial h_s(r,t)}{\partial r}=-\frac{\partial h_w(r,t)}{\partial r}$$

Without going through all the minutiae, the following equation is obtained:

$$\frac{\partial e(r,t)}{\partial t}=-e(r,t)\frac{\partial(V_w(r,t)-V_s(r,t))}{\partial r}-\frac{r\dot{h}}{2h(t)}\frac{\partial e(r,t)}{\partial r}$$
$$-(V_w(r,t)-V_s(r,t))\frac{1-e(r,t)}{1+e(r,t)}\frac{\partial e(r,t)}{\partial r}-e(r,t)\frac{(V_w(r,t)-V_s(r,t))}{r} \qquad [3]$$

For an initially homogeneous sample at $t=0$, $\frac{\partial e(r,t)}{\partial r}=0$.

If the compression speed is high enough to ensure that no water migration occurs then Vw(r,t)–Vs(r,t) = 0 and e(r,t) stays constant through the test. But, if the compression speed is low enough, there is a water migration and the term Vw(r,t)–Vs(r,t) must be studied and expressed in terms of the void ratio and its spatial derivatives.

4.2. *Liquid-solid interaction*

Darcy's law reads:

$$V_w(r,t) - V_s(r,t) = -K_p(r,t)\frac{\partial U(r,t)}{\partial r} \qquad [4]$$

where Kp(r,t) is the interaction coefficient and U(r,t) the pore pressure. The interaction coefficient can be expressed in terms of the void ratio.

[KOZ 27] gives a relation between interaction coefficient and void ratio:

$$K_p(r,t) = K_{p0}\frac{e(r,t)^3}{(1+e(r,t))} \qquad [5]$$

As the interaction constant Kp0 is a complex combination of different soil properties including particle size and shape, density, mineralogy etc., it's necessary to calibrate the model by identifying Kp0 using, for instance, a permeameter test.

4.3. *Total and effective stress, pore pressure*

The Terzaghi principle reads [TER 25]:

$$U(r,t) + p'(r,t) = p(r,t)$$

where p'(r,t) is the effective pressure and p(r,t) the total pressure.

If the studied material is associated with a Cam Clay model [ROS 58], once the test has begun, the skeleton is supposed to have reached and maintained a plasticity state.

The effective pressure writes:

$$p'(r,t) = \frac{2K_i(r,t)}{M} \qquad [6]$$

where M is a parameter linked to the internal friction angle φ.

If the sample is not homogeneous, [1] becomes:

$$\frac{\partial p(r,t)}{\partial r} = -\frac{2}{h(t)}K_i(r,t) + \frac{\partial \sigma_r^{(d)}(r,t)}{\partial r}$$

Thus, the pore pressure gradient reads:

$$\frac{\partial U(r,t)}{\partial r} = -\frac{2}{h(t)}K_i(r,t) + \left(\frac{1}{\sqrt{3}} - \frac{2}{M}\right)\frac{\partial K_i(r,t)}{\partial r} \qquad [7]$$

Ki(r,t) is then be expressed in terms of the void ratio.

The Cam Clay model gives:

$$p'(r,t) = p'_0 \mathrm{Exp}\left(\frac{e_0 - e(r,t)}{\lambda}\right)$$

where p'0 is the effective pressure when e(r,t) = e0.

Using relation [6], the expression of the plastic yield value reads:

$$K_i(r,t) = K_{i0}\mathrm{Exp}\left(\frac{e_0 - e(r,t)}{\lambda}\right) \qquad [8]$$

where Ki0 is the plastic yield value when e(r, t) = e0.

If two plastometer tests are made with compression speeds so that there is no water migration (undrained tests) for at least two different void ratios, Ki0, e0 and λ can then be identified using the method described by [LAN 99].

4.4. *Final differential equation*

By substituting the relations [4], [5], [7] and [8] in relation [3], a differential equation linking the spatial and temporal derivatives of the void ratio with the drainage parameters and skeleton characteristics is obtained using symbolic calculus tools.

5. Theoretical and experimental results for kaolin clay pastes

5.1. *Homogeneous sample*

5.1.1. *Compression test results*

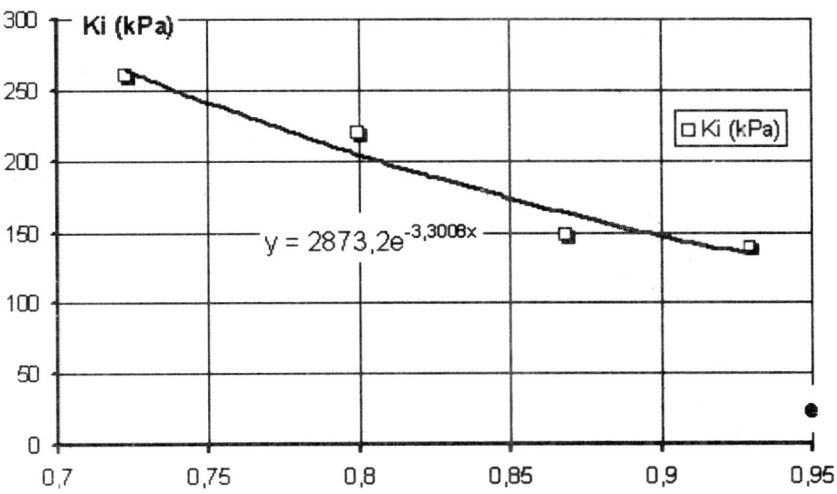

Figure 2. *Plastic and friction experimental yield values versus void ratio*

The tests are realised with a compression speed high enough to ensure that no water migration occurs for four different void ratios in pure Kaolin clay paste sample (Figure 2). The plastic yield parameters Ki0, e0 and λ can then be estimated.

5.1.2. *Permeability measurements*

The interaction coefficient Kp0 is determined through several permeameter tests, Kp0 = 3,6.10-13 m2/s/Pa for the tested Kaolin clay paste.

5.2. *Induced heterogeneity*

Experimental and theoretical simple compression test results

The differential equation is solved using a finite differences algorithm. The initial void ratio of the homogeneous sample is e(r,t) = 0.78 at t = 0, its initial height h0 = 0.03m. The compression speed is 0.001mm.s-1 and the radius of the plates is 0.05m. The local water content and thus the void ratio are experimentally measured for different radius by samplings drying. The results are plotted on Figure 3.

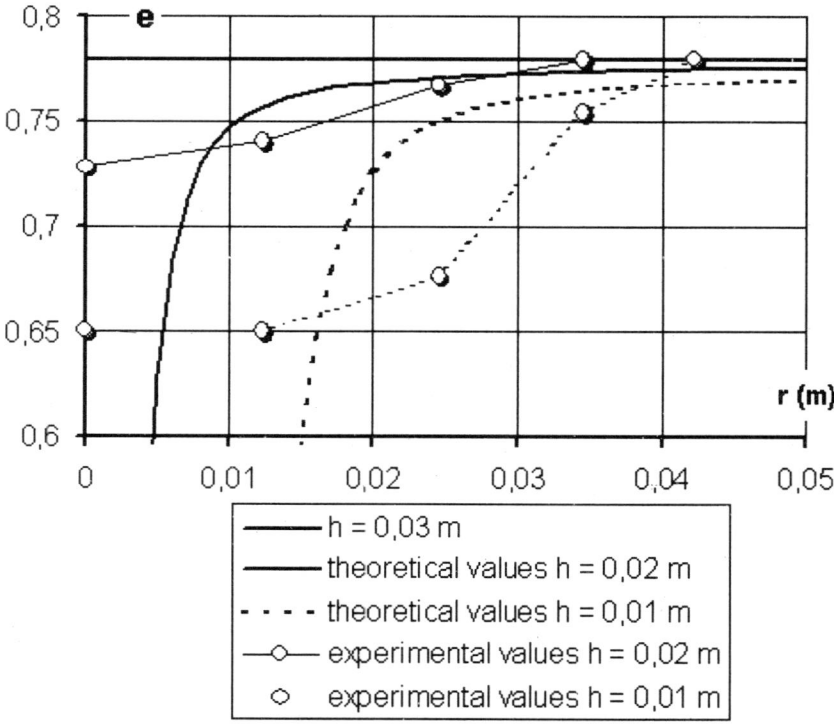

Figure 3. *Experimental and theoretical void ratio values versus h and r (Kozeny's permeability model)*

The experimental and the analytical results are in the same order of magnitude except in the central part of the apparatus. In the central part of the plastometer, the theoretical value of the void ratio becomes fairly low compared to the experimental values especially for small sample heights.

The Kozeny's relation deals with the possibility for a fluid to cross the sample but the phenomena studied here concern as well the ability of a granular material to let all the contained water migrate. It is assumed that a minimal water amount stays linked to the flowing grains. Such consideration is valid knowing surface forces can be particularly high for small clay particles with high specific surface. The stresses created by the flow are not sufficient to break those links. Thus, the "dynamic" permeability does not equal zero when the void ratio equals zero but when what could be called the "free" water has migrated. The model can then be modified to take in account this limit by introducing a modified relation between permeability and void ratio as:

$$K_p(r,t) = K_{p0} \frac{(e(r,t) - e_r)^3}{(1 + (e(r,t) - e_r))(2(e(r,t) - e_r) + 1)}$$

where er is the void ratio for which the sample contains the minimal water amount. For the tested material, er = 0.62 and the associated value for Kp0 is 1,9.10-12 m2/s/Pa. The comparison with the experimental results is plotted Figure 4.

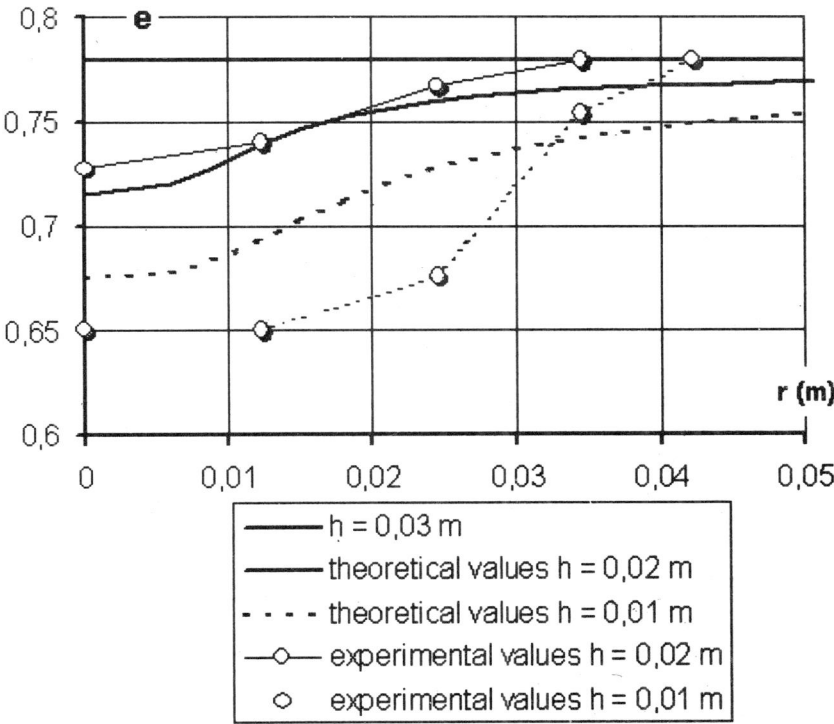

Figure 4. *Experimental and theoretical void ratio values versus h and r (modified Kozeny's permeability model)*

6. Conclusion

The modified model gives a good estimation of the sample drying during the compression test. The understanding of the hardening process allows a better treatment of the plastometer test data.

Most of the model parameters can be identified by using the plastometer test itself which makes it a powerful and complete tool to study the water content influence on the particle/fluid mixture global behaviour.

The limits of Kozeny's expression of the liquid-solid interaction coefficient have been demonstrated in the particular case of sheared small particles but an optimised expression is yet to be found.

7. Bibliography

[COV 77] Covey GH., Application of the parallel plate plastometer to brown coal rheometry, Thèse de doctorat, Université de Melbourne, Australia, 1977.

[KOZ 27] Kozeny J., Sitzber. Akad. Wiss. Math.-naturw. Klasse, 136, Wien, Austria, 1927.

[LAN 99] Lanos C, "Caractérisation du comportement rhéologique des pâtes minérales", *2ème Congrès Universitaire de Génie Civil AUGC*, Poitiers, France, 6–7 mai 1999, Vol. 1, pp. 183–190.

[LAN 00] Lanos C., "Reverse identification method associate to compression test", *Proceedings of the 13th international congress on rheology*, Cambridge, England, 20–25th August 2000, 2, pp. 312–314.

[ROS 58] Roscoe K.H., Schofield A.N., Wroth C.P., "On the yielding of soils", *Geotechnique*, 8, 1958, pp. 22–53.

[ROU 01] Roussel N., Analyse des écoulements de fluides homogènes complexes et plastiques diphasiques: application à l'essai de compression simple, Thèse de doctorat, I.N.S.A. Rennes, 2001.

[TER 25] Terzaghi K., *Erdbaumechanick*, Franz Deuticke, Wien, Austria, 1925.

Chapter 28

A Priori Model Reduction Method for Thermo-mechanical Simulations

David Ryckelynck
Laboratoire de Mécanique des Systèmes et des Procédes, Ecole Nationale Supérieure d'Arts et Métiers, Paris, France

1. Introduction

Thanks to the Finite Element Method, we are able to describe accurately the thermo-mechanical transformations that occur during processes like casting, forming and cutting. It is quite easy to construct a very complex finite element model. It is more difficult to forecast the state of the studied system during the process, by solving a time-dependant and non-linear thermo-mechanical problem. Generally a good knowledge of what are the main significant phenomena allows one to simplify the finite element model. Obviously, the simpler the finite element model is, the faster we forecast the state of the system.

An interesting approach to simplify a finite element model is to build a reduced-order model, thanks to an algorithm. Finite element modelling uses generally a large number of variables to describe the state of the studied system. The aim of model reduction is to define a transformation matrix to get few basis functions, from the Finite Element shape functions, to describe the spatially distributed state. In the framework of non-linear time-dependant problems we can distinguish an a posteriori approach based on Karhunen-Loève expansion [SIR 91] [PAR 01] and an a priori approach based on the Krylov subspace [KNO 95] [WEI 95] [SUN 01]. The first kind of approach is based on the knowledge of the evolution of the state variables of the system, the second one is not based on this knowledge. In the Karhunen-Loève method, state snapshots, taken at various time instants, span a small subspace. The principal directions of this subspace are chosen as basis functions. Hence, the Karhunen-Loève expansion allows one to extract the most significant part of the evolution of the state snapshots. The Krylov subspace allows one to forecast the state evolution of the system. Using this subspace, only few unknowns are needed during the incremental computation of the state. At a given time step, the basic functions are built with the residual of the non-linear equations of the finite element problem and a matrix defined to correct the approximate state forecast at the end of the time step. The basis functions can be used to simplify the computation of the state during several time increments [KNO 95]. In practice, the same Krylov subspace is used only for few time increments. This sub-space had to be adapted if there is large evolution of the state variables after the time step when it was constructed. In this paper we propose an a priori reduction method because we want to use it as a computational strategy.

The purpose of our approach is to define residuals over a time interval involving a large state evolution. To do so, we need to improve the computation of an approximate state over the same time interval, thanks to an iterative algorithm. Obviously this approximate state is described thanks to the basis functions of the reduced-order model. A simple reduced-order model is supposed to be known and several correction stages are used to improve the basis functions thanks to the residuals and to improve also the approximate state over the entire time interval. So, this computational strategy is necessarily a non-incremental one. In order to improve the approximate state at each stage by a linear problem, we use the LATIN method [LAD 85] [LAD 96] to define our algorithm. Because the correction stage is a linear stage, we need basis functions for both displacements and stresses. The system Σ is split into sub-domains. Basis functions are defined for each sub-domain Ω of Σ. Moreover, in the framework of the LATIN method, the basis functions for stresses satisfy some equilibrium conditions. Therefore we need in fact couples of basis functions to describe stresses inside each sub-domain Ω and normal stresses over the boundary of Ω. Even if the approximate solution is far from the solution of the finite element problem, the residuals give efficient indications to construct the reduced-order model. Maybe it is due to equilibrium conditions satisfied when the corrections are computed. Several correction stages of the approximate state must be considered in order to obtain residuals that span a sufficient large subspace to define the reduced-order model. The subspace spanned by the residuals is called the LATIN subspace. During the correction stages the basis functions of the reduced-order model are extracted from the LATIN subspace thanks to the Karhunen-Loève expansion, by selecting the most significant patterns of the known approximate state. The subspace associated to these basis functions is called the LATIN-KL subspace. A 2D example illustrates the capability of the model reduction technique. It corresponds to a kind of casting problem. The part material is a kind of aluminium and the mould material is a kind of sand. We obtain a reduced-order model defined for large non-linear evolutions, during all the cooling of the part. Non-linear couplings are due to the thermal contraction of the part and thermal expansion of the mould, which modify the thermo-mechanical contact conditions between the part and the mould. An accurate approximate state evolution is obtained with few basis functions obtained during 100 correction stages.

2. The Finite Element model

In order to define the LATIN algorithm, we introduce here a convenient manner to describe the finite element model. The considered thermo-mechanical problem involves small strains, small displacements, elasticity, and contact producing thermo-mechanical coupling. Equations are given for each sub-domain Ω. Boundary conditions are given for interfaces between a sub-domain Ω and a sub-domain Ω', and for interfaces between a sub-domain Ω and the boundary $\partial\Sigma$ of the system. A mesh is used to describe the displacement field $\underline{U}$ and the temperature field T by introducing matrix of shape functions $\mathbf{N}_U$ and $\mathbf{N}_T$, and Finite Element degrees of freedom q and θ respectively. The shape functions $\mathbf{N}_U$ and $\mathbf{N}_T$ define two subspaces $\mathcal{U}$ and $\mathcal{T}$ respectively. We assume that these subspaces are large enough to represent accurately the evolution of the gradients of $\underline{U}$ and T. The time interval $]0, t_f]$ is divided into time steps $]t_j, t_{j+1}]$. The forward Euler scheme is used to solve the time differential equations.

2.1. *Equations defined inside* $\Omega \in \Sigma$

Inside each sub-domain Ω the following conditions must be satisfied. There are initial conditions for temperatures and stresses σ. Stresses, normal stresses $\underline{F}$ over $\partial\Omega$, temperatures, heat transfer $\underline{g}$, internal energy e normal heat transfer ϕ over $\partial\Omega$

328 Prediction of Defects in Material Processing

and heat dissipation r, must satisfy equilibrium conditions at any time instant t_{j+1} (forward Euler scheme):

$$\int_\Omega \text{Tr}\left[\varepsilon\left(\underline{U}^*\right)\,\sigma\right] d\Omega = \int_{\partial\Omega} \underline{U}^* \cdot \underline{F}\, dS \quad \forall \underline{U}^* \in \mathcal{U} \qquad [1]$$

$$\int_\Omega T^* \dot{e}\, d\Omega - \int_\Omega \underline{\text{grad}}(T^*)\, \underline{g}\, d\Omega = $$
$$-\int_{\partial\Omega} T^* \phi\, dS + \int_\Omega T^* r\, d\Omega \quad \forall T^* \in \mathcal{T} \qquad [2]$$

The thermo-mechanical constitutive relations concern the elastic strain rate $\dot{\varepsilon}^e$, the thermal expansion $\dot{\varepsilon}^{th}$, the heat dissipation, the internal energy and the heat transfer. The heat dissipation is a fraction of the irreversible mechanical rate of work.

$$\varepsilon\left(\underline{\dot{U}}\right) = \dot{\varepsilon}^e + \dot{\varepsilon}^{th} \qquad [3]$$

$$\dot{\sigma} = \mathbf{K}(T)\, \dot{\varepsilon}^e \qquad \dot{\varepsilon}^{th} = \alpha(T)\, \dot{T} \qquad [4]$$

$$\dot{e} = \rho\, c(T)\, \dot{T} \qquad \underline{g} = -k(T)\, \underline{\text{grad}}(T) \qquad [5]$$

2.2. Boundary conditions over $\partial\Omega$

On each integration point on the boundary of Ω there is an interface between an other sub-domain Ω' or the considered point belongs to the boundary of Σ. Two kinds of interfaces between Ω and Ω' are considered : a perfect interface and a contact interface. In the case of perfect interface the displacement and the temperature must be continuous and some equilibrium conditions must be satisfied. We assume that heat conduction decreases with the gap j such that: $h(j) = h_o\, e^{-\gamma j}$.

3. The non-incremental approach to construct the reduced-order model

The purpose of the LATIN method is to build an algorithm to compute corrections of an approximate state thanks to stages, both global and linear, defined over the entire time interval. The residuals we want to define, must characterise what could be the mean value over $]0, t_f]$ of the corrections by using the Finite Element subspaces $\mathcal{U}$ and $\mathcal{T}$. But, to avoid expensive computations we do not want to compute these corrections by using the Finite Element subspaces. In order to define the residuals, we present the LATIN algorithm as if we used the Finite Element subspaces. Hence we modify the formulation of the linear correction stage to employ the reduced-order model for the computation of the corrections.

3.1. *A LATIN algorithm*

The Finite Element problem presented in section 2 is both global and non-linear. But, we chose a formulation such that the global equations are linear and the non-linear equations are local equations. The LATIN method is based on three principles: P1, P2 and P3 [LAD 85] [LAD 96]:

– P1 : in order to split the difficulties (global and non-linear equations), two groups of equations are created from the reference problem : a group of local equations and a group of linear equations;

– P2 : the algorithm is an iterative procedure that provides a solution for each group of equations at each iteration, these solutions are defined over the structure and over the entire time interval;

– P3 : an appropriate space-time representation has to be used to solve the global equations.

The reference problem is the Finite Element problem. The linear equations are: the initial conditions, the equilibrium conditions (1) (2) and the strain definition (3). Necessarily, the other equations belong to the group of the local equations.

The third principle of the LATIN method allows us to exploit various kinds of time and space representation of the variables ($\underline{U}$, σ, $\underline{E}$, T, ϕ). In our approach we use basis functions associated to the reduced-order model. Different approaches were proposed to define a convenient representation of the corrections ($\underline{U} - \underline{U}_n$, $\sigma - \sigma_n$, ...), expressed linearly in terms of several space functions [LAD 96] [ART 92]. P. Ladevèze has proposed in [LAD 96] an approach very closed to the Karhunen-Loève expansion to construct these space functions at each correction stage. To employ the previously computed space functions during a given correction stage, P. Bussy [BUS 90] proposed to compute a part of the corrections while using the known space functions as basis functions. In practice, this last approach is efficient enough to provide the entire correction of the state variables [PEL 00]. The proposed approach can be understood as an extension of the approach proposed by P. Bussy, in order to construct a reduced-order model of a process, and not only to simplify the computation of the correction stage of a LATIN algorithm. The most simple approximate state ($\underline{U}_o$, σ_o, $\underline{E}_o$, T_o, ϕ_o) is obtained from the initial state ($\underline{U}_{ini}$, σ_{ini}, $\underline{E}_{ini}$, T_{ini}, ϕ_{ini}).

3.2. *The Finite Element formulation of the correction stage*

In accordance with the principle P2, the purpose of the correction stage is to construct a solution of the linear equations (1), (2) and (3). To do so, we have to replace the local equations by linear equations defined thanks to search directions. Let's consider that an approximate state ($\underline{U}_n$, σ_n, $\underline{E}_n$, T_n, ϕ_n) satisfying (1), (2) and (3), is known. A fixed-point approach allows to replace the local equations defined over Ω by:

$$\dot{\sigma} = \mathbf{K}(T_n)\,\dot{\varepsilon}^e \qquad \dot{\varepsilon}^{th} = \alpha(T_n)\,\dot{T}_n \qquad [6]$$

$$r = \mathrm{Tr}\left[\left(\varepsilon\left(\underline{\dot{U}}_n\right) - \dot{\varepsilon}_n^e\right)\sigma_n\right] \qquad [7]$$

$$\dot{e} = \rho\,c(T_n)\,\dot{T} \qquad \underline{g} = -k(T_n)\,\underline{\mathrm{grad}}(T) \qquad [8]$$

On the boundary of each sub-domain, search directions have been defined in [CHA 96] for mechanical problems involving contact conditions. Similar search directions are used at the correction stage, for interfaces between Ω and Ω':

$$\underline{\dot{F}} = \underline{\dot{F}}_{n+1/2} - \widehat{k}^{me}\left(\underline{\dot{U}} - \underline{\dot{U}}_{n+1/2}\right) \qquad [9]$$

$$\phi = \phi_{n+1/2} + \widehat{h}^{th}\left(T - T_{n+1/2}\right) \qquad [10]$$

where $\underline{F}_{n+1/2}$, $\underline{U}_{n+1/2}$, $\underline{F}'_{n+1/2}$, $\underline{U}'_{n+1/2}$, $T_{n+1/2}$, $T'_{n+1/2}$, $\phi_{n+1/2}$ and $\phi'_{n+1/2}$ are defined by the contact conditions or the perfect interface conditions and these equations:

$$\underline{\dot{F}}_{n+1/2} = \underline{\dot{F}}_n + \widehat{k}^{me}\left(\underline{\dot{U}}_{n+1/2} - \underline{\dot{U}}_n\right)$$
$$\underline{\dot{F}}'_{n+1/2} = \underline{\dot{F}}'_n + \widehat{k}^{me}\left(\underline{\dot{U}}'_{n+1/2} - \underline{\dot{U}}'_n\right) \qquad [11]$$

$$\phi_{n+1/2} = \phi_n - \widehat{h}^{th}\left(T_{n+1/2} - T_n\right)$$
$$\phi'_{n+1/2} = \phi'_n - \widehat{h}^{th}\cdot\left(T'_{n+1/2} - T'_n\right) \qquad [12]$$

To find the variables $\underline{F}_{n+1/2}$, $\underline{U}_{n+1/2}$, $\underline{F}'_{n+1/2}$, $\underline{U}'_{n+1/2}$, $T_{n+1/2}$, $T'_{n+1/2}$, $\phi_{n+1/2}$ and $\phi'_{n+1/2}$, in case of contact condition, on an integration point, we assume that there is no contact, if it's not correct hence there is contact and we assume there is adherence. At last, if it's not correct there is sliding contact. In case of an integration point on $\partial\Sigma$, the variables $\underline{F}_{n+1/2}$, $\underline{U}_{n+1/2}$, $T_{n+1/2}$ and $\phi_{n+1/2}$ must satisfy the boundary conditions. In practice, the Young modulus E and a characteristic length d of Ω are used to define the parameters of the algorithm such that:

$$\widehat{k}^{me} = \frac{E(T_n)}{d} \qquad \widehat{h}^{th} = \frac{k(T_n)}{d}$$

3.3. The LATIN subspace

Basis functions have to be defined for $\underline{U}$, σ, $\underline{F}$, T and ϕ. Let's note respectively $\underline{W}_U$, $\mathbf{W}_\sigma$, $\underline{W}_F$, W_T and W_ϕ the residuals defining these basis functions. The three residuals $\underline{W}_U$, $\mathbf{W}_\sigma$ and $\underline{W}_F$ must characterise respectively $\underline{U} - \underline{U}_n$, $\sigma - \sigma_n$ and $\underline{F} - \underline{F}_n$, over $]0, t_f]$. To avoid several resolutions of Finite Element problems during a correction stage, we introduce two given scalar time functions λ^{me} and λ^{th} such that:

$$\underline{U} = \underline{U}_n + \lambda^{me}(t)\, \underline{W}_U$$

$$\sigma = \sigma_n + \lambda^{me}(t)\, \mathbf{W}_\sigma$$

$$\underline{F} = \underline{F}_n + \lambda^{me}(t)\, \underline{W}_F$$

$$T = T_n + \lambda^{th}(t)\, W_T$$

$$\phi = \phi_n + \lambda^{th}(t)\, W_\phi$$

The couple $(\mathbf{W}_\sigma, \underline{W}_F)$ must satisfy the equilibrium conditions (1). Because the residuals are space functions, only a weak form of the equations of the correction stage can be satisfied. The main equations of this weak form are:

$$\int_0^{t_f} \dot{\lambda}^{me}\, \dot{\sigma}\, d\tau = \int_0^{t_f} \dot{\lambda}^{me}\, \mathbf{K}(T_n)\left(\dot{\varepsilon} - \alpha(T_n)\, \dot{T}_n\right) d\tau \qquad [13]$$

$$\int_0^{t_f} \dot{\lambda}^{me}\, \underline{\dot{F}}\, d\tau = \int_0^{t_f} \dot{\lambda}^{me}\, \underline{\dot{F}}_{n+1/2} - \widehat{k}^{me}\left(\underline{\dot{U}} - \underline{\dot{U}}_{n+1/2}\right) d\tau \qquad [14]$$

Hence $\underline{W}_U$ is computed to satisfy a linear system:

$$\underline{W}_U = \mathbf{N}_U\, \underline{q}_W$$

$$\mathbf{K}_W^{me}\, \underline{q}_W = \underline{F}_W^{me}(t_f)$$

The residuals $\mathbf{W}_\sigma$ and $\underline{W}_F$ are deduced from $\underline{W}_U$ by using the equations (13) and (14). The sign of $\dot{\lambda}^{me}$ is chosen such that $\|\underline{F}_W^{me}(t)\|$ is an increasing time function, like it has been proposed in [PEL 00].

The same approach is used to compute W_T and W_ϕ:

$$W_T = \mathbf{N}_T\, \underline{\theta}_W$$

$$\mathbf{K}_W^{th}\, \underline{\theta}_W = \underline{F}_W^{th}(t_f)$$

$$W_\phi = \int_0^{t_f} \dot{\lambda}^{th}\, \mathbf{N}_T^T \left(\phi_{n+1/2} - \phi_n + \widehat{h}^{th}\left(T_n - T_{n+1/2}\right)\right) d\tau$$

The residuals can be used as basis functions only if they have significant values. If it is the case they are normalised.

3.4. The LATIN-KL subspace

As the number of correction stages increases, the size of the LATIN subspace increases also. Moreover the fields $\underline{W}_U$, $\mathbf{W}_\sigma$, $\underline{W}_F$, W_T and W_ϕ do not have any physical meaning. Since we know an approximate state evolution ($\underline{U}_n$, σ_n, $\underline{F}_n$, T_n, ϕ_n) we can use the Karhunen-Loève expansion [SIR 91] to define a subspace with a better physical sense. Therefore we propose to use the Karhunen-Loève method to extract a subspace from the LATIN subspace. This subspace is the LATIN-KL subspace. For each variable $\underline{x}$ represented with the basis functions of the LATIN subspace, we use the coefficients of the linear combination $\widetilde{\underline{x}}(t)$ of basis functions placed in a matrix $\mathbf{X}$ and a particular state evolution $\underline{x}_o$:

$$\underline{x}(t) = \mathbf{X}\,\widetilde{\underline{x}}(t) + \underline{x}_o$$

An approximate value of each variable is known at the end of each time step. Hence we know $\dot{\widetilde{\underline{x}}}$ over the entire time interval. To construct the LATIN-KL subspace, we apply the Karhunen-Loeve expansion to the different variables $\dot{\widetilde{\underline{x}}}(t)$. The values of $\dot{\widetilde{\underline{x}}}$ on each time step define a smaller subspace than the LATIN subspace. The size of this subspace cannot be higher than the number of time steps. For example, if there is only one time step the size of the LATIN-KL subspace is one. The Karhunen-Loève expansion is used to choose the principal directions $\widetilde{\underline{v}}_i$ of the subspace built with the values $\dot{\widetilde{\underline{x}}}(t)$. These principal directions are defined by the following eigensystem:

$$\mathbf{M}\cdot\widetilde{\underline{v}}_i = \mu_i\,\widetilde{\underline{v}}_i \quad \text{with} \quad \mathbf{M} = \int_0^{t_f} \dot{\widetilde{\underline{x}}}(t)\,\dot{\widetilde{\underline{x}}}^T(t)\,dt \qquad [15]$$

Let's note μ_{max} the highest eigenvalue of $\mathbf{M}$. Only a part of the principal vectors is saved in a matrix $\mathbf{V}$. These vectors are corresponding to the eigenvalues greater than $10^{-6}\,\mu_{max}$. Hence we obtain a matrix $\mathbf{X}_v$ containing the basis functions of the LATIN-KL subspace such that:

$$\mathbf{X}_v = \mathbf{X}\,\mathbf{V} \qquad [16]$$

In practice, only the basis functions of the LATIN-KL subspace are saved. For each sub-domain Ω, the different matrix of basis functions are: $\mathbf{A}_U$ for the displacements, $\mathbf{A}_\sigma$ for the stresses, $\mathbf{A}_F$ for the normal stresses over $\partial\Omega$, $\mathbf{A}_T$ for the temperatures and $\mathbf{A}_\phi$ for the normal heat transfer over $\partial\Omega$. Therefore the known approximate state is represented with few degrees of freedom $\underline{a}_{U\,n}$, $\underline{a}_{\sigma\,n}$, $\underline{a}_{T\,n}$ and $\underline{a}_{\phi\,n}$, to define the linear combinations:

$$\underline{U}_n = \mathbf{A}_U \, \underline{a}_{U\,n}(t) + \underline{U}_o \quad \sigma_n = \mathbf{A}_\sigma \, \underline{a}_{\sigma\,n}(t) + \sigma_o \quad \underline{F}_n = \mathbf{A}_F \, \underline{a}_{\sigma\,n}(t) + \underline{F}_o$$

$$T_n = \mathbf{A}_T \, \underline{a}_{T\,n}(t) + T_o \quad \phi_n = \mathbf{A}_\phi \, \underline{a}_{\phi\,n}(t) + \phi_o$$

The same degrees of freedom $\underline{a}_{\sigma\,n}(t)$ are used for the stresses and the normal stresses on the boundary in order to obtain a couple $(\sigma_n, \underline{F}_n)$ that satisfy the equilibrium conditions (1) at any time instant. So, for stresses, the matrix $\mathbf{V}$ in equation (16) is obtained from $\underline{a}_{\sigma\,n}$. But this matrix modifies both $\mathbf{A}_\sigma$ and $\mathbf{A}_F$. After each correction stage, $\underline{W}_U$, $\mathbf{W}_\sigma$, $\underline{W}_F$, W_T and W_ϕ are added to the convenient matrix $\mathbf{A}_x$, if we keep a set of independent space functions. The Karhunen-Loève simplification is done only once after three correction stages, before adding the last computed vectors W_x. When the basis functions have been modified, we need to adapt their linear combination. To do so, we use the Moore-Penrose inverse of $\mathbf{V}$:

$$\widetilde{\underline{x}}_v(t) = \left(\mathbf{V}^T \mathbf{V}\right)^{-1} \mathbf{V}^T \widetilde{\underline{x}}(t) \quad \forall\, t \in [0, t_f]$$

3.5. *Improvement of the approximate state over* $]0, t_f]$

To find a correction of the approximate state with the basis functions of the reduced-order model we use a classical weak formulation of the equations of the correction stage. The main variables of the corrections stages are the displacements and the temperatures. From the corrections of these variables we deduce the corrections of the stresses and the heat transfer. During the correction stage, the correction $(\delta \underline{U}, \delta\sigma, \delta\underline{F}, \delta T, \delta\phi)$ belong to the known LATIN-KL subspace. So the few variables of this global stage are $\delta\underline{a}_U$, $\delta\underline{a}_\sigma$, $\delta\underline{a}_T$, and $\delta\underline{a}_\phi$ such that:

$$\delta\underline{U} = \mathbf{A}_U \, \delta\underline{a}_U(t) \quad \delta\sigma = \mathbf{A}_\sigma \, \delta\underline{a}_\sigma(t) \quad \delta\underline{F} = \mathbf{A}_F \, \delta\underline{a}_\sigma(t)$$

$$\delta T = \mathbf{A}_T \, \delta\underline{a}_T(t) \quad \delta\phi = \mathbf{A}_\phi \, \delta\underline{a}_\phi(t)$$

A weak form of the equations of the correction stage provides the four following incremental problems:

$$\begin{cases} \delta\underline{a}_U(0) = 0 \\ \mathbf{G}_U^{me}(t_{j+1}) \, \delta\underline{a}_U(t_{j+1}) = \underline{N}_U^{me}(t_{j+1}) \quad \forall t_{j+1} \end{cases}$$

$$\begin{cases} \delta\underline{a}_T(0) = 0 \\ \mathbf{G}_T^{th}(t_{j+1}) \, \delta\underline{a}_T(t_{j+1}) = \underline{N}_T^{th}(t_{j+1}) \quad \forall t_{j+1} \end{cases}$$

$$\begin{cases} \delta\underline{a}_\sigma(0) = 0 \\ \mathbf{G}_\sigma^{me}(t_{j+1}) \, \delta\underline{a}_\sigma(t_{j+1}) = \underline{N}_\sigma^{me}(t_{j+1}) \quad \forall t_{j+1} \end{cases}$$

$$\begin{cases} \delta \underline{a}_\phi(0) = 0 \\ \mathbf{G}^{th}_\phi(t_{j+1})\, \delta \underline{a}_\phi(t_{j+1}) = \underline{N}^{th}_\phi(t_{j+1}) \quad \forall t_{j+1} \end{cases}$$

3.6. *Algorithm of the linear correction stage*

The local computation of $(\underline{U}_{n+1/2},\, \underline{F}_{n+1/2},\, T_{n+1/2},\, \phi_{n+1/2})$ is both necessary for the computation of the residuals and for the computation of the correction of the approximate state. More over, this local computation and the global computations are incremental ones. Hence we use the following algorithm:

- For each sub-domain Ω : we do a copy of the boundary variables
$(\underline{U}'_n,\, \underline{F}'_n,\, T'_n,\, \phi'_n)$ of the neighbour sub-domain Ω',
over the entire time interval $[0, t_f]$
- For each sub-domain Ω
  - Initial conditions
  - For each time step $]t_j, t_{j+1}]$
    For each integration point inside Ω : local contributions to
$\mathbf{G}^{me}_U(t_{j+1}),\, \underline{N}^{me}_U(t_{j+1}),\, \mathbf{G}^{me}_\sigma(t_{j+1}),\, \underline{N}^{me}_\sigma(t_{j+1}),\, \mathbf{G}^{th}_T(t_{j+1}),$
$\underline{N}^{th}_T(t_{j+1}),\, \overline{\sigma},\, \mathbf{K}^{me}_W,\, \mathbf{K}^{th}_W$
    For each integration point on $\partial\Omega$: local contributions to
$\mathbf{G}^{me}_U(t_{j+1}),\, \underline{N}^{me}_U(t_{j+1}),\, \mathbf{G}^{me}_\sigma(t_{j+1}),\, \underline{N}^{me}_\sigma(t_{j+1}),\, \mathbf{G}^{th}_T(t_{j+1}),$
$\underline{N}^{th}_T(t_{j+1}),\, \mathbf{G}^{th}_\phi(t_{j+1})\, \underline{N}^{th}_\phi(t_{j+1})\, \overline{F},\, \mathbf{K}^{me}_W,\, \mathbf{K}^{th}_W$
    Computation of : $\delta \underline{a}_U(t_{j+1}),\, \delta \underline{a}_T(t_{j+1})$
    For each integration point inside Ω : local contributions to
$\mathbf{G}^{me}_\sigma(t_{j+1}),\, \underline{N}^{me}_\sigma(t_{j+1})$
    For each integration point on $\partial\Omega$: local contributions to
$\mathbf{G}^{th}_\phi(t_{j+1}),\, \underline{N}^{th}_\phi(t_{j+1})$
    Computation of : $\delta \underline{a}_\sigma(t_{j+1}),\, \delta \underline{a}_\phi(t_{j+1})$
    Choice of the sign of $\dot\lambda^{me}$ and λ^{th} for the contributions to
$F^{me}_W(t_{j+1}),\, \underline{F}^{th}_W(t_{j+1}),\, W_\phi$
  - Computation of new space functions $\underline{W}_U,\, \mathbf{W}_\sigma,\, \underline{W}_F$ and W_T
- For each sub-domain Ω : update the approximate state $\underline{a}_{x\,n+1} = \underline{a}_{x\,n} + \delta \underline{a}_x$
and eventually update the LATIN-KL subspace

There is no time synchronisation for the computation of the corrections on each sub-domain. This algorithm could be easily used on a parallel computer.

4. 2D examples

4.1. *A first casting problem*

Let us consider the Finite Element model in Figure 1. 40 time steps are used to describe the evolution of the state variables. The two sub-domains Ω_1 and Ω_2 are elastic bodies, respectively in aluminium and sand. There is no initial gap between the two sub-domains. This is a plan strain problem.

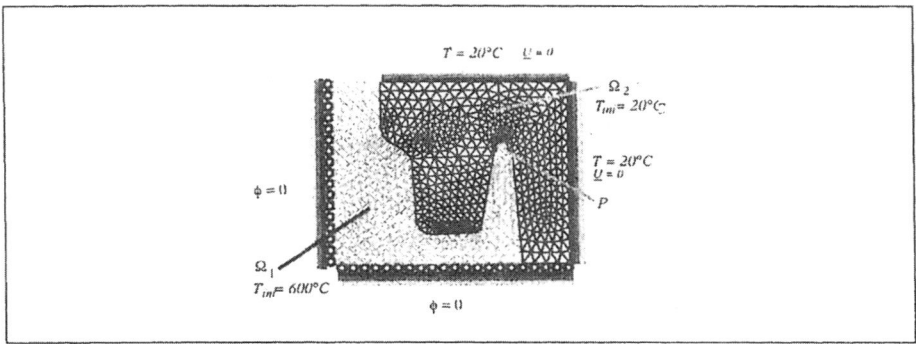

Figure 1. *Meshes on each sub-domain*

The way Ω_1 is cooled down depends on the contact conditions between the two sub-domains. Moreover these contact conditions are defined by the thermal expansion of each sub-domain. The thermo-mechanical coupling is strong and similar to the one encountered in casting problems.

4.2. *Few stages to obtain the main part of the state evolution*

To forecast the defaults in a process with a simulation, it is interesting to obtain a quick overview of the evolution of the state variables. After 20 correction stages a good overview of the displacements and temperature fields is obtained (Figure 2).

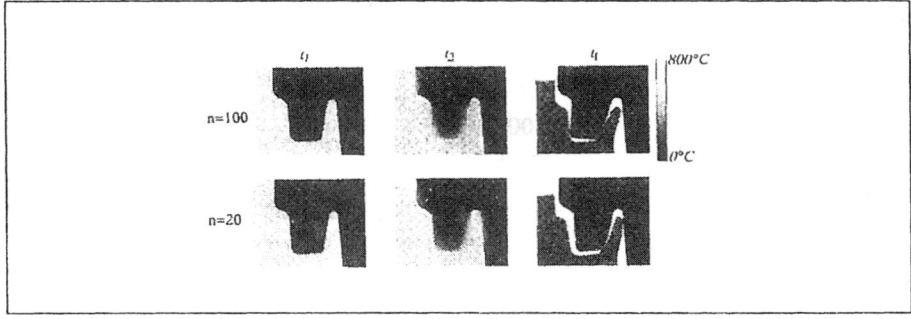

Figure 2. *An interesting overview of the displacement and temperature field after 20 correction stages.*

Even accurate local results are quickly available, with few correction stages. In Figure 3 there is the example of the temperature evolution (Figure 1) on the point P obtained after different correction stages.

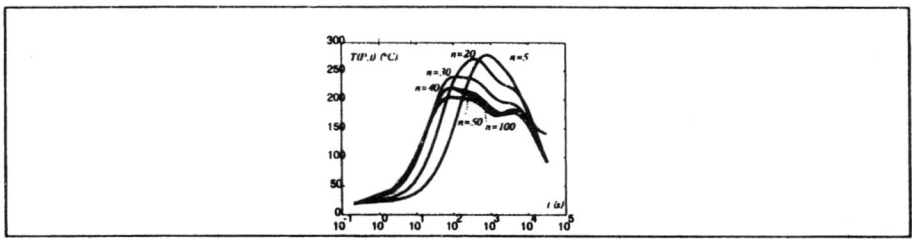

Figure 3. *Different local evolutions of the temperature at the point P.*

4.3. Discussion about the reduced-order model and the LATIN-KL subspace

The LATIN subspace is very close to the subspace proposed by P. Bussy to represent the corrections of the state variables. The Karhunen-Loève expansion allows a very nice reduction of the number of basis functions of the reduced-order model. In Figure 4 we can observe, for the displacement defined over Ω_2, the difference between the size of the LATIN-KL subspace and the size of the LATIN subspace, during the correction stages.

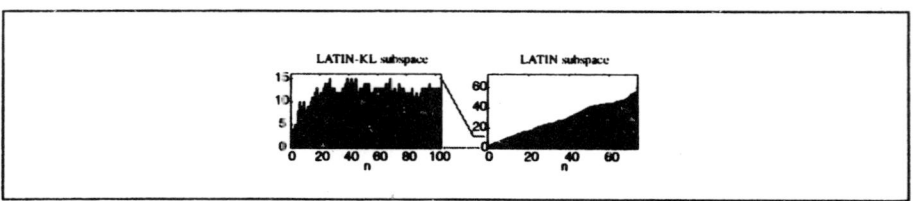

Figure 4. *Sizes of the LATIN subspace and LATIN-KL subspace during the correction stages.*

In Figure 5 there is the first basis functions of A_T and the corresponding values of the degree of freedom $\underline{a}_{T\,n}$ after 100 correction stages (n = 100). These basis functions allow to represent 80% of the temperature field over the entire time interval.

In Table 1, we summarise the number of unknowns of the reduced-order model obtained after 100 correction stages, and the number of unknowns of the Finite Element model.

	$\underline{U}$	σ	T	ϕ
Ω_1	7 / 636	8 / 1647	10 / 318	6 / 47
Ω_2	12 / 926	9 / 2532	11 / 463	6 / 47

Table 1. *Number of unknowns for each variable over each sub-domain, for the reduced-order model (n=100)/for the Finite Element model in Figure 1*

Reduction Method for Thermo-mechanical Simulations 337

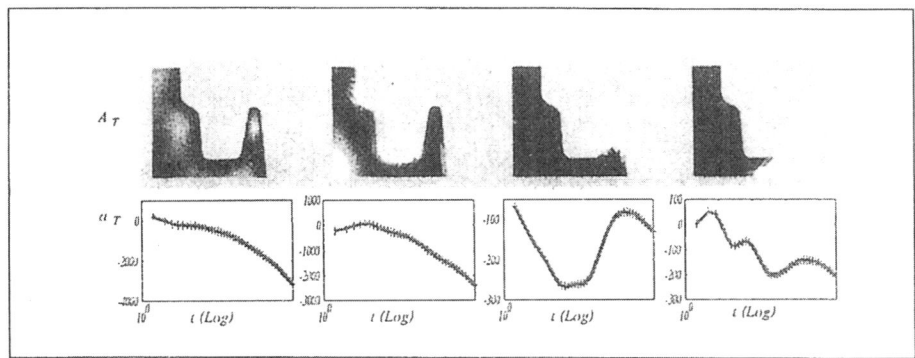

Figure 5. *Basis functions for the temperature fields and corresponding degrees of freedom (n=100).*

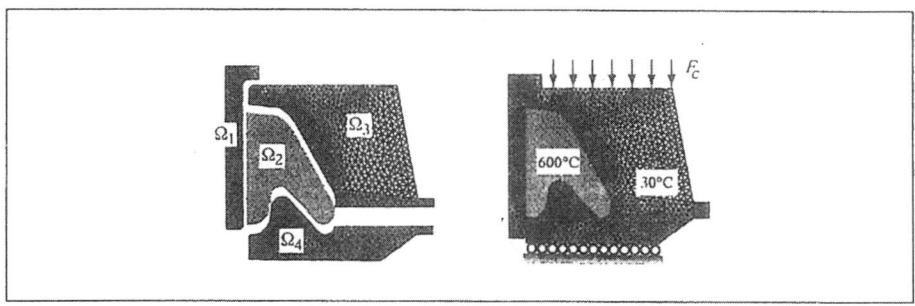

Figure 6. *A more complex problem : Meshes and sub-domains*

A more complex geometry has been studied for the same kind of problem. The system is split into four sub-domains (Figure 6). The numbers of unknowns of the reduced-order model for each sub-domain are given in Table 2.

	U	σ	T	ϕ
Ω_1	15 / 720	30 / 2468	20 / 360	19 / 94
Ω_2	26 / 1422	32 / 5200	17 / 711	22 / 242
Ω_3	16 / 1162	32 / 4144	23 / 583	11 / 154
Ω_4	19 / 884	30 / 3084	26 / 442	15 / 134

Table 2. *Number of unknowns for each variable over each sub-domain, for the reduced-order model (n=100)/for the Finite Element model in Figure 6*

Despite a representation of the displacement and the stresses needed for our approach, only few unknowns are necessary to obtain an accurate approximation of the state evolution.

5. Conclusion

Thanks to the proposed reduction method we can quickly obtain an approximate evolution of the state of the studied system. Therefore, if there is any important default on the process, we can see it very quickly without any specific reflection to construct a simple model to describe what is happening in the case studied. The main phenomena are described automatically by the reduction method. So we suggest that very fine Finite Element model should be realised to describe what could happen during the transformation. The reduction method will automatically extract the basis functions to describe the evolution of the state with few unknowns. To perform accurate simulations, we just have to continue correction stages as long as it is necessary. This approach should be interesting to study processes involving multiphysic coupling which are difficult to describe with a simple model.

6. References

[ART 92] ARTZ M., COGNARD J.-Y., LADEVÈZE P., "An efficient computational method for complex loading histories", *Comp. Plas. Fund. and Appl.*, Editions Owen, Onate and Hinton, Barcelona, 1992, p. 225-236.

[BUS 90] BUSSY P., ROUGÉE P., VAUCHEZ P., "The Large Time INcrement method for numerical simulation of metal forming processes", *Proc. NUMETA*, Editions Elsevier, 1990, p. 102-109.

[CHA 96] CHAMPANEY L., COGNARD J.-Y., DUREISSEIX D., LADEVÈZE P., "Numerical experimentations of parallel strategies in structural non-linear analysis", *Calc. parallèles.*, vol. 8, num. 2, 1996, p. 245-249.

[KNO 95] KNOLL D.-A., MCHUGH P.-R.,"Newton-Krylov methods applied to a system of convection-diffusion-reaction equations", *Comput. phys. commun.*, vol. 88 , 1995, p. 141-160.

[LAD 85] LADEVÈZE P., "Sur une famille d'algorithmes en mécanique des structures", *CRAS, Paris, 300, série II*, vol. 2, 1985, p. 41-44.

[LAD 96] LADEVÈZE P., *Mécanique non linéaire des Stuctures*, Etudes en mécanique des matériaux et des structures, Editions Hermès, Paris, 1996.

[OCA 89] O'CALLAHAN J., AVITABILE P., RIEMER R., "System Equivalent Reduction Expansion Process", *7pInternational Modal Analysis Conference*, Las Vegas, 1989, p. 29-37.

[PAR 01] PARK H.-M., JUNG W.-S., "The Karhunen-Loève Galerkin method for the inverse natural convection problems", *Int. j. heat mass transfer*, vol. 44, 2001, p. 155-167.

[PEL 00] PELLE J-P., RYCKELYNCK D., "An Efficient Adaptive Strategy to Master the Global Quality of Viscoplastic Analysis", *Computers & Structures*, Editions Elsevier Science, vol. 78, num. 1-3, 2000, p. 169-184.

[RYC 00] RYCKELYNCK D., "A LATIN Approach to Solve Thermomechanical Problems", *Techniques and Developments,* Editions B.-H.-V. Topping, Civil-Comp Press, 2000, p. 237-248.

[SIR 91] SIROVICH L., "Empirical Eigen functions and low dimensional systems", *New Perspectives in Turbuence*, vol. 5, num. 139, 1991.

[SUN 01] SUNDAR S., BHAGAVAN B.-K., PRASAD S., "Newton-Preconditioned Krylov Subspace Solvers for System of Nonlinear Equations : A Numerical Experiment", *Applied Mathematics Letters*, vol. 14, 2001, p. 195-200.

[WEI 95] WEISS R., "A theorical overview of Krylov subspace methods", *Appl. numer. math.*, vol. 19, 1995, p. 207-233.

Chapter 29

Fatigue Analysis of Materials and Structures using a Continuum Damage Model

Omar Salomón, Sergio Oller and Eugenio Oñate
International Center for Numerical Methods in Engineering (CIMNE) and Universitat Politècnica de Catalunya, ETS Ingenieros de Caminos, Barcelona, Spain

1. Introduction

Fatigue can be defined as the process of permanent, progressive and localised structural change which occurs to a material point subjected to strains and stresses of variable amplitudes which produces cracks leading to total failure after a certain number of cycles. progressive loss of the material strength occurs as a function of the number of stress/strain cycles, reversion index, load amplitude, etc. Loss of strength may be interpreted as micro-cracking followed by crack coalescence leading to the final collapse of structural parts. Failure typically takes place for stress levels well below the strength limit of the material obtained in static tests.

Structural materials are, in general and at less from a microscopic point of view, non-homogeneous materials. In the case of metals, they include discontinuities such as grain boundaries, microscopic pores, and other particles as impurities and carbon. All of them are commonly considered as defects. This does not denote that those materials are defective and consequently useless. Defects are a normal feature of metallic material microstructure. However, defects are also source of local stress concentrations inducing local plasticity and/or damage and therefore they can turn into fatigue crack initiation sites.

According to their location, defects can be classified as internal or external. External or surface defects can be introduced through component assembly, impact of foreign objects or material processing. They can be detectable by optical inspection. Internal defects, in contrast, are introduced in the structure during manufacturing. The precise size, location and type of defects cannot be predicted. Only post-manufacturing inspections *(i.e.:* radiography and computed tomography) can provide accurate information about the defect population present in structures.

Internal defects in metals include porosity, inclusions as oxide films or carbon layers and, in case of welded parts, lack of fusion and lack of penetration of the welded joint. As discontinuities in geometry, defects cause stress concentration and are therefore prone to fatigue crack initiation.

The classical and simplest approach to take into account the effects of internal defects on the fatigue strength of structures has been to consider the fatigue stress limit of the material reduced by a size factor. The larger the size of a part, the more chance there is for internal defects to be present.

This paper extends the previous work of the authors (Oller *et al.*, 2001; Salomón *et al.*, 2002, 1999), on the application of the theoretical framework of continuum mechanics to the study of non linear fatigue problems, accounting for the influence of internal defects in the resulting fatigue prediction model. As in previous work, restriction is made here to small deformations and isotropic damage.

2. Internal defects in metallic structures

From a detailed study (DARCAST, 00) conducted on aluminium alloy samples, AlSi8.5Cu3.5Fe, at last two basic conclusions arise on the influence of internal defects:

– Porosity significantly lowers the static properties of the material (strength and strain to failure).

– The reduction is almost linear with the porosity level (classified as A0, A2 and A4 according to ASTM E505 standard).

A significant reduction in the fatigue strength was also found when the degree of porosity was increased from A0 to A2. The decrease was distinctly lower from A2 to A4. The ratio fatigue-strength/static-strength was 0.54, 0.44 and 0.51 for materials A0, A2 and A4, respectively. As fatigue data suffer of large scattering this observation is not conclusive.

A real industrial component, an alternator support made of the same material, was also experimentally tested and numerically analysed. The alternator support was selected among the components of the supplier actual production. Two batches were provided (about 50 components) for the experimental testing: one batch of non-porous components (components with an acceptable level of porosity) and one batch of porous components; the distinction being made by the supplier based on radioscopy examination. Subsequent laboratory examination showed that a scattered porosity was also present in the non-porous components.

The fatigue testing showed that for both components (porous and non-porous) four different regions of failure could be identified. At regions of failure not highly stressed, pores of significant dimensions, oxide films and defects in the microstructure were observed on the fracture surfaces. The difference in the fatigue limit, as found for the porous and non-porous components, was observed to be statistically non significant.

Numerical simulations of the aluminium support were done and results of the analysis were compared with those from experimental testing. Figure 1 compares experimental regions of failure with stress distribution obtained from the numerical simulation. Regions of failure 2, 3 and 4 are in coincidence with extended highly stressed areas. Figure 1 also displays the correlation between the damage areas due to fatigue and experimental regions of failure. Damage is observed in concordance with regions of failure 3 and 4. Region of failure 2 (upper part) correspond to a highly compressed area what dismiss the fatigue effects. For region of failure 1 the numerical analysis only yielded a small area of stress concentration due to local geometry conditions.

Life prediction of the aluminium support based on numerical analyses (material A2) agreed with experimental tests: Failure around 900.000 cycles for a maximum load of 10.5 kN (50% probability of survival, porous and non-porous components.

Non-porous components indicate acceptable level of porosity). A different material characterization (as A0 or A4) in the analysis would bring a different lifespan, but the region where failure occurs would always be the same if the material is considerer homogeneous.

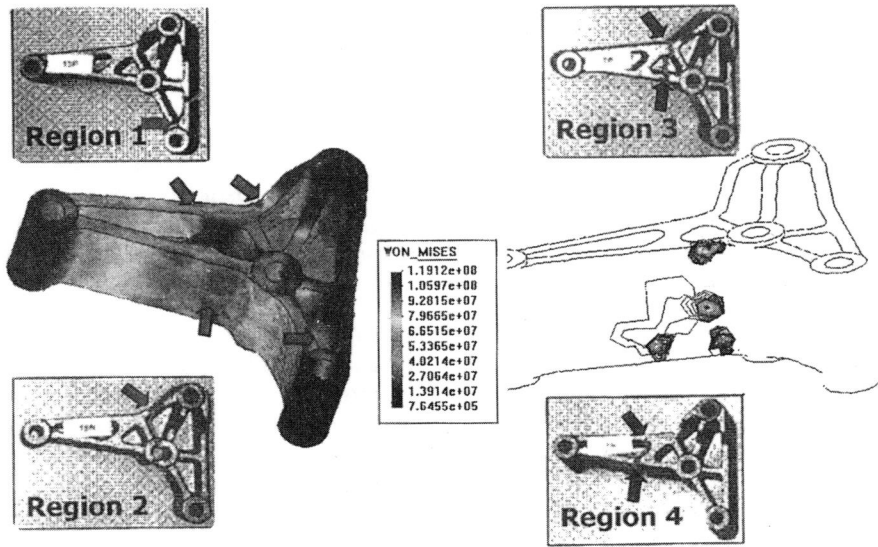

Figure 1. *Stress and fatigue damage distribution from numerical simulation and experimental regions of failure*

Internal defects, mainly porosity, can be regarded as responsible for the fatigue failure of the component at four different regions. Therefore, a first approach to simulate this behaviour is to consider a random distribution of such defects. As it was mentioned before, porosity reduces, in an almost linear way, the static material strength. Consequently, the value of material strength is randomly assigned, material is not homogenous any more. Maximum and minimum values will be the material strength experimentally obtained for samples with porosities A0 and A4, respectively.

Concerning fatigue characterization of the material, porosity not only causes a reduction in the fatigue strength, but eventually also a change in the slope of the Stress-$N°$ of cycles curves. In this paper a master S-N curve is proposed, based in experimental results on rotating bending fatigue, for the whole structure to be analysed. That master curve is scaled down according to the randomly assigned material strength of each point.

3. Fatigue analysis using continuum mechanics

The theoretical structure of continuum mechanics is suitable for the study of non-linear fatigue problems (Osgood, 1982; Suresh, 1998). In previous works of the authors (Oller et al., 2001; Salomón et al., 2002, 1999) a deterministic fatigue prediction model based on a continuum mechanics formulation, considering coupling between elasticity, damage and plasticity, was developed.

The formulation assumes that each point of the solid follows a damage-elasto-plastic constitutive law with the stress (S) evolution depending on the free elastic strain variable (E^e) and a set of internal plastic and damage variables $q = \{\alpha^p, d\} = \{E^p, \kappa^p, d = \kappa^d\}$, were E^p and $\kappa^{ini} \le (\kappa = \kappa^p + \kappa^d) \le 1$ represent the plastic part of the strain and a unit normalized dissipation composed by the plastic plus damage parts, respectively. The initial defect level defines the initial threshold of the normalized dissipation $\kappa^{ini} \equiv d^{ini}$ in the initial damage form. The free energy for isothermal, isentropic and adiabatic processes and for small elastic strains and large plastic strains is written in the reference configuration, accepting the additivity of its elastic Ψ^e and plastic Ψ^p parts, as

$$\Psi = \Psi^e(E^e_{ij}, d) + \Psi^p(\alpha^p_i) = (1-d)\frac{1}{2m^o}\left[E^e_{ij} C^o_{ijkl} E^e_{kl}\right] + \Psi^p(\alpha^p_i) \quad [1]$$

where E^e is elastic Green strain tensor, m^o is the material density, $d = \kappa^d$ is the internal mechanical damage variable for the damage processes $d^{ini} \le (d = \kappa^d) \le 1$ with the initial value $d^{ini} \equiv \kappa^{ini}$ provided by the defect level, and C^0_{ijkl} the initial constitutive tensor. The stress tensor in the reference configuration S_{ij} can be expressed as

$$S_{ij} = m^o \frac{\partial \Psi}{\partial E^e_{ij}} = (1-d) C^o_{ijkl} E^e_{kl} \quad [2]$$

For the plastic behaviour, the general forms of the yield F and potential G plastic functions take into account the influence of the current stress state, the internal plastic variables, and other variables such as the number of cycles N:

$$F(S_{ij}, \kappa, \theta) = f(S_{ij}) - K(S_{ij}, \kappa, N)$$
$$G(S_{ij}) = g(S_{ij}) = \text{constant.} \quad [3]$$

where $f(S_{ij})$ and $g(S_{ij})$ are the uniaxial equivalent stress functions, $K(S_{ij}, \kappa, N)$ is the strength threshold (see Figure 2). All the internal variables at current time t are obtained by means of an integration process $\alpha_i^p = \int_0^t \dot{\alpha}_i^p dt$, starting from its evolution law $\dot{\alpha}_i^p = \dot{\lambda} H_i^p(S_{kl}, \alpha_k^p)$, where λ is the plastic consistent factor.

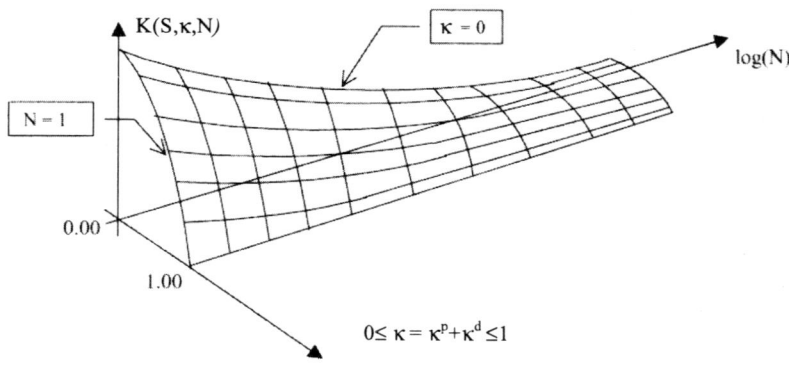

Figure 2. *Uniaxial strength threshold for a symbolic material*

The damage function is defined as

$$G^D(S_{ij}, \kappa, \theta) = \bar{S}(S_{ij}) - K(S_{ij}, \kappa, N) \qquad [4]$$

where $\bar{S}(S_{ij})$ is the uniaxial equivalent stress function in the undamaged space, $K(S_{ij}, \kappa, N)$ is the same strength threshold as in [3] and $\kappa^d = d = \int_0^t \dot{d} \, dt$ the damage internal variable with an evolution defined as $\dot{d} = \dot{\mu} H^D(S_{kl}, d)$, where μ is the consistency damage factor. In both, [3] and [4], the normalized dissipation is defined as $\kappa = \kappa^p + \kappa^d = (\Xi^p + \Xi^d)/\Xi^{max}$, where Ξ^p, Ξ^d, Ξ^{max} are the Clausius-Duhem dissipation for the current plastic, damage process and its maximum capacity of the solid dissipation at each point, respectively.

The effect of the number of cycles on the plastic and/or damage consistency conditions ($\dot{F} = 0$, $\dot{G}^D = 0$) is introduced as follows,

$$f(S_{ij}) - \underbrace{K(S_{ij}, \kappa) \cdot f_{red}(N, S_{med}, R)}_{K(S_{ij}, R, N)} = 0 \qquad [5]$$

$$\overline{S}(S_{ij}) - \underbrace{K(S_{ij}, \kappa) \cdot f_{red}(N, S_{med}, R)}_{K(S_{ij}, R, N)} = 0 \qquad [6]$$

where $0 \le f_{red} \le 1$ represent the unit normalized reduction part of the strength threshold K – plastic and/or damage strength evolution – by load cyclic effect

4. S-N Curves

Stress-N° of cycles (S-N) curves are experimentally obtained by subjecting identical smooth specimens to cyclic harmonic stresses and establishing their life span measured in number of cycles. The curves depend on the level of the maximum applied stress and on the ratio between the lowest and the highest stresses ($R = S_{min} / S_{max}$). In previous works (Oller et al., 2001; Salomón et al., 2002) an exponential function to approach steel and aluminium experimental S-N curves was proposed. This function depends on and is capable of dealing with any value of the ratio between minimum and maximum stress. However, it is a bit difficult to adjust its parameters to obtain a good approximation to experimental curves, which are usually not defined over the whole life-span of the material. Instead of using such an exponential function, here the experimental data is introduced by points in a table-like mode.

Usually, S-N curves are obtained for a fully reversed stress state ($R = S_{min} / S_{max} = -1$) by rotating bending fatigue tests. This zero mean stress is however not typical of real industrial components working under cyclic loads. Based on the actual value of the R ratio and a basic value of the endurance stress Se (for $R = -1$) the model proposed postulates a threshold stress S_{th}. The meaning of S_{th} is that of an endurance stress limit for a given value of $R = S_{min} / S_{max}$; if the actual value of R is $R = -1$ then, $S_{th} = S_e$.

$$\begin{aligned} S_{th} &= S_e + (S_u - S_e) \cdot (0.5 + 0.5 \cdot R)^{STHR1} \longrightarrow abs(R) \le 1 \\ S_{th} &= S_e + (S_u - S_e) \cdot (0.5 + 0.5 / R)^{STHR2} \longrightarrow abs(R) \ge 1 \end{aligned} \qquad [7]$$

STHR1 and STHR2 are material parameters that need to be adjusted according to experimental tests.

5. Cyclic strength reduction function

The S-N curves proposed in the previous section are fatigue life estimators for a material point with a fixed maximum stress and a given ratio R. If, after a number of cycles lower than the cycles to failure, the constant amplitude cyclic loads giving that maximum stress S_{max} (and ratio R) are removed, some change in S_u is expected due to accumulation of fatigue cycles. In order to describe the variation of S_u the following function is proposed:

$$fred(R, Ncycles) = \exp(-B0 \cdot (\log 10(Ncycles))^{BETAF})$$
$$B0 = -\log 10(S_{max}/S_u)/(\log 10(N_F))^{BETAF}$$
[8]

BETAF is a material parameter and N_F the number of cycles to failure.

6. Time advancing strategy

An advantage of the methodology presented consists in the way the loading is applied. In a mechanical problem each load is applied in two intervals, in the following order (see Figure 3),

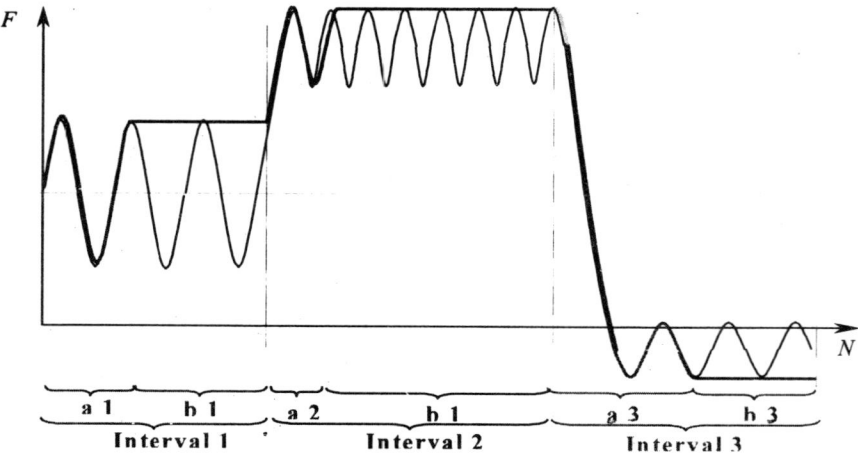

Figure 3. *Schematic representation of the time advancing strategy*

– **Tracing load**, (described by "ai" periods on Figure 3). It is used to obtain the stress ratio $R = S_{min}/S_{max}$ at each integration point, following the load path during several cycles until the R relationship tends to a constant value. This occurs when the following norm is satisfied,

$$\eta = \sum_{GP} \left\| \frac{R_{GP}^{i+1} - R_{GP}^i}{R_{GP}^{i+1}} \right\| \to 0$$
[9]

where $R_{GP}^i = S_{min}/S_{max}\big|_{GP}^i$ is computed at each Gauss interpolation point for the load increment "i".

– **Enveloping load**, (described by "bi" periods on Figure 3). After the first tracing load interval (ai), the number of cycles N is increased while keeping constant the maximum applied load (thick line in Figure 3) and the stress ratio R. In this new load interval, the variable is not the load level (kept constant) but the number of cycles.

This two-stages strategy allows a very fast advance in the time loading. A new interval with the two stages explained should be added for each change in the loading level.

7. Numerical application

A real industrial component, the aluminium alloy support of Figure 1, was numerically analyzed to test the potential of the proposed formulation.

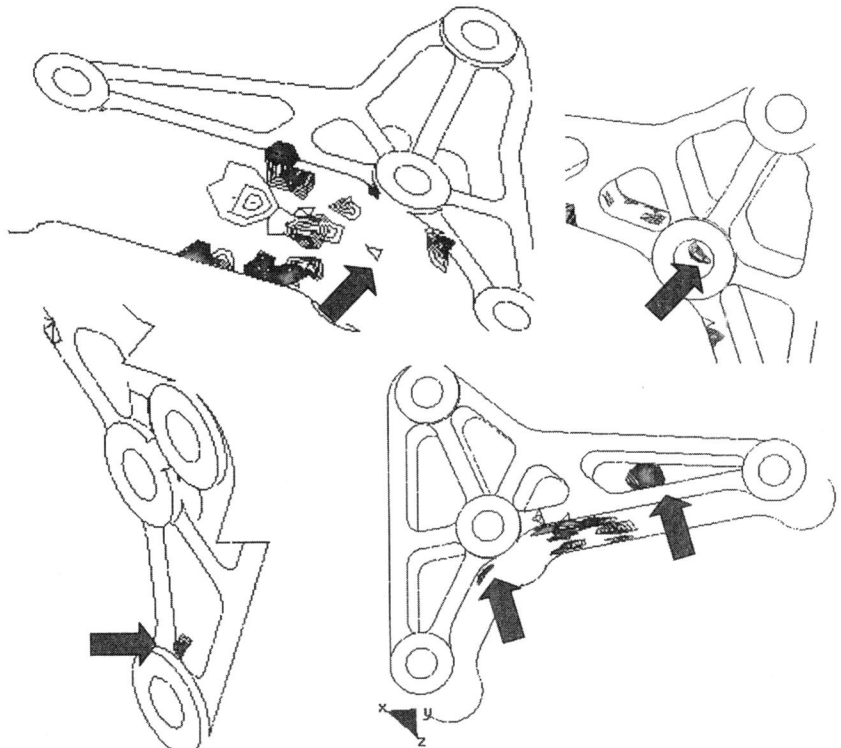

Figure 4. *Fatigue damage localization obtained in different numerical analysis with a random material characterization*

This formulation has been implemented into a general-purpose thermo-mechanical finite element code (COMET, 00). The material strength is randomly assigned to each Gauss point using a normal distribution with mean value corresponding to porosity A2 material strength (see Section 2). Maximum and minimum values of material strength matching porosity A0 and A4, respectively. A S-N master curve, corresponding to experimental values for rotating bending fatigue of A2 material, was scaled down (or up) according to the randomly assigned material strength of each point. The results of fatigue damage localization obtained during different tests are summarized in Figure 4. The arrows show regions of damage different of the ones obtained previously with material characterized as homogenous and corresponding to experimental regions of failure (see Figure 1).

8. Conclusions

A fatigue model based in continuum mechanics accounting for elasto-plastic-damage constitutive equations, previously developed by the authors, has been extended to include the influence of material internal defects. In this stress life approach the material is considered to be non-homogenous. The material strength is randomly assigned and Stress-N° of cycles curves are scaled down according to material strength of each point. Limit values for material strength are based on experimental tests of samples with different degree of porosity.

Preliminary results from the application of the formulation presented to the analysis of an aluminium industrial component, indicate that the approach is adequate for the fatigue analysis of metallic structures with scattered porosity and/or other internal defects.

Acknowledgements

The authors thank the financial support provided by European Commission under the CRAFT Project BES2-5637, DARCAST. This support as well as the experimental results obtained by Politecnico de Torino on industrial components supplied by Fonderie 2A, Italy, are gratefully acknowledged.

9. References

COMET Coupled Mechanics and Thermal Analysis. Data Input Manual. Technical Report n° 308, CIMNE, Barcelona, January 2000.

DARCAST Enhanced Design and Manufacturing of High Resistance Casted Parts. Final Technical Report. Craft Project N°: BES2-5637 funded by the European Community, December 2000.

Oller S., Salomón O., Oñate E., "Thermo-mechanical fatigue analysis using generalized continuum damage mechanics and the finite element method", Sent for publication at: *Int. J. Numer. Meth. Engng*, 2001.

Osgood C., *Fatigue Design*, Pergamon Press, 1982.

Salomón O., Oller S., Oñate E., "Fatigue damage modelling and finite elements analysis methodology: Continuum basis and applications", to be presented at *FATIGUE 2002, 8th International Fatigue Congress*, Stockholm, Sweden, June 2002.

Salomón O., Oller S., Car E., Oñate E., "Thermo-mechanical fatigue analysis based on continuum mechanics", *Congreso Argentino de Mecánica Computacional, MECOM'99*, Mendoza, Argentina, 1999.

Suresh S., *Fatigue of Materials*, Cambridge Univ. Press, 2nd edition, 1998.

Chapter 30

Influence of Initial and Induced Hardening on the Formability in Sheet Metal Forming

Sébastien Thibaud and Jean-Claude Gelin
Applied Mechanics Laboratory, CNRS, University of Besançon, ENSMM, France

1. Introduction

As new materials are now appearing for sheet metal components used in automotive industry (Moussy, 1999), this requires careful analysis of both the formability and possible occurrence of defects (necking, springback) as well as in-use properties as dent resistance, fatigue and crash resistance. As in the last ten years, FEM is more and more extensively used in sheet metal industry (Gelin and Picart, 1999; Makinouchi, 2001), this requires accurate material description in order to get with accuracy not only the shape of the sheet metal components after stamping, but also the shape after tool removing accounting springback (Xia, 2001) and possible occurrence of defects as necking during processing (Boudeau and Gelin, 2000). A large amount of studies in sheet metal forming rely on the determination of plastic yield surfaces and hardening curves, including anisotropy due to rolling processes (Banabic, 2001).

The paper underlines other aspects of material behaviour that are associated to the evolution of the elastic coefficients in relation with the initial hardening resulting from cold rolling rate and with induced hardening associated to sheet metal stamping. A new methodology is proposed to accurately measure the elastic coefficients, both associated to initial hardening and to the induced one.

In the second section of the paper one describes a new methodology that combines vibrometric measurements for the determination of the Young modulus and tensile tests that are used for the determination of the Poisson's ratio. The measurements are performed both on sheet metal strips at different rolling rates and on strips strained from tensile tests specimens at different elongations.

The third section of the paper concerns one of a major shape defects occurring in sheet metal stamping that is related to springback after tools removal and shows the effects of a correct characterization of the elastic parameters on the amount of springback.

The last section of the paper concerns necking that is another important defect occurring in sheet metal stamping. The approach that is used to investigate necking is the linearized perturbations technique. Different analysis are carried out accounting the influence of elastic parameters simply accounting their effects on elastic deformations, or accounting them through a coupled damage model (Lemaitre, 1985; 1992) that affects as well as elastic properties as plastic ones.

2. Vibrometric identification method for the determination of elastic parameters

2.1. *Presentation of the method*

2.1.1. *Framework of the method and theoretical aspects*

The determination of elastic properties is often considered as an easy operation. The most widely used method is certainly the tensile test. It consists in the determination of the linear part of the true stress – true strain curve to obtain Young modulus. Nevertheless if theoretically it is easily to determine the slope of a straight line, from experiments the linear part is not so clearly defined for a great number of materials and the identification error is quite important. The *vibrometric identification method* can be used alternatively.

Sheet metal strips that are obtained by rolling can be assimilated to plates. In these plates, specimens are cut in order to measure the Young modulus. These specimens are cut with an EDM process and their geometry can be assimilated to beams or plates depending on the width of the specimen relatively to thickness. Consequently, beams and plates vibration theory can be used to determine the Young modulus and the Poisson's ratio. The framework of the method is the Bernoulli beam theory and Kirchhoff plate vibration theory. The natural frequencies are determined by solving partial differential equations associated to the structured vibration problem.

A major problem occurring in dynamic measurements is related to the boundary conditions especially the embedded ones. To overcome this problem, free-free boundary conditions have been chosen to carry out the experiments because they do not induce uncontrolled behaviour due to contact problems or the uncertain stiffness of the clamping conditions. However, under three-dimensional consideration, six rigid body modes appear and must be taken into account in the experimental analysis. These considerations permit to obtain the analytical expression of the first natural bending frequency which is a function of the Young modulus the Poisson's ratio in the case of plate theory and only of the Young modulus in the case of beam theory.

The expression of the first bending frequency for plate and beam are respectively:

For plate:

$$f_{1,plate} = \frac{1.506^2}{2\pi d^2}\sqrt{\frac{Eh^2}{12\rho(1-v^2)}}$$ [1]

For beam:

$$f_{1,beam} = \frac{4.73^2}{2\pi d^2}\sqrt{\frac{Eh^2}{12\rho}}$$ [2]

where l is the length of the plate or beam, h the thickness of the plate or beam and where E, v, ρ are respectively the Young modulus, the Poisson's ratio and the density.

The Young modulus can be evaluated with experiments conducted in free-free conditions. In the case of plates, the determination of Poisson's ratio has to be carried out by first measurements of the Young modulus on beam or by other techniques.

2.1.2. *The vibrometric identification method – complete procedure*

The vibrometric identification method that is proposed to get the elastic coefficients first consists in measuring the first bending natural frequency of a specimen and then to introduce the result in equation [2] to get the value of the Young modulus. To access boundary conditions closed to the free ones, the strip is fixed on a hanged rubber band as shown in Figure 1. Then the apparatus is excited with a random electrical signal through a coil. For excitation and measurements, we use the commercial data acquisition system SIGLAB®. This system generates a random signal and treats electrical response generated by a capacitive displacement sensor. Comparing the amplitude of the excitation and the response signal, software built up the Frequency Response Function (FRF) of the plate or beam. On the FRF, the pics in amplitude correspond to natural frequencies of the structure.

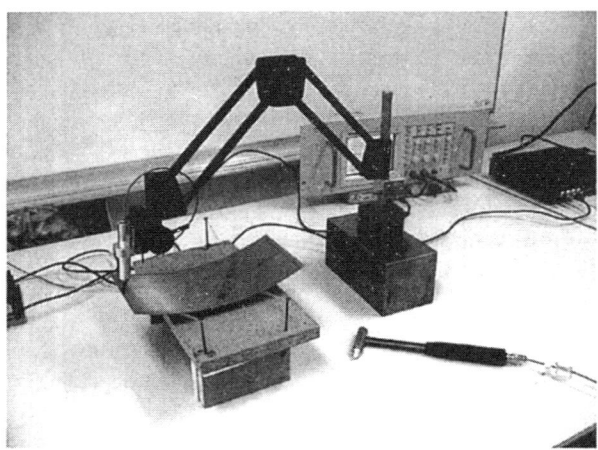

Figure 1. *Photograph of the vibrometric measurements with the specimen fixed on a hanged plate*

The main interest of the dynamic method is that within a short measurement time and without consequent preparation of the specimens, it is possible to determine the Young modulus with a very high accuracy, avoiding the destruction of the specimen.

The accuracy depends on the accuracy of the evaluation of the dimensions of the specimen and relative error calculations have been done in this sense, that lead to the following expressions:

Young Modulus error

$$\frac{\Delta E}{E} = \frac{\Delta \rho}{\rho} + 4\frac{\Delta l}{l} + 2\frac{\Delta h}{h}$$ [3]

Poisson ratio error

$$\frac{\Delta v}{v} = \frac{3}{2}\pi^2 |D|\frac{\Delta \rho}{\rho} + \frac{3}{2}\pi^2 |D|\frac{\Delta E}{E} + 3\pi^2 |D|\frac{\Delta h}{h} + 6\pi^2 |D|\frac{\Delta l}{l}$$ [4]

where $D = \frac{\lambda^4 E h^2}{C l^4 \rho f_{1,plate}^2}$ and $C = 144 - \frac{3\pi^2 \lambda^4 E h^2}{\rho l^4 f_{1,plate}^2}$.

These relationships underline the large influence of geometrical dimensions on the determination of Young modulus and Poisson's ratio.

Unfortunately, the Poisson ratio cannot be determine in an accurate way with this method (the error is approximately 40% for Poisson ratio and less than 5% for Young modulus). So one have chosen to determine the Poisson's ratio from classical tensile tests.

2.2. Characterization of the hardening on Young modulus

The proposed vibrometric method has been applied on TRIP steel. Experiments are conducted on several strips with the same material with different rolling reductions. The evolution of the Young modulus vs. the rolling reduction rate is illustrated in Figure 2 where the rolling reduction rate is expressed as $\ln\left(\frac{h_{initial}}{h_{current}}\right)$.

Figure 2 reveals a linear decreasing of the Young modulus vs. the rolling reduction rate which is in good agreement with literature (EFB-Forschungsbericht, 1998). Figure 2 clearly shows that the Young modulus for a sheet metal part can be very different from the one that can be measured on a bulk testing specimen. The Young modulus looses 12% when the thickness is reduced from 1.44 to 1.10 by rolling. That means also that high rolling reduction rates lead to sheet metal parts more sensitive to springback and that exhibit strongly different dynamic behaviour. One can also conclude that in numerical simulations, the use of the classical value for the Young modulus (meaning 210 GPa) leads to underestimate springback occurring in sheet metal forming processes.

358   Prediction of Defects in Material Processing

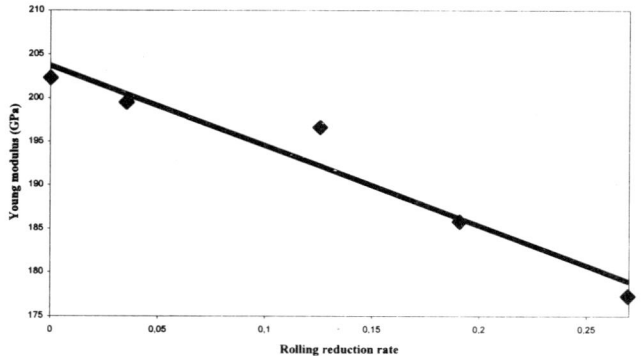

Figure 2. *Young modulus evolution vs. initial hardening characterized by rolling reduction rate (TRIP steel)*

2.3. *Young modulus variation with induced hardening*

In the previous section, we have investigated the influence of the initial hardening on the Young modulus and the necessity to account this phenomenon. Now, one investigates the effect of induced hardening on Young modulus variations. To take into account the evolution of the Young modulus with the plastic deformation, the Young modulus has to be defined as a function of the plastic strain as it has already underlined in literature (Lems, 1963; Morestin *et al.*, 1996).

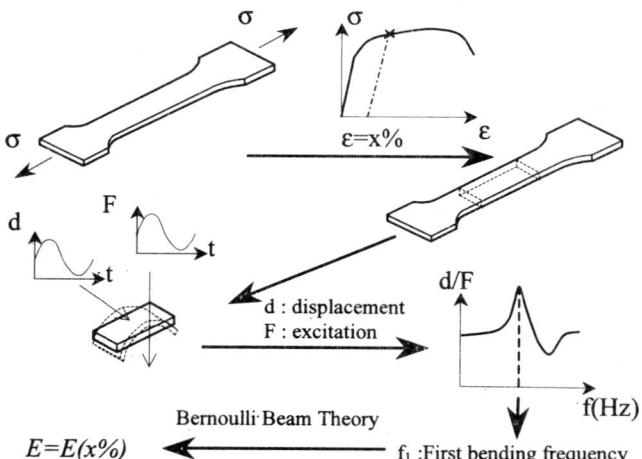

Figure 3. *Schematic illustration of the tensile test combined with vibrometric measurements method for the determination of the induced hardening effects on the Young modulus*

To determine the variation of Young modulus with straining, it is proposed to combine the use of classical tensile tests and vibrometric measurements. Tensile specimens are cut by EDM process in strips obtained from rolling at different reduction rates. Then tensile test are carried out for different tensile elongations and new specimens are cut in the extended tensile specimens. And finally the vibrometric method is used to determine the new Young modulus and to evaluate its evolution within the induced hardening. The method is summarized in the Figure 3.

Figure 4 shows the evolution of Young modulus vs. induced plastic straining obtained from the vibrometric identification method.

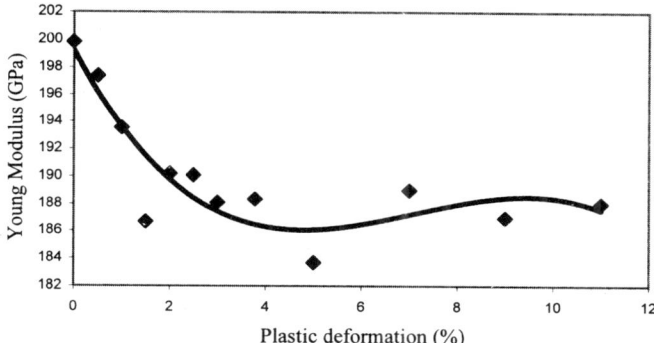

Figure 4. *Young modulus vs. plastic deformation obtained by coupling tensile straining and vibrometric analysis*

3. Effects of the variation of elastic coefficients on sheet metal forming

3.1. *Effects on springback after stamping or deep drawing*

One of the major defects occurring in sheet metal forming is the springback after tools removal. To investigate the effects of elastic parameters on springback the benchmark proposed for Numisheet 2002 (Numisheet, 2002) has been chosen. The initial geometry corresponds to a sheet metal strip (120x30x1 mm) and the drawing is carried out with a semi-circular punch with a radius equal 23.5 mm as indicated in Figure 5. The material that is chosen for doing the analysis corresponds is the same as in section 2. The Young modulus contours after deep drawing is given in Figure 6. It is clearly shown that the fact to consider a constant Young modulus underestimates the amount of springback.

360   Prediction of Defects in Material Processing

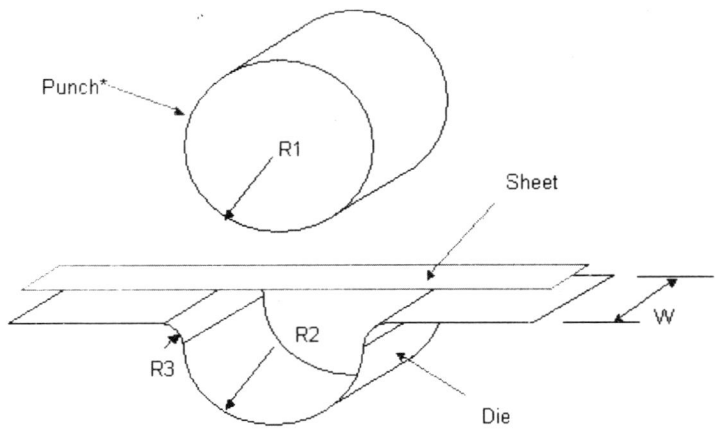

Figure 5. *Geometry used for the springback simulation (R1 = 23.5 mm, R2 = 25 mm, R3 = 4 mm, W = 50 mm)*

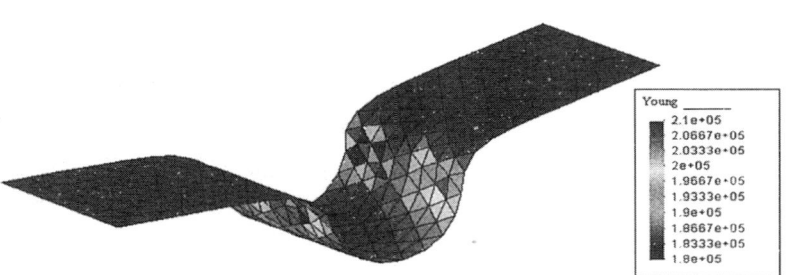

Figure 6. *Young modulus contours resulting from the deep-drawing of a semi-cylindrical strip*

In order to illustrate the effect of induced hardening on the variation of elastic coefficients resulting from stamping process, one consider the stamping of a part damping cup used in car construction.

Figure 7 shows the contours of Young modulus in such a part after deep drawing. It can be seen that the Young modulus varies in a range approximately equal to 8%. As example, the elastic coefficients as well as the thickness contours plays a crucial role on dynamic behaviour after forming as it discussed in (Thibaud et al., 2002).

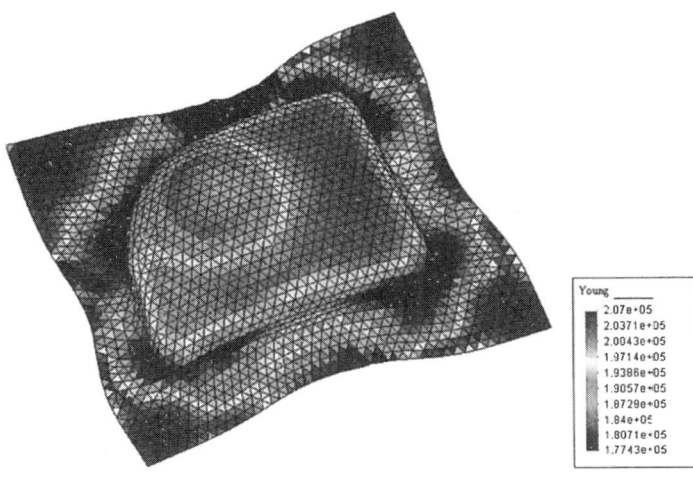

Figure 7. *Young modulus contours on a car component after deep drawing*

3.2. *Influence of elastic properties variations on forming limit diagrams and damage accounting*

In this section, one investigates the effects of the variation of elastic properties on the limit strains at necking. The linearized perturbations method is used for necking simulation and the variation of elastic properties are considered by two ways: the first one consists only to account variation of elastic properties with straining, the second one consists to introduce a full damage model.

3.2.1. *The linearized perturbations technique*

The formability of sheet metal is generally characterized by Forming Limit Diagrams (FLD) that represent the upper straining limits before necking. A lot of experiments on various materials have been realized to build such FLDs. For the prediction of necking, there are also several theoretical approaches. Among these approaches, the two zones model initially introduced in (Marciniak and Kuckzynski, 1967; Marciniak and Duncan, 1992) is well known but necessitates to perform the analysis overall the straining path. It was the reason for that the authors (Boudeau and Gelin, 2000) prefer to use the Linearized Perturbations Technique first introduced for necking in (Molinari, 1990) and further developed in (Boudeau *et al.*, 1998).

The material behaviour is assumed to be isotropic or orthotropic. The local equations defining the complete local problem to solve are expressed as following.

The mass conservation resulting in the plastic incompressibility is expressed as:

$$D_{ii} = 0 \qquad [5]$$

The compatibility equations are expressed with:

$$D_{ij,kl} + D_{kl,ij} = D_{jl,ik} + D_{ik,jl} \qquad [6]$$

The hardening law depend on the material behaviour and can be expressed as

$$\overline{\sigma} = K(\varepsilon_0 + \overline{\varepsilon})^n (\dot{\overline{\varepsilon}})^m \qquad [7]$$

The conservation of momentum is expressed as:

$$(l\sigma_{ij})_{,j} = 0 \qquad [8]$$

where l stands for the length along which the stress is acting,

Concerning the plastic yield locus, one consider here the quadratic Hill yield locus expressed as:

$$\overline{\sigma} = \underline{\underline{H}} : \underline{\sigma} : \underline{\sigma} \qquad [9]$$

where H is the Hill48 quadratic tensor of fourth order for orthotropic materials and deviatoric projector for isotropic materials.

From the previous relations, the constitutive plastic law becomes:

$$D^p = \frac{\dot{\overline{\varepsilon}}}{\overline{\sigma}} \underline{\underline{H}} : \underline{\sigma} \qquad [10]$$

The set of equations [5], [6], [7], [9], [10] can be written as:

$$\mathbf{A}(\mathbf{U}) = 0 \qquad [11]$$

where $\mathbf{A}$ is a non-linear operator and $\mathbf{U}$ represents the variable vector expressed on the following form:

$$\mathbf{U} = \{\sigma_{11}, \sigma_{22}, \sigma_{33}, \sigma_{23}, \sigma_{13}, \sigma_{12}, \overline{\sigma}, D_{11}, D_{22}, D_{33}, D_{23}, D_{13}, D_{12}, \dot{\overline{\varepsilon}}\}^T$$

The perturbation technique consists in the introduction of a small perturbation in the local equilibrium and to check if instability can develop along a certain plane. Let $\vec{n}$ be the normal of the instability plane. This plane is defined by the angles φ and ψ as described in Figure 8.

To predict necking, a perturbation is introduced in the homogeneous solution of the local equilibrium (equation [11]). Let $\mathbf{U}_0$ be this solution and let $\delta \mathbf{U}$ be the perturbation vector $\delta \mathbf{U}$ defined by:

$$\delta U = \delta U_0 \exp(\eta t) \exp(i\xi \, \vec{x}.\vec{n}) \qquad [12]$$

where δU_0 represents the amplitude of the perturbation, η and ξ are respectively the temporal and the spatial part of the perturbation. $\vec{x}$ corresponds to the spatial location where necking could occur.

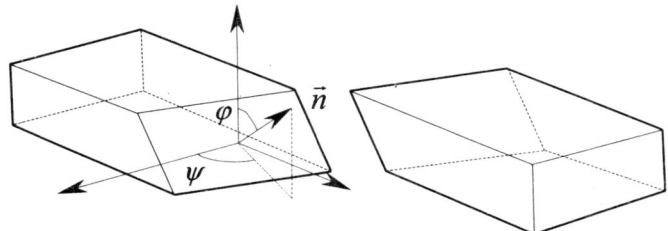

Figure 8. *Geometric description of necking plane*

The perturbed vector $U = U_0 + \delta U_0$ has to satisfy equation [11]:

$$A(U_0 + \delta U_0) = 0 \qquad [13]$$

As the perturbation in the initial stage is smaller than the regular solution, the non-linear system (Eq.[13]) is linearized by neglecting the second order terms resulting from the perturbation. This leads to a linear system expressed as:

$$M(U_0, \eta, \varphi, \psi)\delta U_0 = 0 \qquad [14]$$

A regular solution of equation [14] is obtained when:

$$\det[M(U_0, \eta, \varphi, \psi)] = 0 \qquad [15]$$

that results in:

$$\eta^2 [a\eta^2 + b\eta + c] = 0 \qquad [16]$$

where a, b, c depend on the angles (φ, ψ), the material parameters and the strain- and stress-state at the equilibrium.

The instability occurs when:

$$Re(\eta) > 0 \qquad [17]$$

This criterion is generally too severe, so a positive threshold κ is generally used as:

$$Re(\eta) > \kappa \qquad [18]$$

Forming Limit Diagrams are built from the necking analysis consisting in looking for instability occurrence corresponding to different strain paths. These diagrams have been established to compare the formability of the sheet metals used in deep-drawing (Keeler, 1965) (Goodwin, 1968).

3.2.2. Linear Perturbations Technique accounting elastic coefficients variation with hardening

A simple way to introduce the variation of elastic coefficients vs. hardening in the LPT analysis is to consider that the plastic strain is the total strain rate minus the elastic one that is related to the elastic coefficients by:

$$\varepsilon^e = \left[C^e(\overline{\varepsilon}) \right]^{-1} : \sigma \qquad [19]$$

where C^e is the elastic tensor modified by the evolution of elastic coefficients with the accumulated plastic strain.

Such a modification is trivial and does not change the set of equation presented in the previous section.

3.2.3. Introduction of damage in the Linearized Perturbations Method

As previously shown in section 2, the elastic modules are variable when straining increase. A well known way to account for elastic modules evolution with straining is to include damage in the Kachanov's definition through the concept of effective stress.

In this section, one introduced the well known Lemaitre damage model in the linearized perturbations method with the perspective to detect the fracture in forming limit diagrams. The Lemaitre model is done under an effective stress and strain equivalence principle consideration (Lemaitre, 1992):

$$\underline{\widetilde{\sigma}} = \frac{\sigma}{1-D} \qquad [20]$$

where D is the damage variable introduced by Kachanov (Kachanov, 1958) and $\underline{\widetilde{\sigma}}$ is the effective tensor stress.

The plastic yield locus (equation [9]) become

$$\overline{\sigma} = \underline{\underline{H}} : \underline{\widetilde{\sigma}} : \underline{\widetilde{\sigma}} \qquad [21]$$

Then damage evolution is obtained by a power law

$$\dot{D} = \left(-\frac{Y}{S_0} \right)^{s_0} \dot{\overline{\varepsilon}} \qquad [22]$$

where S_0 and s_0 are material parameters and obtained by tensile test. The authors has chosen the inverse FEM based identification method proposed in (Ghouati et al., 1994) to determine S_0 and s_0 from experiments previously related.

$\dot{\bar{\varepsilon}}$ is the equivalent plastic strain rate and Y is the strain energy density release rate defined by

$$Y = -\frac{\bar{\sigma}}{2(1-D)^2 E}\left(\frac{2}{3}(1+\nu) + 3(1-2\nu)\left(\frac{\sigma_m}{\bar{\sigma}}\right)^2\right) \qquad [23]$$

where E and ν are respectively the Young modulus and Poisson coefficient. D is the damage variable with $0 \leq D \leq 1$.

$\frac{\sigma_m}{\bar{\sigma}} = \frac{tr(\sigma)}{3\bar{\sigma}}$ is the triaxiality ratio which plays a very important role in the rupture of materials.

In the Linearized Perturbations Technique, equations. [5], [6], [7] are not changed. The local equilibrium is now realized by equations [5], [6], [7], [21] and equations [9], [10] where the stress tensor is substituted by the effective stress one. The perturbation function is unchanged and defined by [13]. It is shown in (Lejeune, 2002) that the solution of the homogeneous system is independent of Y and D. So, this equation is identically to equation [14] for a virgin material but where stress is substituted by the effective stress. We recover the equivalent strain principle. Therefore, the solution is done by

$$\det \widetilde{M}(\psi, \varphi, \eta, U_0, D_0) = 0 \qquad [24]$$

Resulting in

$$\eta^2 \left[a\eta^2 + b\eta + c\right] = 0 \qquad [25]$$

where a, b, c now depend on the angles (φ, ψ), the material parameters, the damage state, the strain- and stress-state at the equilibrium. Instability always occurs when $Re(\eta) > \kappa$.

3.3. Comparison of the different approaches

FLDs have been built with the perturbation method and are illustrated in Figure 9. One can observe a decreasing formability in the left hand-side of the FLD with a decreasing thickness. Formability is also diminished in the expansion domain (right hand-side of the figure) when the thickness decreases. In case where one account the effects of the variation of elastic coefficients on FLDs and where damage evolution with straining is also accounted, the FLDs are the same in the left hand side but in the expansion range (right hand side), these approaches permit to take in account the decreasing of the level of necking as expressed in (Marciniak et al., 1992). But one has to notice that the damage model results in a fracture point that is inferior to the hardening coefficient.

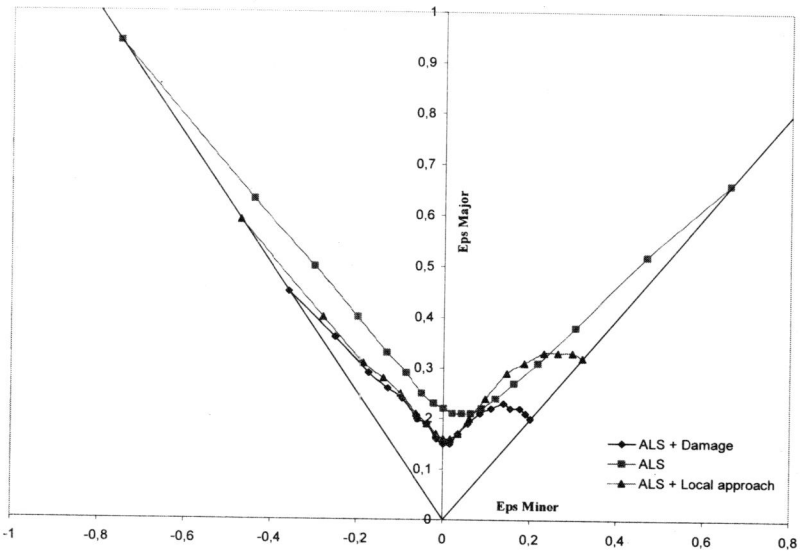

Figure 9. *Forming Limit Diagrams built with LPT, (i) considering constant elastic coefficients, (ii) elastic coefficients variable with plastic straining, (iii) and accounting damage evolution*

3.4. Simulations on a bulging test

A bulging test has been carried out with a circular cavity (radius = 50 mm) for the TRIP steel material under consideration. The results are summarized in table 1, considering the simulations are corresponding to the standard LPT approach, the LPT accounting the effects of hardening on elastic modules and the LPT accounting damage evolution.

Table 1. *Displacement of the spherical cap when necking appeared*

	Standard LPT	LPT + Damage	LPT + Local approach
Necking Detection	12.9 mm	11.2 mm	8.9 mm

4. Conclusions

The paper underlines the effects of initial and induced hardening on the modifications of elastic coefficients that strongly affect the apparition and the amount of defects resulting in sheet metal forming as springback and necking. The methodology that is proposed to access to the variation of elastic coefficients, based on vibrometric analyses, is accurate and reliable, and doesn't need long preparation for experiments, resulting that it can be effectively used in industry.

The Young modulus variation law is quite important in the simulation of springback and the softening of the steel that has been investigated results in the increasing of the amount of springback as it has been through the numerical investigations.

Concerning the necking phenomena, the fact to account variation of elastic coefficients in the LPT considerately increases the quality of the results whereas the fact to account elastic coefficients variation through damage evolution underestimates the limit strains at fracture in the expansion range.

5. References

Alberti N. Barcellona A., Cannizzaro L., Micari F., "Prediction of ductile fractures in metalforming processes: an approach based on the damage mechanics", *Annals of the CIRP*, 43/1, 1994, pp. 207–210.

Banabic D. *et al.*, *Formability of Metallic Materials*, Springer Verlag Publ., 2000.

Boudeau N., Gelin J.C., Salhi S., "Computational prediction of the localized necking in sheet forming based on microstructural material aspects", *Computational Materials Science*, Vol. 11, 1998, pp. 45–64.

Boudeau N., Gelin J.C., "Necking in sheet metal forming. Influence of macroscopic and microscopic properties of materials", *Int. J. Mech. Sci.*, Vol. 42, 2000, pp. 2209–2232.

Boudeau N., Lejeune A., Gelin J.C., "Damage in sheet metal forming: prediction of necking phenomenon", NUMEDAM'00, *Rev. Eur. Eléments Finis*, Vol. 10, No. 2–3–4, 2001, pp. 295–310.

Dudzynski D., Molinari A., "Modélisation et prevision des instabilités plastiques en emboutissage", *Physique et Mécanique de la Mise en Forme des Métaux, Presses du CNRS*, Paris, 1990, pp. 444–460.

EFB-Forschungsbericht, Rückfederungs verhalten beim Umformen von Feinblechen,Nr., 1998, pp.109.

Gelin J.C., Picart Ph., "Proceedings of Numisheet'99", 4th *Int. Conf. And workshop on Numerical simulation of 3D Sheet Forming Processes*, 1999, Burs Publ.

Ghouati O., Identification et modélisation numérique directe et inverse du comportement viscoplastique des alliages d'aluminium, PhD Thesis, Université de Franche Comté, 1994.

Goodwin G.M., "Application of the strain analysis to steel metal forming in press shop", *La Metallurgia Italania*, Vol. 8, 1968, pp. 767–772.

Keeler S.P., "Determination of the forming limits in automotive stamping", *Sheet metal industries*, Vol. 42/461, 1965, pp. 683–691.

Lejeune A., Boudeau N., Gelin J.C., "Prediction of localized necking in sheet metal forming processes for 3D-stress-state", *Proc. of the 4th Int. ESAFORM Conf. On Material Forming*, Liège, 2001, pp. 249–252.

Lejeune A., Modélisation et simulation des défauts d'aspects lors d'un procédé de mise en forme, PhD Thesis in preparation, 2002.

Lemaitre J., Chaboche J.L., *Mécanique des matériaux solides*, Paris, Dunod Publ., 1985.

Lemaitre J., *A course on damage mechanics*, Berlin, Springer-Verlag Publ., 1992.

Lems W., The change of Young's modulus after deformation at low temperature and its recovery, Ph. D Thesis, Delft, 1964.

Makinouchi A., "Recent developments in sheet metal forming simulation", *Simulation of Materials Processing – Theory, Methods and applications*, Ed. by K.I. Mori, Balkema Publ., 2001, pp. 3–10.

Marciniak Z., Duncan J., *Mechanics of sheet metal forming*, London, Edward Arnold Editions, 1992.

Marciniack Z., Kuckzynski, K., "Limit strains in the processes of stretch-forming sheet metal", *Int. J. Mech. Sci.*, 9, 1967, pp. 609–620.

Morestin F., Boivin M., "On the necessity of taking into account the variation in the Young modulus with plastic strain in elastic-plastic software", *Nuclear Engineering and Design*, 162, 1996, pp. 107–116.

Morestin F., Boivin M., Silva C., "Elasto plastic formulation using a kinematic hardening model for springback analysis in sheet metal forming", *J. of Materials Processing Technology*, 56, 1996, pp. 619–630.

Moussi F., "Future trends in materials for automotive body: driving and restricting factors", *Proc of Numisheet '99*, Ed. by J.C. Gelin and P. Picart, 1999, Burs Publ., 1.12.

Thibaud S., Boudeau N., Gelin J.C., Greisert C., Ghouati O., "On the influence of the young modulus evolution on the dynamic behaviour and springback of a sheet metal forming component", submitted to Numisheet 2002.

Xia Z.C., "A parametric study of springback behavior", *Simulation of materials Processing – Theory, Method and Applications*, Ed. by K.I. Mori, Balkema Publ., 2001, pp. 711–716.

Chapter 31

Computational Characterization of Micro- to Macroscopic Mechanical Behaviour and Damage of Polymer containing Second-phase Particles

Yoshihiro Tomita and Wei Lu
Graduate School of Science and Technology, Kobe University, Japan

1. Introduction

In polymers containing soft rubber particles, blended polymers, the onset of cavitation in the rubber particles relaxes the high triaxiality stress state and suppresses the onset of crazing in the polymer (Parker et al., 1990). As a result, large plastic deformation is substantially promoted compared with single-phase polymers (Bucknall, 1977).

The studies associated with blended polymers are generally classified as those for polymers containing noncavitating rubber particles (Tanaka et al., 2000; Lu et al., 2001; Tomita et al., 2002), those containing voids (Steenbrink et al., 1997; Socrate et al., 2000; Pijnenburg et al., 2001) which represent the polymers containing cavitated rubber particles, and those containing rubber particles (Steenbrink et al., 1999) which cavitate with the deformation. Either case employs the unit cell or the RVE (representative volume element) to analyze the micro- to macroscopic deformation behavior under macroscopically uniform deformations. Those studies without consideration of the onset of cavitation aimed to characterize the deformation localization and its propagation in the polymer, the normal stress on the interface between polymer and rubber particles which governs the onset of debonding on the interface, the maximum mean stress in the rubber particles which dominates the onset of cavitation in the rubber particles and the maximum mean stress in the polymer which is closely related to the onset of crazing. On the other hand, the study (Steenbrink et al., 1999) associated with the onset of cavitation under macroscopically homogeneous deformation parallel to the unit cell suggested that an average normal stress remains on the interface of the polymer and rubber particles after cavitation and affects the deformation behavior of the polymer.

In order to clarify the effect of macroscopic deformation behavior on the cavitation of rubber particles and the deformation behavior of the polymer under general macroscopical deformation, a polymer blended with soft rubber particles is modeled with a plane-strain unit cell containing cylindrical rubber particles with different diameters. This model is the very fundamental one which has been used by many researchers and has provided much useful information with restricted computer resources. The responses of the rubber particles and the polymer are respectively modeled by employing the generalized neo-Hooke's strain energy function (Knowles, 1977) and the constitutive equation (Tomita et al., 1997) established based on the nonaffine molecular chain network model. Although the volume strain energy and the surface energy stored in a particle are related to the cavitation (Bucknall, 1994), the effect of the surface energy on cavitation decreases as the size of a particle increases to $100nm$ (Fond, 2001). The present investigation aims to evaluate the pre- to post-cavitation behavior of rubber particles over $1\mu m$ in size. Therefore, the cavitation is evaluated by using rubber particles with small holes (Ball, 1982). The unit cell model together with the homogenization method approximated by the finite element method provides the computational tools (Higa et al., 1999).

2. Constitutive equation

The complete constitutive equation for polymer employed in this investigation is given by Tomita et al., 1997; here we provide a brief explanation of the constitutive equation. The total strain rate is assumed to be decomposed into elastic strain rate and plastic strain rate. Elastic strain rate is expressed by Hooke's law and plastic strain rate is modeled by using the nonaffine eight-chain model (Tomita et al., 1997). The final constitutive equation that relates the rate of Kirchhoff stress $\dot{S}_{ij}$ and strain rate $\dot{\varepsilon}_{kl}$ becomes

$$\dot{S}_{ij} = L_{ijkl}\dot{\varepsilon}_{kl} - P'_{ij}, \quad L_{ijkl} = D^e_{ijkl} - F_{ijkl},$$

$$F_{ijkl} = \frac{1}{2}(\sigma_{ik}\delta_{jl} + \sigma_{il}\delta_{jk} + \sigma_{jl}\delta_{ik} + \sigma_{jk}\delta_{il}),$$

$$P'_{ij} = D^e_{ijkl}\frac{\tilde{\sigma}'_{kl}}{\sqrt{2\tilde{\tau}}}\dot{\gamma}^p, \quad \tilde{\tau} = \frac{1}{2}\tilde{\sigma}'_{ij}\tilde{\sigma}'_{ij}, \quad \tilde{\sigma}_{ij} = \sigma_{ij} - B_{ij}, \qquad [1]$$

where D^e_{ijkl} is the elastic stiffness tensor and σ_{ij} is the Cauchy stress. The shear strain rate $\dot{\gamma}^p$ in Eq. [1] is related to the applied shear stress $\tilde{\tau}$ as (Argon, 1973)

$$\dot{\gamma}^p = \dot{\gamma}_0 \exp\left\{-\frac{A\tilde{s}}{T}\left[1 - \left(\frac{\tilde{\tau}}{\tilde{s}}\right)^{5/6}\right]\right\}, \qquad [2]$$

where $\dot{\gamma}_0$ and A are constants, T is the absolute temperature, $\tilde{s} = s + \alpha p$ indicates shear strength (Boyce et al., 1988), s is the shear strength which changes with plastic strain from the athermal shear strength $s_0 = 0.077\mu/(1-\nu)$ to a stable value s_{ss}, p is the pressure, α is a pressure-dependent coefficient, μ is the elastic shear modulus and ν is Poisson's ratio. Since s depends on strain rate, the evolution equation of s can be expressed as $\dot{s} = h\{1 - (s/s_{ss})\}\dot{\gamma}^p$ where h is the rate of resistance with respect to plastic strain. Furthermore, B_{ij} in Eq.[1] is the back-stress tensor of which principal components are expressed by employing the eight-chain model as (Arruda et al., 1993)

$$B_i = C^R \frac{\sqrt{N}}{3} \frac{V_i^2 - \lambda^2}{\lambda} \mathcal{L}^{-1}\left(\frac{\lambda}{\sqrt{N}}\right),$$

$$\mathcal{L}(x) = \coth x - \frac{1}{x}, \quad \lambda^2 = \frac{1}{3}(V_1^2 + V_2^2 + V_3^2), \qquad [3]$$

where V_i is the principal plastic stretch, N is the average number of segments in a single chain, $C^R = nkT$ is a constant, n is the number of chains per unit volume, k is Boltzmann's constant, and $\mathcal{L}$ is the Langevin function. In the nonaffine eight-chain model, the number of entangled points, in other words, the average number of segments N may change depending on the temperature change and distortion of an orthogonal frame which represents the local deformation of polymeric material.

The response of rubber particles is modeled by employing the generalized neo-Hooke's strain energy function (Knowles, 1977) and can be expressed as

$$\sigma_{ij} = -p\delta_{ij} + \mu[1 + \frac{b}{n}(I_1 - 3)]^{n-1}A_{ij} \qquad [4]$$

where I_1 is the first invariant of the left Cauchy-Green tensor A_{ij}. The constants b and n are determined experimentally. The rate type expression of the constitutive equation [4] becomes

$$\dot{S}_{ij} = -\dot{p}\delta_{ij} + \mu[1 + \frac{b}{n}(A_{mm} - 3)]^{n-1}$$

$$\{\frac{b(n-1)}{n}[1 + \frac{b}{n}(A_{mm} - 3)]^{-1}A_{ij}A_{kl} + \delta_{ik}A_{jl} + A_{ik}\delta_{jl}\}\dot{\varepsilon}_{kl} - F_{ijkl}\dot{\varepsilon}_{kl}$$

$$= -\dot{p}\delta_{ij} + (R_{ijkl} - F_{ijkl})\dot{\varepsilon}_{kl}. \quad [5]$$

Furthermore, the penalty method is employed to approximately satisfy the volume constant deformation. The corresponding pressure rate is expressed as $\dot{p} = -K\dot{\varepsilon}_{mm}$, where $K = \psi\max(R_{ijkl})$, and the value ψ is set to 1000 (Steenbrink *et al.*, 1999).

3. Computational method

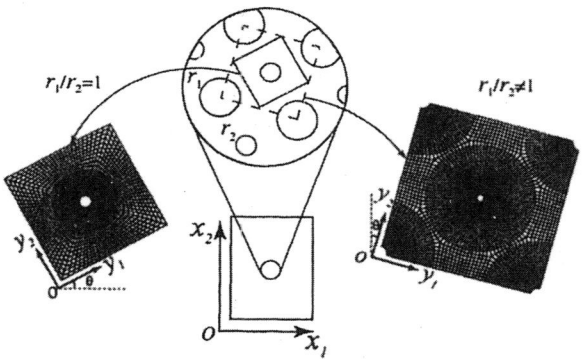

Figure 1. *Computational model*

Figure 1 illustrates the computational model in which heterogeneous particles are assumed to be distributed periodically and we employ the homogenization method and the finite element method to establish a computational tool. In order to reproduce the cavitation behavior of the rubber particles by employing the computational simulation, the unit cell model in which the particles are assumed to be circular cylinders with radii of curvature r_1 and r_2, containing small holes with radii of curvature c_1 and c_2 respectively, is used. Through the precise discussions associated with the effect of radius ratio c_i/r_i on the cavitation behavior and the complementation by the analytical solution (Hou *et al.*, 1990), we employ the ratio $c_i/r_i = 0.2$ in order to investigate the onset of caviation and the subsequent deformation behavior. The macroscopically homogeneous deformations are applied by prescribing the average strain rates $\dot{\Gamma}_1$, $\dot{\Gamma}_2$ or stress rates $\dot{\Sigma}_1$, $\dot{\Sigma}_2$ with respect to the coordinate directions x_1, x_2, respectively. In order to characterize the direction of the unit cell, y_1 or y_2, with respect to the coordinate axis x_1 or x_2, angle θ shown in Figure 1 is introduced. The coherent boundary condition is assumed to be applied to the particle and polymer interface. Furthermore, the elasticity modulus ratio of particles to polymer is defined by E_i / E_m. Macroscopic equivalent stress and strain are defined as $\Sigma_e = (\frac{3}{2}\Sigma_i'\Sigma_i')^{\frac{1}{2}}$ and $\Gamma_e = (\frac{2}{3}\Gamma_i'\Gamma_i')^{\frac{1}{2}}$, respectively.

Although the mechanisms of plasticity, cavitation and crazing are strongly related to the material, the applied stress sate, the strain rate and the temperature (Bucknall, 1977), in this study, we restrict our attention to the deformation behavior of rubber-blended polycarbonate under low strain rate at room temperature. For a typical unit cell, which is the microscopic element of the blended material, a macroscopically homogeneous strain rate $\dot{\Gamma}_2 = \dot{\varepsilon}_0 = 10^{-5}/s$ is applied. The material parameters for the polymer employed are $E_m/s_0 = 23.7$, $s_{ss}/s_0 = 0.79$, $h/s_0 = 5.15$, $As_0/T = 78.6$, $\alpha = 0.08$, $\dot{\gamma}_0 = 2.0 \times 10^{15}/s$, $s_0 = 97$MPa, $T = 296$K. For rubber particles, $\mu = E_i/3$, $b = 0.1$ and $n = 0.54$ are introduced.

4. Results and discussion

4.1. *Effect of stiffness of particle and direction of macroscopic loading on the deformation behavior*

A series of simulations have been performed for the unit cells containing homogeneous sized particles having different stiffness ratios, elasticity modulus ratios, of particles and polymer $E_i/E_m = 0.05, 0.005$ and volume fraction $f_0 = 20\%$, deformed under the strain ratios $\Gamma_1/\Gamma_2 = 0.0$. The direction θ of the unit cell defined in Fig.1 changes according to the values $0°$ and $30°$.

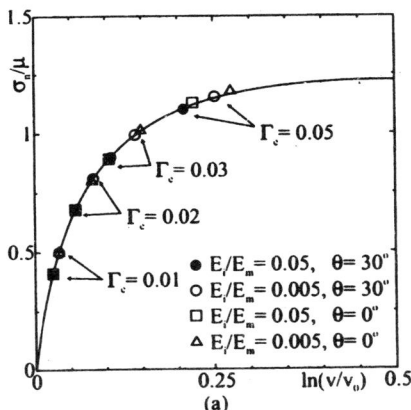
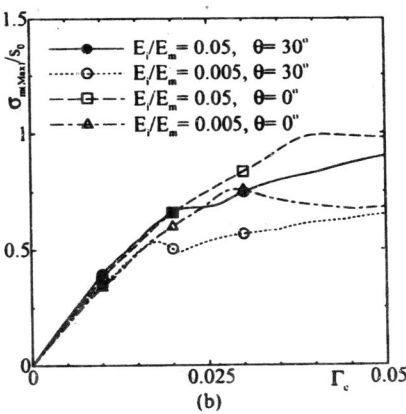

Figure 2. *Relations : (a)Average normal stress on the interface vs. logarithmic volumetric strain of the particle ; (b)Maximum mean stress in the polymer vs. macroscopic equivalent strain for $f_0 = 20\%$, $\Gamma_1/\Gamma_2 = 0.0$, $r_1/r_2 = 1$*

Figure 2a indicates the relationships between average normal stress on the interface and the volumetric strain of the particle. The deformations corresponding to the macroscopic equivalent strains of $\Gamma_e = 0.01, 0.02, 0.03, 0.05$ are indicated by arrows. The results confirm that regardless of the magnitude of stiffness of the particles or the direction of the unit cell, the relationships relating to the average normal stress and the volumetric strain of the particles are cast into a characteristic single line, which is the more general indication of the characteristics of the deformation behavior of the hole in the axisymmetric particles (Steenbrink et al., 1999). Furthermore, we also confirmed similar characteristics for the different stress conditions and the heterogeneous particles. These results suggest that the evolution of the hole in the particles is governed by the average normal stress on the particles and is independent of the formation and propagation of shear bands in the unit cell. In other words, the soft rubber particles behave almost as a compressible fluid.

Figures 2b and 3 respectively show the maximum mean stress in the polymer with respect to the macroscopic equivalent strain and the distribution of equivalent plastic strain rate. The change of the maximum mean stress governing the onset of crazing shown in Figure 2b indicates that growth of the hole in the particles is promoted in the polymer containing soft rubber particles, which lowers the maximum value of the mean stress in the polymer. At the point with the change in the upward trend of maximum mean stress shown in Figure 2b, the formation of a strong shear band is observed as depicted in Figure 3 and subsequently the maximum mean stress gradually changes.

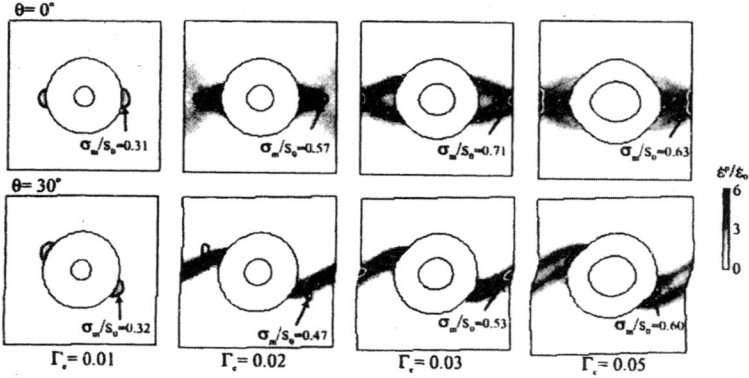

Figure 3. *Equivalent plastic strain rate distributions and location of maximum mean stress in the polymer for* $f_0 = 20\%$, $\Gamma_1/\Gamma_2 = 0.0$, $E_i/E_m = 0.005$, $r_1/r_2 = 1$

4.2. Effect of macroscopic strain ratio on the deformation behavior

Figure 4a indicates the macroscopic equivalent stress versus strain relations for unit cells under the macroscopic strain ratio $\Gamma_1/\Gamma_2 = 0.0, 0.4, 0.8$ and the stress ratio $\Sigma_1/\Sigma_2 = 0.0$. With the increase of the strain ratio ($\Sigma_1/\Sigma_2 = 0.0 \rightarrow \Gamma_1/\Gamma_2 = 0.4$), the growth of the particles becomes more remarkable, which promotes plastic deformation in the polymer and results in the onset of macroscopic yielding at an earlier stage of deformation and decreases the macroscopic yield stress. With further increase of the strain ratio ($\Gamma_1/\Gamma_2 = 0.4 \rightarrow \Gamma_1/\Gamma_2 = 0.8$), the main shear bands connecting the particles are formed almost simultaneously, which leads to the onset of macroscopic yielding with similar level in stress. The mean stress in the polymer shown in Figure 4b suggests that the maximal value of mean stress increases with strain ratio. The evolution of volume change of the particles promotes plastic deformation and therefore, the maximum mean stress changes to decrease in the early stage of deformation and then gradually increases. However, due to the high triaxiality of the macroscopic stress, the mean stress in the polymer increases in the case when the strain ratio increases.

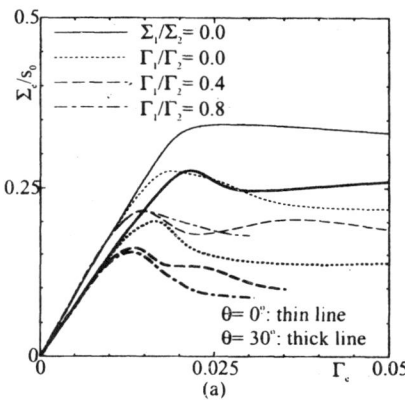
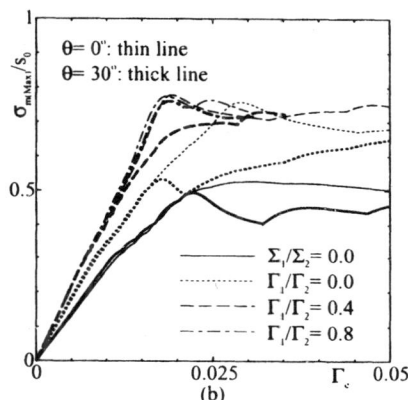

Figure 4. *Relations : (a)Macroscopic equivalent stress vs. macroscopic equivalent strain; (b)Maximum mean stress in polymer vs. macroscopic equivalent strain for* $f_0 = 20\%$, $E_i/E_m = 0.005$, $r_1/r_2 = 1$

4.3. Effect of heterogeneity of rubber particles on deformation behavior

Figures 5a and 5b respectively indicate macroscopic deformation behavior and the distribution of equivalent strain rate and mean stress in the polymer. Similar to the results reported in a previous paper (Tomita et al., 2002), the macroscopic equivalent stress versus strain relations are rather insensitive to the change of strain ratio when the sizes of particles are moderately different. Figure 5c suggests that the volumetric strain of the small particles grows first whereas that of the large particles is suppressed. Therefore, as can be seen in Figure 5b, the formation of a high-strain-rate region and its propagation are indicated near the small particles. On the other hand, the shear bands connecting the particles onset almost simultaneously and propagate with preservation of their configuration. Consequently, the macroscopic responses shown in Figure 5a are almost insensitive to the strain ratio. In this study, the suitable length scale such as size of the particles is not considered through the introduction of volume strain energy as well as surface energy (Bucknall, 1994). Its effect on the interaction of heterogeneous particles requires further study.

Figure 5d depicts the maximum mean stress with respect to the macroscopic equivalent strain. For the case with small strain ratio, the small particles contribute to increasing the maximum mean stress in the polymer, while for the case with a large

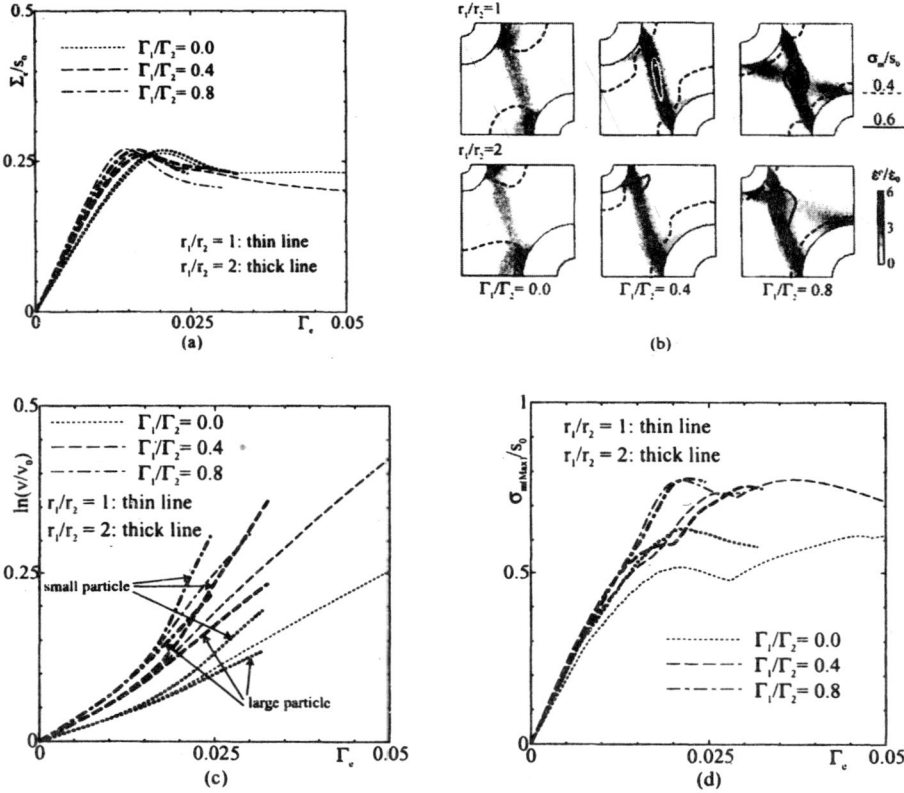

Figure 5. *Relations : (a)Macroscopic equivalent stress vs. macroscopic equivalent strain; (b)Plastic strain rate and mean stress distributions in polymer at* $\Gamma_e = 0.02$; *(c)Logarithmic volumetric strain of the particle vs. macroscopic equivalent strain; (d)Maximum mean stress in polymer vs. macroscopic equivalent strain for* $f_0 = 20\%$, $E_i/E_m = 0.005$, $\theta = 0°$

strain ratio, regardless of the size of the particles, the maximum mean stress is kept almost unchanged. This is attributable to the change of location at the point where the maximum mean stress appears. As shown in Figure 5b, the location of the maximum mean stress point is the front of a shear band or the inflection point of the inclination of the shear band depending on the growth of the shear bands. The latter often appears at the point of intersection of the two shear bands. A similar tendency can be found in the mean stress distribution along the slip line in rigid plastic materials deformed under plane strain. The mean stress changes in proportion to the change of inclination angle of the slip line (Hill, 1950). With reference to Figure 5b, although the location of the maximum mean stress point is different depending on the size of the particles, the profile of the shear bands is almost unchanged within the present range of deformation. Therefore, the maximum values of mean stress approximately maintains their values during the deformation for the cases of $\Gamma_1/\Gamma_2 = 0.4$ and 0.8. The location of the maximum mean stress point moves depending on the deformation near the particles for the case of $\Gamma_1/\Gamma_2 = 0.0$.

5. Conclusions

Computational simulations employing a plane-strain unit cell model containing soft rubber particles clarified the effect of stiffness and size of rubber particles, macroscopic strain ratio, and the direction of the unit cell, on the cavitation behavior in the rubber particles and the onset of shear bands and their propagations in the polymer. Also, the relationships between these phenomena and the maximum mean stress in the polymer that directly affects the onset of crazing was investigated. The following is a summary of the results.

1) Regardless of the changes of the parameters in the present investigation, the relations related to the average normal stress versus volume change of the particles are cast into a characteristic single line.

2) The maximum mean stress appears near the front of the shear band for insufficient growth of the shear band connecting the particles and it appears at the inflection point of the inclination of the shear band for the sufficiently developed shear band connecting the particles. The latter point corresponds to the intersection point of the shear bands.

3) Soft rubber particles promote the onset of shear bands connecting the particles and the growth of holes in the particles that lower the maximum value of the mean stress in the polymer.

4) The increase of macroscopic strain ratio results in the increase of volumetric strain of particles, which promotes the plastic deformation of the polymer and the onset of macroscopic yielding with a lower value of yield stress. Due to the high triaxiality of the macroscopic stress, the mean stress in the polymer increases when the strain ratio increases.

5) The volumetric strain of the small particles increases first and the formation of a high-strain-rate region and its propagation are marked near the small particles. The small particles contributes to increase the maximum mean stress in the polymer for the case of a small strain ratio, while regardless of the size of the particles the maximum mean stress is kept almost unchanged for the case with a large strain ratio.

The computational strategy employing the homogenization method and the finite element method provides the homogenized constitutive equation accounting all microscopic deformation behavior in the unit cell subjected to macroscopically uniform deformation so that it is straight forward to introduce thus obtained constitutive equations at each material point for macroscopically nonuniform deformation as observed in the forming processes. However, in the present investigation, the cavitation is modeled by introducing small holes in the particles, therefore, onset and growth of cavity is controlled by the deformation behavior of particles and surrounding polymer. To clarify the absolute size effect on controlling cavitation, a suitable length scale should be considered through the introduction of volume strain energy as well as surface energy (Bucknall, 1994). The size of particles may affect the interaction of heterogeneous particles as well, which requires further study.

Acknowledgements

Financial support from the Ministry of Education, Culture, Sports, Science and Technology of Japan are gratefully acknowledged. We wish to thank Dr. Adachi, Associate Professor of Kobe University, for many useful comments and Dr. Higa, Research Associate of Osaka University, for valuable discussions.

6. References

Argon, A.S. "A theory for low-temperature plastic deformation of glassy polymers", *Philos. Mag.*, vol. 28, n° 39, 1973, pp. 839-865.

Arruda, E.M., Boyce, M.C. "A three-dimensional constitutive model for large stretch behavior of rubber materials", *J. Mech. Phys. Solids*, vol. 41, 1993, pp. 389-412.

Ball, J.M. "Cavitation in nonlinear elasticity", *Phil. Trans. R. Soc. Lond.*, A306, 1982, pp. 557-611.

Boyce, M.C., Parks, D.M., Argon, A.S. "Large inelastic deformation of glassy polymers, Part I : rate dependent constitutive mode", *Mech. Mater.*, vol. 7, 1988, pp. 15-33.

Bucknall, C.B., *Toughened Plastics*, London, Applied Science, 1977.

Bucknall, C.B., Karpodinis, A., Zhang, X.C., "A model for cavitation in rubber-toughened plastics", *J. Mat. Sci.*, vol. 29, 1994, pp. 3377-3383.

Fond, C. "Cavitation criterion for rubber materials : A review of void growth models", *J. Polym. Sci. B : Polym. Phys.*, vol. 39, 2001, pp. 2081-2096.

Higa, Y., Tomita, Y. "Computational prediction of mechanical properties of Nickel-based superalloy with Gamma prime phase precipitates", *Advance Materials and Modeling of Mechanical Behavior*, 1999, pp. 1095-1099.

Hill, R., *The Mathematical Theory of Plasticity*, London, Oxford University Press, 1950.

Hou, H.S., Zhang, Y. "The effect of axial stretch on cavitation in a elastic cylinder", *Int. J. Non-Linear Mechanics*, vol. 25, 1990, pp. 715-722.

Knowles, J.K. "The finite anti-plane shear field near the tip of a crack for a class of incompressible elastic solids", *Int. J. Fracture*, vol. 13, 1977, pp. 611-639.

Lu, W., Tomita, Y. "Estimation of deformation behavior of rubber-blended glassy polymers", *Trans. JSMS*, vol. 50, n° 6, 2001, pp. 578-584, in Japanese.

Parker, D.S., Sue, H.J., Huang, J., Yee, A.F. "Toughening mechanisms in core-shell rubber-modified polycarbonate system", *Polym.*, vol. 31, 1990, pp. 2267-2277.

Pijnenburg, K.G.W., Van der Giessen, E. "Macroscopic yield in cavitated polymer blends", *Int. J. Solids Struct.*, vol. 38, 2001, pp. 3575-3598.

Socrate, S., Boyce, M.C. "Micromehanics of toughened polycarbonate", *J. Mech. Phys. Solids*, vol. 48, 2000, pp. 233-273.

Steenbrink, A.C., Van der Giessen, E., Wu, P.D. "Void growth in glassy polymers", *J. Mech. Phys. Solids*, vol. 45, 1997, pp. 405-437.

Steenbrink, A.C., Van der Giessen, E. "On cavitation, post-cavitation and yield in amorphous polymer-rubber blends", *J. Mech. Phys. Solids*, vol. 47, 1999, pp. 843-876.

Tanaka, S., Tomita, Y., Lu, W. "Computational simulation of deformation behavior of glassy polymer with cylindrical inclusions under tension", *Trans. JSME*, vol. 66, n° 643, 2000, pp. 36-45, in Japanese.

Tomita, Y., Adachi, T., Tanaka, S. "Modelling and application of constitutive equation for glassy polymer based on nonaffine network theory", *Eur. J. Mech. A/Solids*, vol. 16, n° 5, 1997, pp. 745-755.

Tomita, Y., Lu, W. "Computational characterization of micro- to macroscopic mechanical behavior of polymers containing second-phase particles", *Int. J. Damage Mech.*, vol. 11, n° 2, 2002, pp. 129-150.

Index

3D sheet metal stamping processes, shape metal measurement 136

analytical fracture mechanics 53
anisotropic damage model, aluminium 118 et seq
   identification 125
   numerical implementation 123
axisymmetric extrusion, upper bound solutions 5

backward extrusion process, continuum damage model 12
bending process, FEM simulation 168
Bernoulli's theorem
   central bursting defects in extrusion and drawing 1 et seq
blanking
   process, 2D finite element simulation 277
   tension effects 246
   thin sheet, modelling 272 et seq
   Zhou and Wierzbicki tension zone model 243
blow molding polymer crystallization, deformation induced 262 et seq
Brazilian test 157
bulge test 287

cavity growth, 3 dimensional prediction 64 et seq
central bursting
   in extrusion and drawing 2 et seq
   initiation 7
ceramic defects, cavity defects and failure 148 et seq
   progressive damage and failure, analysis 149

continuous damage model, fatigue analysis 342 et seq
cracked body loading 54
crystallization, deformation induced 265
cyclic strength reduction function 347

damage measurements 13
   tensile test 14
damageable media material constants 203
deformation induced crystallization 265 et seq
   tests 267 et seq
ductile damage model with inclusion considerations 64 et seq

edge cracks 56
ellipsoid, transformation of 64
extrusion and drawing, central bursting effects 2 et seq

fatigue analysis, materials and structures, continuous damage model 342 et seq
   continuum mechanics 345
fatigue crack growth
   initial and newly born growth 52
   numerical analysis 57
   theory 55
formability, sheet metal forming, initial and induced hardening 354 et seq
   vibrometric identification 355
forming operations, failure prediction, anisotropic sheet metals 74 et seq
   damage model 74
   damage parameters 78
   necking failure criterion 80

irreversible deforming, dynamic
    process modelling 200 et seq

material defects, prediction in reactive
    polymetric flows 107 et seq
    numerical model 111
    simplified flow model 109
    transport problems 112
metal forming processes, numerical
    simulation of ductile damage 212
    et seq
    applications 226
    Forge2, Forge3 227
    theoretical and numerical aspects
        212 et seq
    various examples 228 et seq
metallic structures, internal defects
    343
micro- and macro-fracture of solids
    and structures 200 et seq

necking initiation during bending of
    metal-rubber profiles 166 et seq
    MK analysis 172
    necking features 167
non spherical voids in deformed
    metals 24 et seq
    various models 24, 25, 26

plastic spin 65
    compressible potential 67
    plastic potential 68
polymer containing 2nd phase
    particles, mechanical behaviour
    and damage 370 et seq
punching, induced residual stresses
    model 240 et seq
    strain mechanisms 240
    stress 241

sheet rolling process, void closure
    behaviour 86 et seq
    with internal void 90
slow shearing, saturated granular
    media, induced non homogeneity
    316

homogeneous plastic material 317
    kaolin clay pastes 321
S-N curves 347
solid state drawing of
    poly(oxymethylene), strain
    hardening and damage 186 et seq
spurt flow instability during extrusion
    of molten HDPE 306 et seq
square boxes, deep drawing of 137
steel, damage in at high temperature,
    microscopic modelling 294 et seq
stress rate 70
structural components, strength of
    156

thermomechanical simulation, model
    reduction method 326 et seq
    finite element method 327
thermoplastic composites forming
    processes, interply porosities 96 et
    seq
    reconsolidation stage 99
    sheet forming simulations 97
    shell element with pinching 101
thin canning sheet materials,
    anisotropy in 282 et seq
    sheet blanking, modelling 273 et
    seq
titanium
    alloys lattice misorientations 38 et
    seq
    deformation textures, variational
        self-consistent model 176 et seq
    globularization 40
    stress response and flow softening
        45
tube hydroforming, wrinkling and
    necking instabilities 250 et seq

upsetting tests 17

void closure behaviour during sheet
    rolling process 86 et seq
    finite element modelling 87
    mesh configuration 88